SpringerWienNewYork

Yearbook on Space Policy

Edited by the
European Space Policy Institute
Director: Kai-Uwe Schrogl

European Space Policy Institute
Kai-Uwe Schrogl, Charlotte Mathieu, Nicolas Peter (eds.)

# Yearbook on Space Policy 2006/2007

## New Impetus for Europe

SpringerWienNewYork

# List of acronyms

**A**

ABM: Anti-ballistic Missile
ACTS: Advanced Crew Transportation System
AEB: Brazilian Space Agency (Agencia Espacial Brasileira)
ALOS: Advanced Land Observing Satellite
ALT: Radar Altimeter
AP-MCSTA: Asia-Pacific Multilateral Cooperation in Space Technology and Applications
APR-SAF: Asia-Pacific Regional Space Agency Forum
APS: Aerosol Polarimeter Sensor
APSCO: Asia-Pacific Space Cooperation Organization
AR4: Fourth Assessment Report
ARTES: Advanced Research in Telecommunications Systems
ASAT: Anti-SAtellite Test
ASI: Italian space agency (Agenzia Spaziale Italiana)
ASTRO: Autonomous Space Transport Robotic Orbiter
ATK: Alliant Techsystems
ATV: Automated Transfer Vehicle

**B**

BERR: Department for Business, Enterprise and Regulatory Reform of UK
BMD: Ballistic Missile Defence
BNSC: British National Space Centre

**C**

CaLV: Cargo Launch Vehicle
CBM: Confidence-Building Measures
CD: Conference on Disarmament
CDI: Center for Defense Information
CEA: Space Conference of the Americas (Conferencia Espacial de las Américas)
CEO: Chief Executive Officer
CEOI: Centre for Earth Observation Instrumentation of UK
CEOS: Committee on Earth Observation Satellites
CEV: Crew Exploration Vehicle
CFSP: Common Foreign and Security Policy

CHAMP: CHAllenging Minisatellite Payload
CLA: Alcantara Launch Centre
CLV: Crew Launch Vehicle
CMEs: Coronal Mass Ejections
CMIS: Conical Scanning Microwave Imager
CNES: French Space Agency (Centre National d'Etudes Spatiales)
CNSA: China National Space Administration
COPUOS: Committee On the Peaceful Uses of Outer Space
COROT: COnvection, ROtation and planetary Transits
COSPAS-SARSAT: International Search and Rescue Satellite
COSTIND: Commission of Science Technology and Industry for National Defense
COTS: Commercial Orbital Transportation Services
CRS: Congressional Research Service
CSA: Canadian Space Agency
CSI: Constellation Services International
CSRS: Counter Surveillance Reconnaissance System

**D**

DARPA: Defense Advanced Research Projects Agency
DARS: Digital Audio Radio Satellite
DBS: Direct Broadcast Services
DCS: Defensive Counterspace operations
DEFRA: Department for Environment, Food and Rural Affairs of UK
DG: Director General
DIUS: Department for Innovation, Universities and Skills of UK
DLR: German Aerospace Center (Deutsches Zentrum für Luft-und Raumfahrt)
DoD: U.S. Department of Defense
DPJ: Democratic Party of Japan
DSP: Defense Support Program

**E**

EBN: European Business & Innovation Centre Network
EC: European Commission
ECS: European Cooperating State
EELV: Evolved Expendable Launch Vehicle
EGNOS: European Geostationary Navigation Overlay Service
EHNWI: Extremely High Net Worth Individuals
EISC: European Interparliamentary Space Conference
EOMD: Earth Observation Market Development program

EPO: European Patent Office
EPS: Eumetsat Polar System
ERBS: Earth Radiation Budget Sensor
ERC: European Research Council
ERCS: Emergency Response Core Service
EROS: Earth Remote Observation Satellite
ESA: European Space Agency
ESCAP: United Nations Economic and Social Commission for Asia and the Pacific
ESDP: European Security and Defence Policy
ESINET: European Space Incubators network
ESMD: Exploration System Mission Directorate
ESNI: European Satellite Navigation Industries
ESP: European space policy
ESPI: European Space Policy Institute
ESTB: EGNOS System Test Bed
EU: European Union
EVA: Extravehicular Activity
EVCA: European Private Equity & Venture Capital Association

**F**

FAA: United States Federal Aviation Administration
FMCT: Fissile Material Cut-off Treaty
FOCAC: Forum On China-Africa Cooperation
FP: Framework Program
FP7: 7$^{th}$ Framework Program
FSS: Fixed Satellite Services
FTC: U.S. Federal Trade Commission
FY-1C: Feng Yun-1C

**G**

GAC: GMES Advisory Council
GBAORD: Government Budget Appropriations or Outlays allocated to Research and Development
GCOS: Global Climate Observing System
GE: General Electric
GEO: Geostationary Earth Orbit
GEO: Group on Earth Observations
GEOSS: Global Earth Observation Systems of Systems
GJU: Galileo Joint Undertaking

Glonass: Russia's Global Navigation Satellite System
GMD: Ground-Based Midcourse Defense System
GMES: Global Monitoring for Environment and Security
GNP: Gross National Product
GOC: Galileo Operating Company
GOSPS: The group on the strategic directions of Defence Space Policy
GPS: Global Positioning System
GRACE: Gravity Recovery and Climate Experiment
GSA: European GNSS Supervisory Authority
GSC: Guiana Space Centre
GSLV: Geosynchronous Satelite Launch Vehicle
GTO: Geostationary Transfer Orbit
GUF: General University Funds

**H**
HSPG: High-level Space Policy Group
HTC: Houston Technology Center

**I**
IADC: Inter-Agency Space Debris Coordination Committee
IASF: Israel Air and Space Force's
IAU: International Astronomical Union's
ICT: Information and Communication Technology
IEOS: U.S. Integrated Earth Observation System
IGC: Intergovernmental Conference
IGOS: Integrated Global Observing Strategy
IGS: Information Gathering Satellites
IGY: International Geophysical Year
IHY: International Heliophysical Year
IJPS: International Joint Polar System
IKI: Russian Space Research Institute
ILS: International Launch Services
INF: Intermediate Range Nuclear Forces
IOV: In-Orbit Validation
IPCC: Intergovernmental Panel on Climate Change
IPY: International Polar Year
IRNSS: Indian Regional Navigation Satellite System
ISAS: Institute of Space and Astronautical Science
ISR: Intelligence, Surveillance and Reconnaissance
ISRO: Indian Space Research Organisation

ISS: International Space Station
ISS: Information Satellite Systems
ISTAR : Intelligence, Surveillance, Target Acquisition, Reconnaissance
IT: Information Technology
ITAR: International Traffic in Arms Regulations
ITU: International Telecommunication Union

**J**
JAXA: Japan Aerospace Exploration Agency
JDA: Japanese Defence Agency
JPO: Japanese Patent Office
JWST: James Webb Space Telescope

**K**
KARI: Korea Aerospace Research Institute

**L**
LCROSS: Lunar Crater Observation and Sensing Satellite
LCT: Laser Communication Terminals
LDC: Least Developed Countries
LDP: Liberal Democrat Party
LEO: Low Earth Orbit
LISA: Laser Interferometer Space Antenna
LLC: Lunar Lander Challenge
LMCLS: Lockheed Martin Commercial Launch Services
LMCS: Land Monitoring Core Service
LRO: Lunar Reconnaissance Orbiter
LTTE: Liberation Tigers of Tamil Eelam

**M**
MAD: Mutual Assured Destruction
MDA: U.S. Missile Defense Agency
METI: Japanese Ministry of Economy, Trade and Industry
MEXT: Japanese Ministry for Education, Culture, Sports, Science and Technology
MFA: Ministry of Foreign Affairs
MHI: Mitsubishi Heavy Industries
MoE: Japanese Ministry of Education and Culture
MoU: Memorandum of Understanding
MRO: Mars Reconnaissance Orbiter

MSG: Meteosat Second Generation
MSS: Mobile Satellite Services
MUSIS: MUltinational Space-based Imaging System for surveillance, reconnaissance and observation

**N**

NABS: Nomenclature for the Analysis and Comparison of Scientific Programmes and Budgets
NAL: National Aerospace Laboratory of Japan
NASA: U.S. National Aeronautics and Space Administration
NATO: North Atlantic Treaty Organization
NDU: National Defense University
NEC: Network-Enabled Capabilities
NEO: Near-Earth objects
NERC: Natural Environment Research Council of UK
NFIRE: Near Field Infrared Experiment
NGA: U.S. National Geospatial-Intelligence Agency
NGO: Non-Governmental Organisations
NMD: National Missile Defense
NOAA: National Oceanic and Atmospheric Administration
NPOESS: U.S. National Polar-orbiting Operational Environment Satellite System
NRO: U.S. National Reconnaissance Office

**O**

OCS: Offensive Counterspace operations
OECD: Organisation for Economic Cooperation and Development
OMPS: Ozone Mapping and Profiler Suite
OPECST: Parliamentary Office for Scientific and Technological Assessment of the Assembly and Senate (French acronym)
OSI: Office of Science and Innovation of UK
OST: Outer Space Treaty
OSTP: White House Office of Science and Technology Policy

**P**

PAROS: Prevention of an Arms Race in Outer Space
PDA: Personal Digital Assistants
PFI: Private Financing Initiative
PLA: People's Liberation Army
PNT: Position, Navigation and Timing

POES: Polar Operational Environmental Satellite
PPP: Public-Private-Partnership
PRC: People's Republic of China
PRS: Public Regulated Service
PSLV: Polar Satellite Launch Vehicle

## Q

QDR: Quadrennial Defense Review
QMV: Qualified Majority Voting
QZSS: Quazi Zenith Satellite System

## R

RAIDRS: Rapid Attack Identification Detection and Reporting System) made for on-board threat analysis and reporting
RAST: Recent Advances in Space Technologies
R&D: Research and Development
RLV: Reusable Launch Vehicle
RMA: Revolution in the Military Affairs
RpK: RocketplaneKistler

## S

SAC: Japanese Space Activities Committee
SBI: Space-Based Interceptors
SBIR: Small Business Innovation Research
SBIRS: Space-Based Infrared System
SCSD: Special Committee on Space Development of Japan
SELENE: SELenological and ENgineering Explorer
SIPRI: Stockholm International Peace Research Institute
SMESE: SMall Explorer for Solar Eruptions
SOHO: ESA's Solar and Heliospheric Observatory
SOFIA U.S.-German Stratospheric Observatory or Infrared Astronomy
SPIDER: Space-based Information for Disaster Management and Emergency Response
SPM: Summaries for Policymakers
SSA: Space Situational Awareness
SSN: United States Space Surveillance
SSTL: Surrey Satellite Technology Ltd.
S&T: Science and Technology
STA: Science and Technology Agency

STEREO: Solar TErrestrial RElations Observatory

**T**

TCBM: Transparency and Confidence-building Measures
TEL: Transporter-Erector Launcher
THAAD: Terminal High-Altitude Area Defense
TSIS: Total Solar Irradiance Sensor
TTE Council: EU Transport, Telecommunications and Energy Council

**U**

ULA: United Launch Alliance
UK: United Kingdom
UMP: Union pour un Mouvement Populaire
UN: United Nations
UNCoPUOS: United Nations Committee on the Peaceful Uses of Outer Space
UNEP: United Nations Environment Programme
UNESCO: United Nations Educational, Scientific and Cultural Organization
UNFCCC: United Nations Framework Convention on Climate Change
UNGA: United Nations General Assembly
UNISPACE III: Third United Nations Conference on the Exploration and Peaceful Uses of Outer Space
UNOOSA: United Nations Office for Outer Space Affairs
U.S.: United States
USAF: U.S. Air Force
USAT: Ultra Small Aperture Terminals
USPTO: U.S. Patent and Trademark Office
USSSN: U.S. Space Surveillance Network

**V**

VAST Committee: Committee for the Evaluation of Scientific and Technological Options
VSAT: Very Small Aperture Terminals
VTB: VneshTorgBank

**W**

WMO: World Meteorological Organization
WSSD: World Summit on Sustainable Development
WTO: World Trade Organisation
WWW: World Weather Watch

# List of figures and tables

## Figures

### Part 1: The Year in Space 2006/2007

## Part 2: Views and Insights

# PART 1

# THE YEAR IN SPACE 2006/2007

2006 by the Japanese parliament, succeeding Junichiro Koizumi who had been Prime Minister since 2001.

In recent years, Japan has also evolved in recent years towards a bigger role in peacekeeping and security, as demonstrated by its contribution of non-combatant troops to the Iraq War. This involvement in the Iraq war marked the first overseas use of its military since World War II. Japan's shift in policy vis-à-vis international security issues has evolved rapidly following North Korea's outburst on 5 July 2006, when the regime of Kim Jong Il fired off seven missiles, including a long-range ballistic Taepodong-2 into neighbouring seas. After the firings, Japan immediately called an emergency meeting of the United Nations Security Council. This missile crisis as well as the North Korean nuclear test of a sub-kiloton device on 9 October 2006 have accelerated the debates and spurred major internal discussions about Japan's pacific stance and the ability to defend itself in case of attacks. In particular, the on-going internal debate consider a modification of the constitution, as it prohibits the use of military force to wage war against other countries,[4] and thus influences the debate on forthcoming Japanese space policy.

### 1.2.5. China

Since the initiation of economic reforms in 1979, China has become one of the world's fastest-growing economies and is increasingly driving not only economic growth in East Asia, but also that of the global economy.

Already a commercial giant, China has also increased the level of its diplomatic involvement. China is increasingly opening to the world in terms of foreign policy, and in addition to traditional diplomatic visits to foreign countries and to international fora it has been increasingly using its "next superpower status" to become a major actor in all domains of international affairs. In particular, China invited leaders from 48 countries to the first Forum On China-Africa Cooperation (FOCAC) held in Beijing in 3–5 November 2006. This Summit, the first of its kind, aimed to boost the development of bilateral relations and the establishment of a new Chinese-African strategic partnership. China's principal interest in the continent is access to natural resources, but also to find new markets for its expanding industries. While China is seeking the transfer of technology and know-how in its relations with developed countries, China is also pursuing greater "South-South" cooperation in S&T, including space activities. However, in contrast to its relations with developed countries, here China is not a receiver, but a donor of technology and uses this as a tool of foreign policy.

### 1.2.6. India

India's economy has grown by an annual average of about 8% for the past three years and is increasingly establishing itself as a dominant actor of the future. The year 2006 was a major turning point in India's economic development, as Indian companies started to take over major foreign companies such as India's Mittal Steel acquisition of the French/Luxemburg Arcelor.

With the rise of India, major countries are now pressing to enter into relations with it and to reach deals to have access to the world's second-most populated country. For instance, the U.S. recently reversed its longstanding policy of limited cooperation with India following India's nuclear test of 1974. In a major policy shift and a reversal of three decades of U.S. non-proliferation policy, President George W. Bush signed a landmark agreement on cooperation in civilian nuclear activities during his visit to New Delhi in early March 2006 (the first by a U.S. President in six years), illustrating the new leverage gained by India in recent years. Space activities are also included in this bilateral cooperation, as illustrated by the U.S.-Indian cooperation on the first Indian lunar mission, Chandrayaan-1.

## 1.3. International security

2006/2007 was marked by several on-going and new conflicts as well as significant military events threatening world peace and stability with three conflict areas claiming international dimension: the Middle East, Afghanistan and Somalia.

Political situations have become more severe in the Middle East, particularly in Iran, Iraq, Palestine territories, Lebanon and Israel. Prospects for peace and stability dimmed significantly in 2006/2007. Gaza, which in August 2005 had celebrated Israel's military withdrawal, found itself on the brink of economic collapse and involved in a military conflict with Israel in summer 2006. Furthermore, another conflict in the Middle East broke out in July and August 2006 in Lebanon and northern Israel between the Israeli military and Hezbollah paramilitary forces. In June 2007, the Palestinian Civil War between Hamas and Fatah intensified, and by 14 June 2007, the Gaza Strip was completely overtaken by Hamas. Consequently, following Hamas' takeover in Gaza, Palestinian Authority Chairman Mahmoud Abbas of Fatah dismissed Hamas from the government and formed a Cabinet based in the West Bank. The on-going Iraq War that started on 20 March 2003 with the United States-led invasion has evolved as an asymmetric warfare with the Iraqi insurgency and a civil war between Sunni and Shia Iraqis that led to an estimated 60 000 Iraqi civilian death in the course of the four-year's conflict. In January 2006, Iran removed UN seals on uranium enrichment equipment, and in April 2006 it announced that it had successfully enriched uranium. Iran's president, Mahmoud Ahmadinejad, has since refused to curb Iran's nuclear programme, which he views as a source of national pride. After the failure of international negotiations and threats, on 9 April 2007, Iran claimed to be enriching uranium on an "industrial scale" at its underground plant at Natanz.

In Asia, the Taliban continued their resurgence in Afghanistan, making 2006 the deadliest year of fighting since the 2001 war. The Taliban now funds their insurgency principally through the drug trade, and in 2006 Afghanistan's opium harvest reached record levels. Taliban militants are now infiltrating southern Afghanistan and attacking Afghan and U.S. troops passing through the mountainous Afghan-Pakistani border from their bases in remote Pakistani tribal regions. On 9 October 2006, North Korea announced that it had conducted its first nuclear test, which was later confirmed by the United States on 16 October 2006. The blast was less than one kiloton, but triggered major international reactions and complaints over this nuclear test.

The Horn of Africa has also witnessed increasing conflicts with combats in Somalia and an escalation of the three-year old civil war in Sudan leading to widespread population displacement causing major humanitarian crises. It is estimated that more than 200 000 have been killed in Darfur and 3.5 million have become refugees since 2003. Furthermore, piracy activities have increased along African coasts, particularly in the Gulf of Guinea and the Indian Ocean, threatening maritime security of major commercial sea routes.

The aforementioned conflicts and crises consequently led to a major expansion of multilateral peacekeeping missions prompted by the United Nations and other multilateral security organizations. However, following the terrorist attacks of 11 September 2001, international terrorism has become a principal security concern of the West. In spring 2007, a series of terrorist bombings in Morocco and Algeria claimed by local Al Qaeda branches reinforced the threat of terrorist attacks in Europe and North America, as shown by the terrorist plots in London and Scotland in late June 2007.

## 1.4. Major highlights in scientific activities and research

Climate change and particularly global warming have being a major topic of debate in 2006/2007, and are increasingly being perceived as a serious global threat that demands an urgent global response. Consequently, a series of high-level reports has been recently released that relayed partly on space-based observation and measurements.

The "Stern Review on the Economics of Climate Change" on the effect of climate change and global warming on the world economy compiled by the economist Sir Nicholas Stern for the government of the United Kingdom's Treasury was released on 30 October 2006. Stern's Review suggests that climate change threatens to be the greatest and widest-ranging market failure ever seen, and it provides prescriptions including environmental taxes to minimize the economic and social disruptions. Using the results from formal economic models, its main conclusions are that 1% of a global Growth Domestic Product (GDP) is required to be invested in order to mitigate the effects of climate change, and that failure to do so could risk a recession worth up to 20% of global GDP by 2050. The report argues also that without action, up to 200 million people could become "environmental" refugees by the middle of the century, as their homes are hit by drought or flood. The Stern Review attracted a great deal of positive attention, but is also often criticized due, among other things, to the failure to acknowledge the scope for the long-term adaptation to possible global warming.

In early 2007, the Intergovernmental Panel on Climate Change (IPCC)[5] published its new Summaries for Policymakers (SPM) of the IPCC Fourth Assessment Report discussing current climate change science.[6]

Its first SPM released on 2 February 2007, in terms somewhat stronger than its predecessor published in 2001, concluded that it is "very likely" that observed changes in climate are human-caused, rather than just "likely".[7] The report indicates also that the warming of the climate system is "unequivocal". This SPM also gave updated estimates for temperature and sea level changes by the end of the century. In particular, the IPCC's range of predictions of the rise in temperature by 2100 has increased from 1.4 to 5.8 °C in the 2001 report to 1.1–6.4 °C in the 2007 report.[8]

The second SPM report on " Climate Change 2007: Impacts, Adaptation and Vulnerability" was presented on 6 April 2007. It described the specific effects of climate changes and described options for limiting risks. According to the report, some of the changes could be beneficial. Areas that now have cold climates will experience longer growing seasons and a greater variety of crops, as well as becoming more attractive to tourists. Nonetheless, most changes will prove negative in the long run, and most regions are likely to be more harmed than helped by the changes. The damages caused by global warming are expected to range from worsening floods and water shortages in the developing world to damages to rich countries' wild life as well as to infrastructure, such as road and rail networks, water and energy systems, and healthcare.

The SPM on "Mitigation of Climate Change" was released on 4 May 2007. This SPM report focuses on the mitigation of climate change by limiting or preventing greenhouse gas emissions and enhancing activities that remove them from the atmosphere. It analysed mitigation options for the main sectors in the near-term, addressing also cross-sectoral matters such as synergies, co-benefits and trade-offs and also provided information on long-term mitigation strategies for various concentration stabilization levels.

In 2007, two year-long scientific programmes relevant to space affairs started. On 1 March 2007, the international scientific community launched the International Polar Year (IPY) that is a large scientific programme focused on studying both the Arctic and Antarctic from March 2007 to March 2009. This initiative constitutes the most intensive period of research on the Polar Regions in half a century.[9] Because of the remoteness and harshness of these regions, and because infrastructure is sparse, space-based assets are particularly beneficial to support this frontier research. And, for the first time during an IPY, the scientific community will have satellite data on the polar regions at its disposal. Another year-long scientific programme that started in 2007 is the UN-sponsored, but scientifically-driven, International Heliophysical Year (IHY) which is an international programme of scientific cooperation aiming to understand external drivers of planetary environments and universal processes in solar-terrestrial-planetary-heliospheric physics. The IHY has been planned to coincide with the 50$^{th}$ anniversary of the International Geophysical Year (IGY), one of the most successful international science programme of all time that also initiated the "Space Age" with the successful launch of Sputnik I on 4 October 1957.

Health issues and particularly avian flu pandemics were also a major topic of concern in the last months. Since the H5N1[10] strain emerged in China's Guangdong province in 1996, the number of cases among humans is rising, and by the end of 2006, the number of human deaths from the disease had more than doubled in a year, with a noticeably higher mortality rate of almost 60%.[11] In addition, about 300 million birds around the world have died or been culled. Avian flu outbreaks in Europe and Africa, and particularly, the first human deaths from H5N1 outside Asia (in Nigeria) in January 2006 heightened concern about the spread of the disease leading to the fear of a potential pandemic influenza that could be similar to the 1918–1919 Spanish flu.[12] The geographical extension of the avian flu has consequently prompted massive global attention on possible prevention measures, with the U.S., the EU and countries such as China and Japan committing hefty financial and human resources to combating the disease.

Energy also was one of the major issues in 2006/2007. Rising oil and gas prices, increasing global dependence on a few suppliers, the fear of shortages of existing reserves and the urgency of global warming issues have rekindled the debate on the need for alternative sources of energy, particularly so-called "green energy".

## 1.5. Main science and technology indicators relevant to space activities

Several science and technology (S&T) indicators provide useful information on the global space sector and its evolution. The approach taken in this section is to provide recent information on indicators that represent the overall S&T environment and context in which the European space sector operates. In particular, two sets of indicators are looked at: research and development (R&D) spending and patents filing. Commitment to R&D spending is one of the main tools for measuring the input of innovative activity of a country and illustrates how much priority governments are giving to the public funding of innovative activities and thus activities linked directly and indirectly to space. Patents' filing on the other hand, is a tool for measuring the outputs of innovative activities particularly the efficacy of the financial investments. These two S&T indicators therefore provide valuable insights worth looking at to assess the dynamism and sustainability of the space sector.

### 1.5.1. Science and technology inputs

S&T indicators related to R&D spending are of particular relevance for the space sector, as space budgets depend heavily on overall R&D investment.

R&D expenditure in the EU-27 rose by 1.5% in real terms on average per year between 2001 and 2005, compared to 1.7% in the United States and 2% in Japan between 2001 and 2004,[13] and 19% in China since 1995.[14] According to OECD projections in 2006, the United States is predicted to remain the world's leading investor in R&D, spending just over 330 billion U.S. dollars (about 254 billion euros) for R&D activities largely ahead of Europe, whereby the EU-15 is predicted to spend just over 230 billion U.S. dollars[15] (about 177 billion euros) and Japan about 130 billion U.S. dollars (about 100 billion euros).

According to Eurostat preliminary results, R&D intensity (i.e. expenditure as a percentage of GDP) in the EU-27 stood at 1.84% in 2005, the same level as in 2004.[16] R&D intensity remained therefore significantly lower in the EU-27 than in other major economies. In 2004, R&D expenditure was 2.68% of GDP in the United States, 3.18% in Japan, while it reached 1.34% in China in 2005.[17]

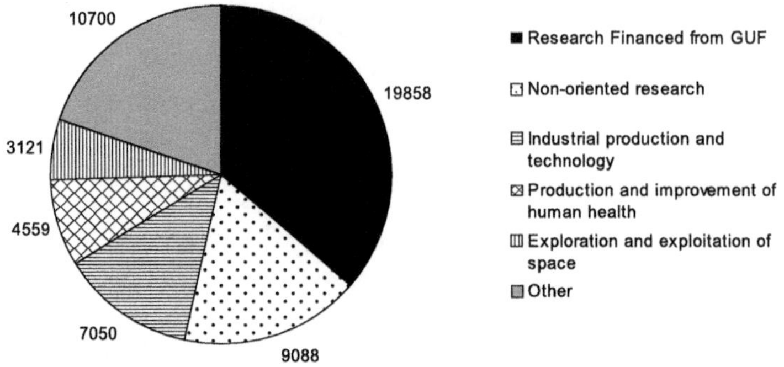


**Fig. 1:** *Distribution of civil GBAORD by main NABS socioeconomic objectives in million constant 1995 PPSS in EU-25 in 2005 (source Eurostat).*

In 2005, Government budget appropriations or outlays allocated to research and development (GBAORD),[18] expressed as a percentage of GDP, amounted to 1.06%, 0.74% and 0.71% for the United States, the EU-25 and Japan respectively.[19,20]

When looking at civil GBAORD broken down by main socio-economic objectives according to the NABS (nomenclature for the analysis and comparison of scientific programmes and budgets), in 2005, the main EU-25 civil socio-economic objective was "Research financed from General University Funds (GUF)"; it represented more than one-third (36%) of the EU-25's total civil GBAORD[21] (Figure 1). This category was followed by "Non-oriented research" and "Industrial production and technology" with respectively 17% and 13% of total civil GBAORD (Figure 1). "Exploration and exploitation of space" accounted for a substantial part of the EU-25's total GBAORD (6%). All other objectives showed a share of less than 5% and are regrouped in the category "Other"[22] (Figure 1).

In 2005, "Research financed from GUF" like in Europe was also the main socio-economic objective in Japan, with the same share as in the EU-25 (36%) (Figure 2). Japan's second main civil socio-economic objective of the government R&D budget was "Production, distribution and rational utilisation of energy" (18%) (Figure 2). As in the EU-25, the objectives "Non-oriented research", "Industrial production and technology" and "Exploration and exploitation of space" were among Japan's main objectives (Figure 2).

The largest part of the civil government budget in the United States in 2005 was devoted to "Protection and improvement of human health" (Figure 3). It represented more than half (53%) the total civil GBAORD.[23] Two other objectives took a share of over 10% of the total civil GBAORD in the United States: 'Exploration and exploitation of space' (18%) and 'Non-oriented research'

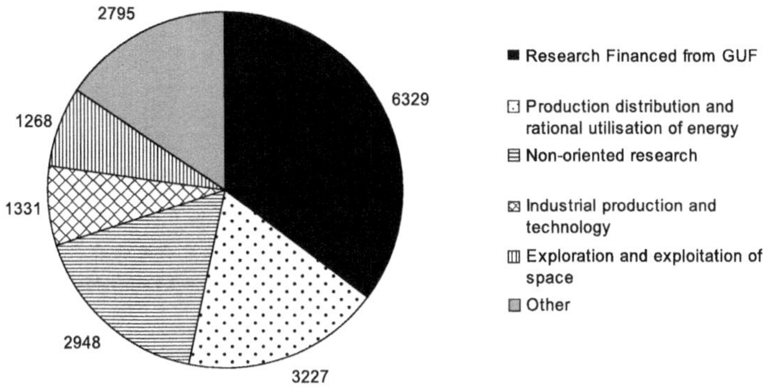


**Fig. 2:** *Distribution of civil GBAORD by main NABS socioeconomic objectives in million constant 1995 PPSS in Japan in 2005 (source Eurostat).*

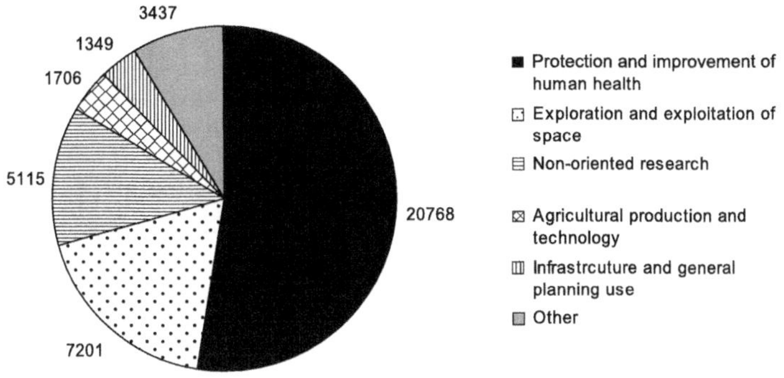


**Fig. 3:** *Distribution of civil GBAORD by main NABS socioeconomic objectives in million constant 1995 PPSS in the United States in 2005 (source Eurostat).*

(13%) (Figure 3). All other objectives had shares of less than 5% of total civil GBAORD. So, overall, government budgets allocated to R&D financed less-diversified research in the United States than in the EU-25 and in Japan.

### 1.5.2. Science and technology outputs

While R&D spending is one of the major inputs to assess the willingness to foster the development of S&T activities, patents filings are a major output metrics for assessing the results of R&D investment and reflect the inventive activity of an entity and its capacity to exploit knowledge and translate it in economic profit.[24] However, not all innovations are subjected to patenting processes (i.e. military

**Tab. 1:** *Patent applications to the EPO in EU-27 and selected countries in 1998, 2003 and Annual Average Growth Rate (source Eurostat).*

| Country | Total Number | | Annual Average Growth Rate 1998–2003 |
|---|---|---|---|
| | 1998 | 2003 | |
| EU-27 | 51194 | 62250 | 4.0 |
| China | 348 | 1898 | 40.4 |
| Canada | 1931 | 2736 | 7.2 |
| Israel | 1088 | 1587 | 7.8 |
| India | 152 | 1003 | 45.8 |
| Japan | 17243 | 27987 | 10.2 |
| South Korea | 1227 | 5400 | 34.5 |
| Russia | 534 | 641 | 3.7 |
| USA | 38345 | 48786 | 4.9 |

research) and therefore certain research fields may be characterized under-representation due to the secrecy of defence research R&D activities (e.g. military space activities).

A common approach in S&T evaluation is to calculate patent indicators based on information coming from a particular patent office. Table 2 gives an overview of patent applications to the European Patent Office (EPO) in 2003 from all EU Member States and several other countries. When looking at absolute EPO patent applications, 62250 patent applications were filed in 2003 by EU Member States, compared to 48786 from the United States. and 27987 from Japan, with China and India developing rapidly (Table 1).[25]

Patent applications from the EU-27 to the EPO grew at an annual average growth rate of 4% between 1998 and 2003, slightly less that the U.S. (almost 5%). Over the same period, very high growth rates were observed in applications from Asian countries, ranging from 10% for Japan to almost 46% for India (Table 1).

When looking at the U.S. Patent and Trademark Office (USPTO) data, 87116 patents granted by the USPTO came from inventors residing in the United States, 32178 from Japanese residents and 24733 from European residents.[26] These figures, like in the case of the EPO data clearly show that there is a "home country advantage bias" where domestic applicants tend to file more patents in their home country.

Triadic patent families that are the patents taken at the EPO, the Japanese Patent Office (JPO), and the USPTO are consequently a good reference of the invention potential of a country outside its domestic advantage. Therefore, they allow the improvement the international comparability of patent-based indicators, as only patents applied for in the same set of countries are looked at. Furthermore, these triadic patents families are often of higher value, because extending the protection of innovation in other patent offices incurs additional costs and time. Therefore, when using the triadic patent families as an indicator, it appears that the United States led in 2000, but Japan took over second place from EU-25 which followed very close behind.[27] The figures in Table 2 show that U.S.'s share of triadic patent families was stable over the period 1990–2000 and that Japan's share grew over the last few years, whereas the EU-25 lost ground.

When looking at the space sector, according to the 2006 OECD Compendium of Patent Statistics, 2267 space-related patent applications published by the EPO between 1980 and 2004 were identified, as well as 2573 patents granted by the USPTO.[29] The majority of inventions filed at these patent offices (about 98%) originated from inventors based in OECD countries with the United States,

**Tab. 2:** *Distribution of triadic patent families, EU-25, Japan and United States, by priority years 1990, 1995, 2000 as % of total (source Eurostat).*[28]

| | 1990 | 1995 | 2000 |
|---|---|---|---|
| EU-25 | 30.5 | 32.5 | 27.1 |
| Japan | 30.5 | 26.9 | 32.2 |
| USA | 34.0 | 34.4 | 34.0 |

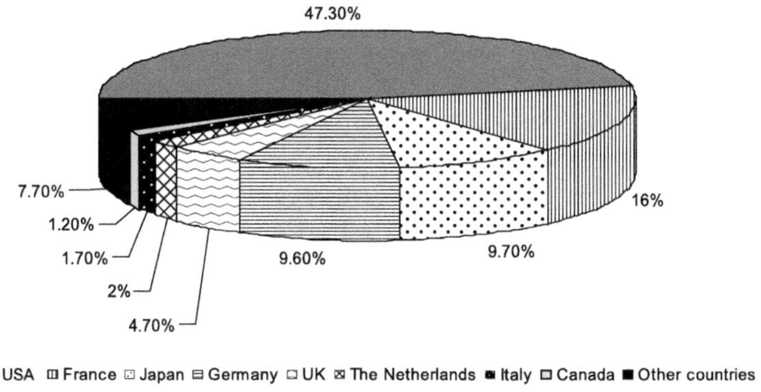


**Fig. 4:** *Share of countries in space-related patents EPO 1980–2004 (source OECD).*

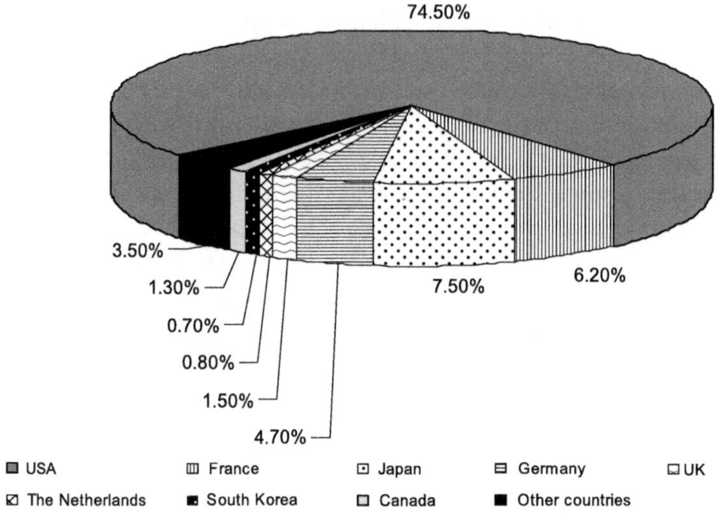


**Fig. 5:** *Share of countries in space-related patents USPTO 1980–2004 (source OECD).*

France, Germany and Japan being the main sources of space-related patents filed with the EPO and the USPTO (Figures 4 and 5).[30]

## 2. Worldwide space policies and strategies

Major space-faring countries have been particularly active in 2006/2007. This period was marked by the increased involvement of China and India in space, a renewal of Russia's space ambitions, a policy-driven transition in Europe and Japan, as well as the release of a new space policy in the United States. The increasing internationalization of space activities was also a striking feature of this period with several new comers laying the foundations of future activities and plans, as well as on-going regional space cooperation plans in Asia and the Americas gaining momentum.

### 2.1. The United Nations system

In 2006, at the 61st plenary session of the United Nations General Assembly (UNGA) two UNGA resolutions pertaining to space security were passed with an overwhelming majority: the annual Prevention of an Arms Race in Outer Space

(PAROS) resolution, A/RES/61/58 and a Russian initiative, "Transparency and confidence-building in outer space activities," A/RES/61/75.[31] Those two resolutions were opposed only by the United States, while Israel abstained from both, signifying therefore the strong international support for those resolutions. On 15 January 2007, the UNGA resolution on "International cooperation in the peaceful uses of outer space" was adopted (A/RES/61/111).

Recognizing that space technology and its applications can play a vital role in supporting disaster relief operations in its resolution (A/RES/61/110) of 14 December 2006, the UNGA agreed to establish the "United Nations Platform for Space-based Information for Disaster Management and Emergency Response" (SPIDER) as a new UN programme. SPIDER's mission statement is: "Ensure that all countries have access to and develop the capacity to use all types of space-based information to support the full disaster management cycle". In contrast to recent initiatives that have contributed to making space technologies available for humanitarian and emergency response, such as the International Charter "Space and Major Disasters",[32] SPIDER aims to ensure access and the use of such solutions during all phases of the disaster, including the risk reduction phase which will significantly contribute to an increasing reduction in loss of lives and property. SPIDER will focus on being a gateway to space information for disaster management support, as well as serving as a bridge to connect the disaster management and space communities, and being a facilitator of capacity-building and institutional strengthening, in particular, for developing countries. It will be implemented as an open network of providers of space-based solutions to support disaster management activities by the United Nations Office for Outer Space Affairs (UNOOSA), as well as SPIDER offices in Beijing and Bonn, and a liaison office in Geneva.

As at 1 January 2007, the status of the five UN treaties on outer space was as follows: the Treaty on Principles Governing the Activities of States in the Exploration and Use of Outer Space, including the Moon and Other Celestial Bodies had 98 States parties and had been signed by 27 additional States; the Agreement on the Rescue of Astronauts, the Return of Astronauts and the Return of Objects Launched into Outer Space had 89 States parties and had been signed by 24 additional States; the Convention on International Liability for Damage Caused by Space Objects had 84 States parties and had been signed by 24 additional States; the Convention on Registration of Objects Launched into Outer Space had 49 States parties and had been signed by 4 additional States; the Agreement Governing the Activities of States on the Moon and Other Celestial Bodies had 13 States parties and had been signed by 4 additional States. In particular, Algeria ratified the 1972 Liability Convention, Brazil acceded to the 1975 Registration Convention, Lebanon acceded to the 1975 Registration

Convention and the 1979 Moon Agreement and Turkey acceded to both the 1968 Rescue Agreement and the 1975 Registration Convention.

### 2.1.1. United Nations General Assembly committees

Two UNGA committees were particularly involved in space affairs in 2006/2007: the First Committee for Disarmament and International Security and the Committee on the Peaceful Uses of Outer Space (COPUOS).[33]

In June 2006 and June 2007 at the 49th and 50th COPUOS sessions, the main agenda items were: ways and means of maintaining outer space for peaceful purposes, the implementation of the recommendations of the Third United Nations Conference on the Exploration and Peaceful Uses of Outer Space (UNISPACE III) as well as the report of the activities of the Scientific and Technical Subcommittee, and the Legal Subcommittee. Review of the benefits of space technology spin-offs as well as the links between space and society, and space and water and the recommendations of the World Summit on the Information Society were also major topics of discussions.

At the First Committee for Disarmament and International Security's session in October 2006 there was overwhelming consensus on the need to preserve outer space for peaceful and cooperative uses. The majority of states recognized that the key threat to preserving outer space is the likelihood of its weaponization and a subsequent arms race. Several states called consequently for further substantive debate and negotiations on a comprehensive, legally-binding PAROS treaty in the Conference on Disarmament (CD) and for the reestablishment of a PAROS Ad Hoc Committee. Two draft resolutions regarding space security issues were presented and adopted on 25 October 2006: the annual draft resolution on PAROS A/C.1/61/L.10/Rev.1 (A/RES/61/58), and a Russian initiative, "Transparency and Confidence-Building Measures (CBMs) in Outer Space Activities" A/C.1/61/L.36 (A/RES/61/75).

### 2.1.2. Other United Nations bodies and organs

Besides the UNGA and its committees, there are other UN programmes, specialized UN agencies and other organs engaged in activities relevant to space primarily in the fields of space applications. In 2006, a new UN body went operational, the International Committee on Global Navigation Satellite Systems (ICG). It was established on a voluntary basis on December 2005 as an informal body for the purpose of promoting cooperation, as appropriate, on matters of mutual interest related to civil satellite-based positioning, navigation, timing, and value-added services, as well as compatibility and interoperability among the GNSS systems. Its first meeting was held on 1–2 November 2006 to review and discuss Global Navigation Satellite Systems and their potential applications.

Several specialized agencies of the UN were also active in space. The two most important were the United Nations Educational, Scientific and Cultural Organization (UNESCO) and the International Telecommunication Union (ITU).[34] The UNESCO presented on July 2006 its Space Education Programme at the 5th Space Conference of the Americas, held in Ecuador. Then, the 2nd International Conference on Remote Sensing in Archaeology sponsored by the UNESCO took place on December 2006. The theme of the conference was "From Space to Place" and focused on the study and conservation of archaeological and ancient landscapes, including UNESCO World Heritage Sites through integrated technologies and virtual reality, as well as remote sensing techniques. Finally, in March 2007, the UNESCO participated in the organisation of the "50 Years of Space Age" celebration (together with IAF, COSPAR, COPUOS, IAA, IISL). The ITU was particularly active in the field of disaster response. In June 2006, the ITU organized together with the UN Office for the Coordination of Humanitarian Affairs the International Conference on Emergency Communications. In December 2006, it also organised together with the United Nations Economic and Social Commission for Asia and the Pacific (ESCAP) a regional workshop on disaster communications were participants discussed technical, policy and institutional issues in the development of networks, systems and possible regional cooperation mechanisms for communications supporting disaster management, with an emphasis on emergency situations for countries in the region.

There is also a formal mechanism to coordinate the activities of all related UN bodies and agencies. The United Nations Coordination of Outer Space Activities convenes on an annual basis to discuss current and future activities, emerging technologies of interest and other related matters through the Inter-Agency Meeting on Outer Space Activities with 27 UN-related organizations taking part.[35] The last meeting (27th session) was held in January 2007 and issued a report on its deliberations for the consideration of the COPUOS, and produced a report on the coordinated space-related activities of the UN system. It covered the current and future plans of common interest, including consideration of how the activities of organizations of the UN system in the area of space science and technology and its applications relate to their mandated programmes. In June 2006, the Inter-

Agency Meeting on Outer Space Activities released also a document entitled "Space Technology and Sustainable Development", which lists all space-related initiatives and programmes carried out by Member States of the COPUOS and within the UN system that respond to specific recommendations contained in the Johannesburg Plan of Implementation of the World Summit on Sustainable Development (WSSD).

## 2.2. The Group on Earth Observations (GEO)

The intergovernmental Group on Earth Observations (GEO) was formally established at the Third Earth Observation Summit in February 2005, hosted by the EC, to carry out the Global Earth Observation Systems of Systems (GEOSS) 10-Year Implementation Plan. GEO membership increased significantly, and now includes 71 member countries, the European Commission and 46 participating organizations.[36]

In 2006, GEO began the implementation of the GEOSS 10-Year Implementation Plan covering nine societal benefit areas: disaster, health, energy, climate, water, weather, agriculture, ecosystems and biodiversity; and five "cross-cutting" transverse areas: user engagement, architecture, data management, capacity building and outreach. The third annual GEO Plenary Meeting that is the official decision-making forum of GEO was held in 28–29 November 2006 Bonn, Germany. On 30 November 2007, the first GEO Ministerial Summit since GEO was formally created will take place in Cape Town, South Africa and will draw on a number of achievements which are specific to developing countries. The Cape Town Ministerial Summit will also allow reporting on the progress made in the implementation of the GEOSS against the GEO 10-Year Implementation Plan, and engage the commitment of Ministers to endorse the forthcoming Declaration of Cape Town.

## 2.3. Regional cooperation in space activities

Today, the dominant form of institutional organization in the space sector is the national space agency.[37] However, in recent years, side-by-side with the domestic agencies, regional or sub-regional organizations have developed based on the European model of a regional space agency, ESA. And, in 2006/2007, several regional space organization initiatives that started in the early 1990s in the Asia-Pacific and in the Americas gained momentum.

### 2.3.1. The Asia-Pacific Space Cooperation Organization

As early as in 1992, China together with Thailand and Pakistan made the proposal for setting up a multinational organization to promote cooperation in space technology and applications in the Asia-Pacific region. This trio proposed an initiative called the Asia-Pacific Multilateral Cooperation in Space Technology and Applications (AP-MCSTA) to promote cooperation to develop space programmes in the region, and organized a Workshop in November of the same year. Asia-Pacific conferences on multilateral cooperation in space technology and applications have since been held regularly to promote regional exchange and promote space cooperation.[38] In December 2002, a workshop on the institutionalization of the AP-MCSTA was held. At this workshop, the Convention on the Establishment of the Asia-Pacific Space Cooperation Organization (APSCO) was introduced.[39] Then, on 28 October 2005, representatives of China, Bangladesh, Indonesia, Iran, Mongolia, Pakistan, Peru, and Thailand signed the APSCO Convention in Beijing as a precursor to establishing an intergovernmental organization facilitating cooperation in space in the region marking a milestone for the official launch of APSCO. In the framework of APSCO, and to make available to its members space technology and applications, in March 2006, China offered a set of weather data-receiving and broadcasting equipment allowing the receipt free of charge of weather information gathered by the Fengyan meteorological satellites to each of the initial seven signatory countries. On 1 June 2006, Turkey signed the APSCO Convention, and became the ninth member of the organization. Other countries are also expected to join APSCO in the years to come. The APSCO Convention entered into force on 12 October 2006 with the fifth ratification of the Convention marking the official establishment of the Organization (Table 3).

**Tab. 3:** *The five ratifications of APSCO convention that established the organisation.*

| Participant | Ratification |
|---|---|
| China | 30 June 2006 |
| Mongolia | 6 July 2006 |
| Bangladesh | 14 September 2006 |
| Pakistan | 9 October 2006 |
| Peru | 12 October 2006 |

### 2.3.2. The Asia-Pacific Regional Space Agency Forum

Following the creation of APSCO, Japan has been trying to revive the Asia-Pacific Regional Space Agency Forum or APR-SAF that was launched in 1993 and aims to enhance the development of its member's space programmes and enable the exchange of views on future cooperation in space activities in the Asia-Pacific region. Japan, through its space agency, JAXA, is aiming to support Asian countries in various applications programmes, and, in particular, in Earth observation and education programmes. APR-SAF meets annually and the last meeting was held in Jakarta, Indonesia on 6 December 2006. A new project that complements the four pre-existing working group (Communication Satellite Applications, Space Education and Awareness, Earth Observation, International Space Station) was also initiated in early 2006. The Disaster Management Support System in the Asia-Pacific Region or "Sentinel Asia" is a "voluntary and best-efforts-basis initiative" led by the APR-SAF to share disaster information in the Asia-Pacific region on the Digital Asia (Web-GIS) platform, and to make the best use of Earth observation satellites data for disaster management in the Asia-Pacific region.[40]

### 2.3.3. Space Conference of the Americas

Another region that has exhibited willingness to develop and support regional space activities is the Americas. Since the early 1990s, a series of pan-American conferences entitled the "Space Conference of the Americas" also know by its Spanish acronym, CEA (Conferencia Espacial de las Américas) have been initiated as an effort to facilitate dialogue and cooperation on space-related activities in the region, particularly in the fields of space science and space technology. The Space Conference of the Americas serves as a forum at which representatives from the region are able to identify specific areas of mutual interest and to formulate coordinated projects where the use of space technology could yield quantifiable benefits in the near to long term. Five conferences have being organized since 1990 that have enabled greater international mutual understanding and have enhanced regional cooperation. These conferences foster discussions and cohesion on regional space activities and entertain the potential of creating a Latin America Regional Space Agency. The Fifth Americas Space Conference was held in Ecuador in July 2006.[41] This conference demonstrated the common interest in creating a pan-American cooperation structure allowing the creation of continental synergy in terms of space capabilities. In particular, at the end of the 2006 Space Conference of the Americas, participants adopted the "Quito Declaration" that reaffirmed their willingness to strengthen international and regional cooperation in developing space technology. The declaration suggested that each country from Latin America set up a national space agency so as to lay the foundation for the establishment of a regional body for space technology cooperation.

## 2.4. European space activities

The period 2006/2007 was particularly dynamic and symbolic for Europe (defined as the EU and the EC, ESA, Eumetsat and their Member States) in the area of space. In particular, the first formalized European Space Policy was adopted in spring 2007. This European Space Policy has a historic value, as it provides for the first time an EU dimension to space policy developed and implemented since 30 years by ESA Member States collectively or individually.[42]

The first "European Space Policy" was presented on 26 April 2007 as a joint Communication from the EC to the Council and the Parliament and as a proposal from the ESA Director General to the ESA Council.[43] It was also accompanied by an EC Staff Working Paper on the "Preliminary elements for a European Space Programme".[44] Drafted in a continuous process of consultation within the High-level Space Policy Group (HSPG),[45] this proposal for a collective space policy is an important milestone for Europe. This document establishes a comprehensive political framework for the development and exploitation of space technologies and systems in Europe and outlines the strategic guidelines for its future activities in space, defining priorities and key actions.[46] The April 2007 proposal for a European Space Policy presents the European vision for space and its related priorities and objectives, including access to space, space technology applications, industrial policy and international relations.[47] The European Space Policy covers five main sections looking at the strategic mission of, space applications, the foundation aspects of space activities, industrial competitiveness, as well as governance issues. This document is also a proposal for a fully functional European Space Programme[48] that will be a common, inclusive and flexible platform encompassing all activities and measures to be developed at the national and the European level in order to achieve the objectives defined in the overall European Space Policy.

While the Communication on the European Space Policy and its associated Space Programme served as a foundation in the process leading to development of the first European Space Policy, the key element in this process was, however, the Resolution on the European Space Policy adopted unanimously by EU/ESA ministers at the Fourth Space Council on 22 May 2007.[49] This Resolution "welcomes and supports" the joint EC-ESA document on European Space Policy and legitimizes the European space policy by the backing of 29 European governments.[50] Specifically, the Resolution has three parts. The first part entitled "Vision for Europe and General Strategy" looks at the overall strategic motives and directions of Europe in space. It highlights the strategic nature of the space sector contributing to the independence, security and economic development of Europe and recognizes the actual and potential contributions from space activities to support EU policies such as the Lisbon Strategy, Europe's Sustainable Development Strategy, as well as the Common Foreign and Security Policy (CFSP), by among others, provides vital information on critical global issues such as climate change and humanitarian aid. The Resolution acknowledges the standing of Europe as a leading space-faring actor and that Europe remains committed to maintaining its position by strengthened intra-European and international cooperation. The second part, "Further Steps – Programmes and Implementation" looks at the various thematic areas introduced in the 2007 Communication. However, the order of priorities is somewhat different with access to space, the International Space Station and exploration, governance being given a higher priority in the Resolution. The new policy also welcomes ESA and EU efforts to implement large user-oriented projects such as the flagship initiatives GMES and Galileo, with greater attention being devoted to the former, and calls for ensuring sustainable funding for space applications. The Resolution deals also prominently with security and defence issues, and while recognizing the intrinsic dual nature of space activities, it affirms the need to set up a "structured dialogue" with the competent bodies of the Member States and within the EU Second and Third Pillars including the European Defence Agency for optimizing synergies between defence and civil space technologies and programmes. Along the same lines, the Resolution does not preclude the use of GMES and Galileo by military users and therefore recognizes the implicit dual-use nature of the future services proposed by those programmes. Issues related to access to space, the International Space Station and exploration, science and technology, governance, industrial policy and international relations are also considered. In overall terms, the Resolution clearly states the strategic importance of space for Europe in demonstrating its independence and its readiness to assume global responsibilities. Finally, the third part is dedicated to key issues to be considered when implementing the European Space Policy.

### 2.4.1. European Space Agency

In 2006/2007, ESA was involved, as aforementioned, in the drafting process of the first European Space Policy. Nonetheless, ESA's Director General (DG) Jean-

Jacques Dordain prepared jointly with the ESA Directors, parallel to the development of the European Space Policy, a plan called Agenda 2011 in which he sets out the objectives for ESA for the coming years. Building on the Agenda 2007, Agenda 2011 released in October 2006 "aims at defining a common framework of strategic action for achieving wide-ranging objectives of ESA Member States and for adapting the Agency to the new environment".[51] It is intended to be an overall roadmap for all ESA stakeholders, and the plan of actions associated with Agenda 2011 is detailed in the ESA Long-Term Plan 2007–2016 that will be the implementing instrument of this Agenda.[52]

Agenda 2011 aims to provide the overall objectives of ESA for the next five years. In particular, ESA's DG wants the Agency to evolve beyond its current core activities to become a model for underpinning the use of space in the world and specifically in the context of Europe's growing needs.[53] Three priorities driving the actions of ESA are identified:

- Consolidation of steps taken at the December 2005 Ministerial Council towards new discoveries and competitiveness.
- Development and promotion of integrated applications (space and non-space) and integration of the security dimension in the European Space Policy.
- Evolution of ESA.

It also looks at current and potential future programmes, including synergies between the civil and defence services, and gives an overall profile of expenditures of its activities. However, the major element of this document is that it acknowledges and takes into account the evolving nature of European space activities, as well as elements such as the development of the first European Space Policy and the accompanying European Space Programme, but also the increasing importance of the EC as a space actor. It consequently indicates that ESA must evolve and prepare "for a situation where the role of ESA will be embedded in a European Space Policy".[54] Several concrete steps are proposed to allow such an evolution. First, amend the ESA Convention at the next Council Meeting in 2008 in order to improve ESA's effectiveness, but also prepare for an ESA enlargement, as well as increasing institutional relationship with the EU. Secondly, to increase the number of ESA Member States to at least 22 by 2011 (particularly EU Member States). Finally, ESA is to become an Agency of the EU by 2014.

To adapt and transform the Agency to the changing European space context, ESA has seen its partnerships evolve in recent months. On 17 February 2006, following Hungary (2003) and the Czech Republic (November 2003), ESA and Romania signed a European Cooperating State (ECS) Agreement allowing Romania to be able to participate in almost all ESA programmes and activities.[55] Romania was followed on 27 April 2007 by Poland that became the fourth ECS. Finally, on 20 June 2007, ESA and Estonia signed an agreement marking closer cooperation. Estonia is the first of the new EU countries to sign a Cooperation Agreement with ESA and in a second step it intends to become an ECS in the years ahead.

ESA accounted for the largest share of European space expenditure in 2006 representing more than half of total European civil spending on space with about 2.9 billions euros.

ESA Council in June 2006 renewed the mandate of Jean-Jacques Dordain as Director General of ESA for a further four-year term to run until July 2011.

### 2.4.2. European Union

In 2006/2007, the Presidency of the Council of the European Union, which is sometimes informally called the "European Presidency" led to the development of a series of milestones at EU-level.[56] In 2006, Austria and Finland assumed the Presidency of the Council of the EU. In the operational programme for 2006 adopted on 22 December 2005, they listed several issues of importance for space affairs. In particular, under the heading "Strengthening Competitiveness" in the section "Innovation and Enterprise", it was stated that both Presidencies "recognise the important role that space policy can play in Europe in terms of industrial and innovation policy. Work on the further development and implementation of an overall European space policy based on the EC/ESA Framework Agreement will be actively taken forward. Further meetings of the "Space Council" in 2006 will ensure continued progress, taking due account of the space research activities proposed under the 7$^{th}$ Framework Programme and of the implementation of Galileo and GMES". Further along in the document under the heading "Information Society, Energy and Transport, Chemical Policy" in the section on "horizontal issues", there is paragraph dedicated to Galileo calling on the two Presidencies to make greater efforts and devote more attention to topic.[57]

In the first half of 2006, following to the decisions of the third meeting of the Space Council in Brussels in November 2005, and the ESA Council at the ministerial level in Berlin (5–6 December 2005), the Austrian Federal Ministry of Transport, Innovation and Technology considered convening a Conference on GMES as a priority of the Austrian EU Council Presidency. Consequently, a conference entitled "A Market for GMES in Europe and its regions – The Graz Dialogue" was held in Graz on 19–20 April 2006 to look at the time frame, funding and governance of GMES. This conference explored the potential market for GMES services and the applications that would be required, and a "Graz Roadmap for GMES service development" was also adopted. This roadmap called for a user-driven approach to ensure the success and long-term sustainability of GMES service development, and for adequate public investments and leadership, as well as a balanced governance structure. It called also for developing additional fast-track services such as atmosphere monitoring, security and cross-cutting thematic areas like mountain regions. Finally, the role of regions in GMES was mentioned as essential to the definition and use of GMES services. This was a major event in the implementation process of GMES, marking the end of a series of sectoral preparatory events.[58]

In the second half of 2006 under the Finnish Presidency, the negotiations on FP7 were brought to a conclusion allocating 1.43 billion euros for the thematic priority "Space" over 7 years out of about 50 billion euros dedicated to the entire FP7. The EU expenditure by the Commission on space-related activities is concentrated mostly on R&D activities in the context of the FPs rather than on operational programmes.[59] An important development in FP6 (2002–2006) was the fact that, the first time, space activities were included under the thematic priority "Aeronautics and Space". Still, FP7 has seen its space emphasis grow thanks to a dedicated "Space" theme, illustrating the EC's willingness to enhance Europe's industrial competitiveness in space activities. The allocation for space activities in FP7 represents a significant advance compared to FP6 (+1.075 billion euros). Nonetheless, about 85% of this has been earmarked for GMES, which leaves "only" about 200 million euros over seven years for launchers, exploration, technological developments and science projects, etc.[60] For 2007, a total of 88.7 million euros is expected to be committed within the scope of the FP that will focus almost exclusively on GMES-related services (Fast Track Services and access to Earth observation data).[61] Throughout the duration of FP7, an average of 205 million euros is planned to be allocated to space through the space thematic alone. Moreover, in addition to the FP, parts of the Trans-European Network funds are also dedicated to space activities, and particularly to the Galileo programme, with 900 million euros over seven years for space infrastructure[62] and another 70 million euros under the heading Competitiveness and Innovation. Over the period 2007–2013, it is therefore estimated that the EC will spend on average some 340 million euros per year on space activities.

Following the Graz Conference organized by the Austrian Presidency in April 2006, a year later on 17 April, 2007, the German Presidency organized a high-level symposium in Munich entitled "The Way to the European Earth Observation System GMES – The Munich Roadmap" to achieve a European consensus on the way forward for the main GMES long-term issues, namely governance and operational funding. This symposium also unveiled the consensus that has emerged among the GMES stakeholders and the GMES Advisory Council (GAC) reflected in the "Munich Roadmap" that is intended to serve as a basis for the upcoming political decisions on GMES. The document released on this occasion details the envisaged overall architecture of European Earth observation services building on the network of existing capacities. The aspects of a governance and financing scheme to guarantee the long-term sustainability of GMES services operations are also raised to ensure in particular an uninterrupted provision of Fast Track services. Regarding Galileo, on 1 January 2007, the European GNSS Supervisory Authority (GSA) officially took over the tasks previously assigned to the Galileo Joint Undertaking (GJU), which was dismantled at the end of 2006.[63] However, the highlight of the German European presidency was the adoption of the aforementioned first European Space Policy on 22 May 2007 jointly drafted by the EC and ESA in consultation with their respective Member States after an extensive debate involving the main European institutional actors for about two years.

### 2.4.3. Other European institutions

Besides the aforementioned institutions of the Union, other bodies and organs linked to parliamentary structures are active and influential in the area of European space activities. The European Interparliamentary Space Conference (EISC) was the most active in space affairs in 2006/2007. It was established in 1999 as a permanent forum to foster cooperation among European national parliaments to develop a continuing dialogue on space policy issues and support the national governments and European institutions in their efforts to achieve a common European space policy for the maximum benefit of European citizens. It is composed of members of parliaments from Belgium, France, Germany, Italy, Spain and the United Kingdom. The Czech Republic joined the EISC as a permanent member in 2006, and Poland and Romania are expected to join in 2007. Besides the members of national parliamentary groups, representatives of the EC, ESA, national space agencies, industry stakeholders and observers from other countries are also participate in the yearly EISC conferences.

In 2006, Belgium held for the second time the presidency of the EISC with the overarching goal to promote the enlargement of the Conference to Member States of ESA and the EU and other European inter-parliamentary assemblies. In particular, it organised the 8$^{th}$ EISC which took place on 12–14 June 2006 in the Belgian Parliament. Members of national parliaments of the ESA and the

EU Member States, as well as the most important space countries outside Europe and representatives of the different competent international bodies and industry attended this event. This ESIC focused on the developments of the soon-to-be adopted European space policy and the associated European space programme, as well as on space application and exploration issues, space and education, and access to space. A new charter for the EISC was also adopted granting the status of member to the national parliament of Russia, among others.

Italy held the presidency of the EISC in 2007. The VAST Committee (Committee for the Evaluation of Scientific and Technological Options) of the Chamber of Deputies, which is responsible of technological and space issues at the parliamentary level animated and organized a series of events such as parliamentary seminars and national initiatives open to the general public (upon invitation) and to the press including the 9th EISC in October 2007.

### 2.4.4. Eumetsat

2006/2007 was a significant period for the European Organisation for the Exploitation of Meteorological Satellites. First, the agency celebrated its 20th anniversary in July 2006. Then, on 1 October 2006, Eumetsat released a document entitled "Eumetsat Strategy: 2030". This 32-page document presents the strategic framework for future activities which envisions making Eumetsat the leading operational satellite agency for European Earth observation programmes that are consistent with its Convention. Consequently, Eumetsat, while maintaining the priority of operational meteorological and climate services, is considering evolving into new areas and developing new services for the environment covering oceans, atmosphere, land and biosphere and natural disasters to the extent that these are linked to meteorology and climate.

Consistent with recent Eumetsat's aspirations, its core mission of providing operational meteorological observations has been expanded to include satellites in Low Earth Orbits (LEOs) following the successful launch of the first European polar-orbiting satellite (Metop-1) in October 2006. This marks the first step outside of Eumetsat's initial perimeter of data provider from its satellite fleet in Geostationary Earth Orbit (GEO). Metop-1 is the first of three satellites developed under a joint programme being carried out by ESA and Eumetsat which are designed to provide meteorological operational data from polar orbit until 2020. Metop-1 is also the inaugural satellite of the space segment of the Eumetsat Polar System (EPS) designed to collect atmospheric and environmental data to complement the hemispheric survey conducted from geostationary orbit by the Meteosat system. EPS will be operated in coordination with the U.S. Polar Operational Environmental Satellite (POES) system managed by the National Oceanic and Atmospheric Administration (NOAA). Completing Eumetsat's push into new activities, the European meteorological agency is broadening its geographical presence and activities. In addition to the cooperation with the United States in the International Joint Polar System (IJPS), the agency has been active in recent years in the Indian Ocean Region, where it has temporarily stationed a Meteosat spacecraft.

Eumetsat also extended its membership, with Estonia becoming the third Baltic State to join Eumetsat as a cooperating state in December 2006, and in July 2006, Croatia became the latest full member of the European meteorological agency. Eumetsat now has 20 Member States[64] (the same as ESA plus Turkey, Croatia and Slovakia) and 10 Cooperating States (Bulgaria, Estonia, Hungary, Iceland, Latvia, Lithuania, Poland, Romania, Slovenia and the Czech Republic[65]). In 2006, Eumetsat received contributions from its Member and Cooperating States of about 251.9 million euros. Of these funds, 11.5% were allocated to the general budget to cover operating costs and to fund activities required in preparing for possible future programmes. Eumetsat therefore had dedicated 222.93 million euros of its budget to programme-related activities in 2006, with EPS representing the biggest earmark for 2006 followed by the Meteosat Second Generation (MSG) programme.

### 2.4.5. National governments

In addition to the activities of the ESA, EU and Eumetsat, a majority of ESA Member States and other EU-27 countries have a dedicated space agency, space office, or funds allocated to domestic space programme. However, there is an important heterogeneity of public support devoted to space activities in Europe. Most European countries funnel the majority of their investment to ESA, and only four Member States (France, Germany, Italy and the United Kingdom) have a wide-ranging spectrum of national activities and invest substantially in domestic programmes.

#### 2.4.5.1. France

A series of high-level policy documents were released in France in the past months illustrating the sustained support of space at the political level. A report issued on 7

February 2007 by the French Parliamentary Office for the Evaluation of Scientific and Technological Choices, also known as the "Cabal-Revol Report" makes a series of proposals to reinvigorate Europe's civil and military space policy. The "Cabal-Revol Report" is of particular significance as it is a report aiming to revigorate both the French and European space sector. The report argues that Europe is loosing ground to the United States, China, India and Russia due to their growing space budgets and that Europe should act soon to avoid falling too far behind the aforementioned countries. It covers almost all space topics and has an exhaustive list of 50 proposals to keep Europe competitive in space in the future.[66] In February 2007 as well, the French Minister for Defence released a document entitled "Let us Make more Space for our Defence. Strategic Guidelines for a Space Defence Policy in France and Europe". The 29-page un-classified report is based on the work of the group on the strategic directions of Defence Space Policy (GOSPS). The project started in October 2003 and was completed in the fall 2004 with the presentation of the GOSPS conclusions to the Minister of Defence and was finally partially released to the public on 15 February 2007. The un-classified document presents a comprehensive range of proposals aimed at strengthening French military space capabilities that will serve as a reference for forthcoming work, and is intended to be used to promote dialogue and strategic analysis between the civilian, military, industrial and the institutional partners in both France and Europe. This policy document advocates that France should increase its annual military space budget to 650 million euros per year. That would represent a 50% increase from the current budget. It also proposes a Europe-wide effort to increase military space capabilities through reciprocal dependence on nationally-owned space-based military assets. The document demonstrates the growing significance of space, at both the military and political level for both France and Europe. It places emphasis on the role that space should play, as a catalyst in enhancing the effectiveness of defence resources and as a unifier in the emergence of a European defence.

France has the largest national civilian budget in Europe with about 1760 million euros in 2006 devoted to its space agency the Centre National d'Etudes spatiales (CNES), with 742 million euros devoted to ESA and the rest to its national programme. Space utilization was the biggest budget item for CNES in 2006 followed by the development of launchers. With a sound financial basis and a stable budget for the period lasting at least until 2010, CNES is looking for new investments in space such as Jason-3. Nonetheless, not everything is going according to plan. Notably, the dual use of the optical Earth observation satellite Pleiades is delayed 18-months, with the launch of the first unit being expected around 2010.

#### 2.4.5.2. Germany

German Chancellor Angela Merkel has endorsed a multiyear increase in German space spending and in 2006 reiterated the government's support for the International Space Station (ISS). On 2 May 2006, speaking at a ceremony marking the end of development of Europe's Columbus orbital module she said that "in the past years, and even decades, technology has not been appreciated as its fair value, especially in a nation like Germany. This needs to be changed"[67] and "space would be one of five priority areas targeted for the extra spending, along with nano- and information technologies".[68] The German government is thus expected to boost the budget for R&D and innovation by 6 billion euros over the next four years in a bid to increase the proportion of Gross National Product (GNP) spent on research from 2.4% to 3% by the end of the decade.[69] In 2006, Germany allocated an estimated 982 million euros to civilian space activities, and therefore represents the second-largest institutional space spender in Europe. Of this total, 71% were spent on Germany's contribution to ESA and Eumetsat, 17% on the German national space programme led by the Deutsches Zentrum für Luft-und Raumfahrt (DLR), and 12% on R&D in DLR's space business.

In 2007, the DLR founded a new Institute of Space Systems in Bremen on 26 January (the ninth DLR location in Germany). The new Institute will focus its work on systems analysis and technology, and their applications for space systems. Finally, as of 1 March 2007, the DLR had a new chairman in the person of Professor Johann-Dietrich Wörner.

#### 2.4.5.3. Italy

Italy is the third European space power ranked by budget with about 794 million euros devoted to its space agency (Agenzia Spaziale Italiana, ASI). In recent months it has shown ambitions to intensify its space effort. In March 2006, the Italian government approved a move to raise space spending by about 8% over the next three years.[70] This major budget increase shows Italy's determination to expand its activities in the space sector, and its expanding budget trend contrasts with most other European agencies. Two-thirds of the spending is earmarked to fund activities in three areas: Earth observation (which will get 29% of the funds), science (21%) and space transportation (18%).[71] The remainder is targeted at telecommunications, navigation and manned space flight. Most of the money will go to existing programmes, such as the Galileo, the GMES network and the Aurora exploration programme,[72] but the new funding will also enable the agency to undertake a number of new projects.

After a 6-month vacancy following the departure of Professor Sergio Vetrella, Professor Giovanni Bignami, has been appointed in April 2007 for a four-year term as the new president of ASI.

#### 2.4.5.4. United Kingdom

2006 was a particularly active year for space policy in the United Kingdom. First, the British Parliament's Science and Technology Committee conducted what it called "a major and wide-ranging inquiry" into aspects of UK space policy in fall 2006.[73] This inquiry has a twofold approach. Firstly, the Committee has been collecting written evidence from interested organizations and individuals on:

- The impact of current levels of investment on space-related activities on the UK's international competitiveness in this sector;
- The benefits and value for money obtained from participation in the European Space Agency and other international programmes;
- The maximisation of commercial benefits and wealth creation from UK space-based technologies through innovation and knowledge transfer;
- The delivery of public benefits from the space-related activities of different government departments, and the co-ordination of these activities; and
- Support for space-related research and the UK skills base.

Secondly, the Committee has been conducting a series of hearing of experts on the aforementioned issues. The results of this wide parliamentary enquiring are expected in the second half of 2007.[74]

The British government – through the British National Space Centre (BNSC)[75] – also started a 12-week survey to determine whether its current space policy should be modified, especially with respect to global exploration effort. The 40-pages document called "A Consultation on the UK Civil Space Strategy 2007–2010" should help to determine which civil space strategy the UK should pursue over the last four years of the decade. The survey, to be published in the fall 2007, will canvass government agencies, academic and research institutions, industry and the general public. The UK Civil Space Strategy proposes three primary objectives:

- Delivering world-class science by exploiting the UK's space activities and expertise;
- Delivering public benefits in partnership with government bodies and institutions to exploit the full potential of space activities;
- Maximizing the potential for wealth creation from space activities by facilitating a progressive business environment.

The United Kingdom is the last member of the quartet of big European space spenders. Through the BNSC, which coordinates civil space activities of its funding partners, in the fiscal year 2005/2006 it spent, £207 million on space programmes or about 305 million euros. About 65% of this budget was the UK's contribution to ESA projects.

Finally, David Williams has been appointed for a four-year term as Director General of the BNSC in May 2006.

## 2.5. United States

The White House Office of Science and Technology Policy (OSTP) rolled out a 10-page unclassified version of the new U.S. "National Space Policy" on its website on 6 October 2006.[76] This document supersedes the September 1996 version of the policy. The new policy supports not only a moon, Mars and beyond exploration agenda, but also responds to a post 9/11 environment of emerging transnational threats and the increased need for intelligence-gathering internally and externally to the United States. The U.S. National Space Policy[77] consists of 13 sections dealing with national security and commercial, civil and international cooperation, as well as U.S. policy goals and principles. It is a document intended to govern the conduct of American space activities by explaining in detail the roles of various agencies in promoting U.S. space efforts, the most important of these being the Department of Defense (DoD) and the National Aeronautics and Space Administration (NASA). The policy enunciates seven primary principles saying that the United States:

- "Is committed to the exploration and use of outer space by all nations by peaceful purposes, and for the benefits of all humanity"
- "Rejects any claims to sovereignty by any nation over outer space or celestial bodies, or any portion thereof, and rejects any limitations on the fundamental right of the United States to operate in and acquire data from space"
- "Will seek to cooperate with other nations in the peaceful use of outer space to extend the benefits of space, enhance space exploration, and to protect and promote freedom around the world"
- "Considers space systems to have the rights of passage through and operations in space without interference"
- "Considers space capabilities – including the ground and space segment and supporting links" vital to its national interests". To safeguard the U.S. to "dissuade or deter others from either impeding those rights or developing,

capabilities intended to do so; take those actions necessary to protect its space, capabilities; respond to interference; and deny, if necessary, adversaries the use of space, capabilities hostile to U.S. national interests"
- "Will oppose the development of new legal regimes or other restrictions that seek to prohibit or limit U.S. access to or use of space"
- "Is committed to encouraging and facilitating a growing and entrepreneurial U.S. commercial space sector"

Major areas of emphasis that were only touched on by previous presidential space directives include: the development of high quality cadre of space professionals, space acquisition reform, the space industrial base, space situational awareness, intelligence collection, space protection, and interagency collaboration. And new subjects include: homeland security, radiofrequency spectrum access and protection, as well as orbital assignments and the space sensitive technology list.

The U.S. government is by far the largest investor in space programmes, with about 53 billion U.S. dollars spent on space activities, which represents about 81% of total institutional space budgets worldwide with two agencies, the DoD and NASA receiving the largest share of U.S. space funds.

Although NASA asked for almost 16.8 billion U.S. dollars (about 12.92 billion euros) for the 2007 fiscal year and planned accordingly, it didn't see a budget increase under the new Democratic congress.[78] For NASA, it means having only around 16.62 billion U.S. dollars (about 12.78 billion euros). The biggest change for NASA concerns the amount of money it plans to spend on science missions. By holding science spending essentially flat, NASA is able to provide its Exploration System Mission Directorate (ESMD) with the funding needed to roll out its exploration infrastructure. Thus, a number of space science programmes would be delayed or deferred because of budget constraints. Consequently, NASA is facing tough choices with little room to manoeuvre and must therefore adapt to new budgetary realities. In particular, the flat funding could wreck NASA's plans for fielding Orion and Ares by 2014. Nonetheless NASA is still by far the biggest civilian space agency in the world with over four times more than the second-largest civil space budget, namely of ESA.

The DoD requested 22.5 billion U.S. dollars in Fiscal Year 2006[79] (about 17.3 billion euros). It is thus the leading governmental agency investing in space worldwide, but also the main driver for the U.S. space industry. In particular, funding for space programmes continued its upward trend in the Uited State Air Force's (USAF) 2007 budget proposal with a 12.5 billion U.S. dollars request (about 7.5 billion euros) for unclassified space activities (+500 million U.S. dollars from last year or about 384 million euros).[80] The intelligence sector allocated an estimated 12.5 billion U.S. dollars (about 9.7 billion euros) through the National Reconnaissance Office (NRO) and the National Geospatial-Intelligence Agency (NGA). The U.S. government also invested about 430 million U.S. dollars in 2006 (about 330 million euros) for developing a layered defence against ballistic missiles through the Missile Defense Agency (MDA) with the biggest unclassified space programmes being the Space Tracking and Surveillance System, a series of experimental satellites that would track missiles during the midcourse phase of flight, at about 228 million U.S. dollars (or about 175 million euros).[81,82] The U.S. military space budget with about 35.5 billion U.S. dollars (about 27 billion euros) allocated to space is therefore by far the largest military space budget in the world.

## 2.6. Russia

Recent years have been marked by an impressive recovery of the Russian space programme as high oil and other natural resources prices have made it possible to balance and grow the Russian institutional budget. In this context, in October 2005, the Russian government adopted a new Federal Space Programme (2006–2015) that comprise a space spending plan to attempt to halt the decline of the country's industrial base and ending years of under-funding. The new research and procurement strategy calls for the civil space programme to receive about 305 billion rubles (about 8.9 billion euros) for space activities in 2006–2015,[83] as well as 182 billion rubles in private investments (about 5.3 billion euros). This growing budget is supporting new ambitions and partnerships in space. Nonetheless, despite the massive investment in space activities, Russia's national space budget remains small in international comparison.[84]

Among the major items in the 10-year-plan are the development, replenishment and maintenance of orbital space constellations in the interests of the country's socio-economic developments, science and national security (communications, TV broadcasting, Earth observation, hydrometeorology, environment monitoring, emergency situations control, fundamental space research, microgravity space research). The development, deployment and subsequent operation of the upcoming ISS Russian segment to carry out fundamental and applied research, the implementation of the long-term applied research programme and experiments planned for ISS are also perceived as important, and also ensure the operational continuity of the Russian segment of the International Search and Rescue Satellite COSPAS-SARSAT. The development of the replacement of the Soyuz space capsule used to support ISS and its associated infrastructure, as well as the maintenance and development of the Baikonur cosmodrome facilities and the development of internationally competitive rocket technology are also major items of the new Federal Space Programme.

On 29 March 2007, the State Council, an advisory body to the Russian President, met in presence of President Putin in Kaluga, hometown of Konstantin Tsiolkovsky, to discuss aerospace issues. This does point out the importance attached to the space sector by the country's leadership. President Putin indicated that the main objective must be to make better use of the findings of space research in navigation, communications, geology, television, radio broadcasting, medicine, ecology, agriculture, education and many other areas.[85]

## 2.7. Japan

Japan's overall space policy is currently in transition, a new bill was submitted on 20 June 2007 to the Lower House of the Diet by the ruling Liberal Democratic Party (LDP) for the establishment of a "Basic Law for Space Activities".[86] The new law pushes three main elements. Firstly, it proposes to set up a new Ministry for Space and Space Development Headquarters (a forum of user ministries with strong authority); this Minister for Space would reside in the Cabinet Office for coordinating space policies governing civil, military and commercial activities of different ministries. Secondly, the "Basic Law for Space Activities" aims to reconsider the assumption of the "exclusively peaceful purpose" clause in the Diet resolution of 1969 to allow the use of space assets by military bodies.[87] Finally, the third element of the new proposal concerns ways and means to increase the competitiveness of the Japanese industry.

Japan transitioned from its second national technology plan (2002–2006) to the third national technology plan (2007–2011) in April 2007. The new plan mentions space as "stem technologies for the nation" (rocket technology and Earth observation programmes) for the first time ever. Consequently, Japan's Aerospace Exploration Agency's (JAXA) budget in fiscal year 2006[88] was of 180 billion yens (1.142 billion euros) up from 177 billion yens in 2005 (1.12 billion euros). Japan's investment in space focuses almost exclusively on civilian space activities and only the Information Gathering Satellites (IGS) programme receives some "non-civilian" funding.

## 2.8. China

On 12 October 2006, China released its new White Paper entitled "China's Space Activities in 2006" that drives its use of space for the next five years.[89] In China, policies are established mainly through White Papers released by the State Council.[90] White Papers are the most important official policy documents and are released by the Information Office of the China State Council on behalf of the government.[91] In contrast to the "U.S. National Space Policy" issued just a week before, the new White Paper has received little media attention outside of China. Furthermore, besides the different treatment of international cooperation by the U.S. and China in their respective policy documents,[92] the Chinese White Paper has avoided any discussion of developing military space capabilities and has sought to portray its space programme as peacefully driven[93] and stresses the benevolence of its space programme.[94] Furthermore, unlike the U.S. policy, " China's Space Activities in 2006" is a much more retrospective document principally concerned with China's space activities during the period of the 10th Five-Year Plan (2001–2005).

The White Paper consists of five sections dealing with the aims and principles of development; progress made in the past five years; development targets and major tasks for the next five years; development policies and measures; and international exchange and cooperation. The overall aims of China's space activities are listed to be "to explore outer space, and enhance understanding of the Earth and the cosmos; to utilize outer space for peaceful purposes, promote human civilization and social progress, and benefit the whole of mankind; to meet the demands of economic construction, scientific and technological development, national security and social progress; and to raise the scientific quality of the Chinese people, protect China's national interests and rights, and build up the comprehensive national strength".[95] Top priorities involve also developing and operating a high-resolution Earth observation system, a polar and geostationary weather satellite network and a system of small disaster protection spacecraft, along with associated satellite, launcher, ground production and operating facilities. Launcher development will focus on a new non-toxic, low-cost, high-performance rocket family capable of lifting 25 ton to low-Earth orbit and 14-ton to geostationary transfer orbit. Extravehicular activity and rendezvous/docking manoeuvres will be the main thrust of manned missions. Sciences will focus, among other things, on the development and launch of "breeding" satellites to expand the application of space technology in the field of agricultural sciences research, but also to strengthen the ability to monitor the space environment and space debris, and set up a space environment monitoring and warning system.[96]

On 12 February 2007, the Commission of Science Technology and Industry for National Defense (COSTIND) issued its "Eleventh Five Years Space Development Plan", which is a blueprint of the Chinese government on future space scientific development.[97] In particular, the COSTIND puts focus on improving the innovation of space science and its continuous development capability and is scheduled to engage in scientific research and exploration in three fields: space astronomy and solar physics, space physics and solar system exploration, microgravity science and space life science.[98] According to the plan, the overall strategic objectives for the next 15 years include: pioneering exploration and the study of key scientific issues, space and astronomical survey and research, solar-terrestrial space environment survey and solar system exploration, capacity-building for microgravity science and space life science research.[99]

China's space programme budget remains relatively opaque. China's White Paper and other official documents on space activities do not provide budgetary figures, though they state the great prominence of the space programme and its vital contribution to achieve the modernization of the country. A number of unofficial statements on various aspects of China's manned space programme have been made but they vary in dimension and scope and thus provide only vague clues about budgetary numbers. But, as indicated in Washington in April 2006 by Luo Ge, one of two Vice Administrators of the China National Space Administration (CNSA), the figures of China's spending in space is difficult to calculate, but said that it is about 500 million U.S. dollars (about 384 million euros) a year, and for this amount China has some 200 000 full-time space workers.[100]

## 2.9. India

India has been particularly active in space affairs in 2006/2007 despite the failure on 10 July 2006 of its Geosynchronous Satellite Launch Vehicle (GSLV) that destroyed the domestically-built Insat 4-C television broadcasting satellite. Indian leadership still supports its space programme financially and politically. India's main space agency, the Indian Space Research Organisation (ISRO), received a 3 billion rupees funding increase for the 2006/2007 fiscal year. The Total ISRO budget for this period was 30 billion rupees (about 535 million euros). Included in the annual plan were 250 million rupees (4.6 million euros) to begin development work on a new generation of communications satellites in the 4-ton class. This larger class of satellites is being developed to meet India's growing domestic needs and possibly for future export on the international market. ISRO is also developing a new vehicle that will be able to lift these new 4-ton satellites to geostationary transfer orbit and 10-ton into low Earth orbit (GSLV Mk3). Another major driver of the spending hike is the Indian Regional Navigation Satellite System (IRNSS), a 7-8-constellation satellite designed to enhance signals of the U.S. GPS satellite throughout India. The Indian government has approved spending 14.2 billion rupees (253.5 million euros) to develop this independent regional satellite navigation system that would be launched starting in 2008.[101] On 28 February 2007, India unveiled a 38.5 billion rupees (or about 690 million euros) space activities budget for the fiscal year beginning on 1 April 2007. The novelty of this 2007/2008 budget is the fact that significant funds (500 millions rupees or about 9 million euros) for a human spaceflight programme haves been allocated.[102] The budget includes plans to develop a new rocket engine using Kerosene/Liquid Oxygen propellants and money for a new observing satellite in GEO.

India has well-developed space capabilities in the field of Earth observation and telecommunications, and has long since demonstrated the ability to build sophisticated launch vehicles and satellites for national development needs. But ISRO is eager to start a human spaceflight programme, and to autonomously launch its first manned flight by 2014–2015 and land an Indian astronaut on the moon by 2020.[103]

## 2.10. Emerging space powers

In addition to the aforementioned traditional space powers, a variety of new actors have increased their space development activities in the past few months.

In 2006/2007, South Korea was particularly active in the space sector, and based on the strength of its information and telecommunications technologies, it is aiming to become a global leader in space technology by 2015.[104] Kim Woo-Sik, Korean Minister of Science and Technology recently vowed to push science-related projects to develop the nation's space capabilities in earnest. The first Korean satellite (Korea Institute of Technology Satellite-1 or KITSAT-1) was only launched in 1992, but since then, Korea has started numerous initiatives with a remarkable success up to now including the launch of a high-resolution optical satellite on 28 July 2006 (Kompsat-2) aboard a rocket vehicle from Russia's Plesetsk Cosmodrome[105] or the launch on 22 August 2006 of its first explicit dual-use satellite, the Koreasat 5 (Mugunghwa 5) communications satellite. The South Korean government is also overseeing launch-related projects including the development of a domestic launch site. Furthermore, on 25 December 2006, the final two runners in the race to become Korea's first astronaut were chosen by the Ministry of Science and Technology and the aeronautics and space agency of South Korea, the Korea Aerospace Research

Institute (KARI). The final candidate set to travel to ISS aboard a Russian Soyuz in April 2008 will be selected later this year. These developments demonstrate South Korea's ambition to become a space power able to cover all of it strategic needs.

In Israel, major developments related to the way Israel manages and operates in space for its military activities occurred in 2006/2007. In particular, in February 2006, the Israeli government ended years of heated debate by announcing that the Israeli Air Force will be given the lead role in all military activities in space, as well as the responsibility for designing and operating the nation's future satellites. Consequently, the Israel Air and Space Force's (IASF) mission will be to operate in the air and space arena for purposes of defence and deterrence. This change is a major reorganization in the way Israel manages and operates in space. The IASF will now provide space capabilities to all Israeli users and particularly to the intelligence agencies. Israel also successfully launched its newest reconnaissance satellite, Ofeq-7 on 11 June 2007 aboard its indigenous Shavit rocket. This launch was much needed to overcome the September 2004 loss of Ofeq-6 due to a failure of a Shavit launcher and consequently replaces the five-year-old Ofeq-5, which is nearing the end of its operational lifespan.

On 31 July 2006, the South African cabinet approved the establishment of South Africa's space agency, which will be tasked with coordinating the use of space technology and local science research.

In 2006, Brazil also entered the exclusive club of countries having Astronauts with the first Brazilian astronaut Marcus Pontes sent into space in March 2006 aboard the Soyuz TMA-8/12S mission for a 9-day trip to ISS. Furthermore, Brazil is increasingly cooperating with India in a similar fashion and it cooperated with China in the 1990s to foster "South–South" cooperation.

Turkey's Air Force is also planning to spend at least 200 million U.S. dollars (including the procurement, as well as the launch and the insurance of the satellite), or about 154 million euros, to buy and launch Turkey's first military satellite by 2011. This satellite will be an optical reconnaissance satellite with a resolution of 80 centimetres under a programme dubbed GOKTURK.[106] As a showcase to demonstrate the advanced development of Turkey, the 3rd International Conference on Recent Advances in Space Technologies (RAST) entitled "Space for a More Secure World" took place in Istanbul from 14 to 16 June 2007. The conference attracted around 300 participants from almost 30 countries.

Finally, Iran completed the conversion of one of its ballistic missiles, the Shahab missile, into a sounding rocket that it tested to fire a research payload into space on 25 February 2007. While the launch of Iran's first space research rocket does not represent a significant development in Iranian rocket technology per se, this event is rather important and is part of a more ambitious Iranian space programme (Iran hopes also to launch five satellites into orbit by 2010). This drive for space is also part of an overall political push of the Iranian leadership to demonstrate its power projection, as well as Iran's S&T capabilities.

## 3. Worldwide space budgets and revenues

The global size of the space sector including both institutional space budgets and commercial space revenues is estimated by the European Space Policy Institute (ESPI) at some 136 billion euros[107] in 2006. The revenues of the space industry are estimated to have reached about 86 billion euros, with the bulk being generated by satellite services while institutional space budgets (including civil and military budgets) accounted for an estimated 50 billion euros. In overall terms, global space budgets and revenues are increasing from year to year due to the higher institutional investments in military space, on the one hand, and the sustained demand for new applications and services, on the other.

### 3.1. Overview of institutional space budgets

Public spending for space programmes and activities at a global level remained robust in 2006 following the expansion of the U.S. space budget as well as continued growth in space expenditure of the space agencies in Asia and a renewed increase of investments in Russian space activities.

World institutional space expenditures are estimated at about 50 billion euros for the year 2006, representing 36% of the overall space sector. Military and intelligence applications represented the biggest portion of public allocations to space activities with an estimated 56% of the world's institutional space budget. The remaining amount was dedicated to civil space programmes, or about 22 billion euros.

In 2006, the trend observed in recent years of increasing public investment in military space activities has been reinforced with the space military/intelligence institutional budget seeing a greater expansion than the space civil sector. However, it has to be stressed that growth of the U.S. military space budget is creating a distortion of the overall space military sector. Nonetheless, it is important to note that the budget allocated to military/intelligence space activities is certainly underestimated due to the secrecy of defence budgets in general, and particularly as regards those of Russia and China.

Although total space military/intelligence budgets are higher than the total budgets of civil space programmes, the latter are more commonly implemented throughout the world. Moreover, the increasing internationalization of space activities witnessed in recent years is leading to an increasing level of public budget allocated to civilian space activities. However, while the number of countries investing in space is growing, the differences in investment volumes among countries remain high, with the major space-faring countries representing an overwhelming majority of the world's institutional expenditure on space activities (and particularly on military activities). ESPI has identified about 60 countries with national space programmes and activities, but only about half of them are estimated as investing substantial amounts (more than 10 millions U.S. dollars or about 7.7 million euros) per year on their domestic activities.

In general, it has been observed that North America, Europe and Asia are the main regions investing in institutional space activities. In 2006, North America invested an estimated 82% of all public funding in space followed by Europe with 10%, Asia with 5% and ex-U.S.S.R. countries about 2%, while the investments of the rest of the world in public space activities were very limited. However, there is a clear difference in the dynamic when comparing public investment in North America driven by the United States, and the major Asian countries and Russia that are increasing their space efforts, while in Europe institutional investment in space has remained constant.

In 2006, three main space powers dominated the institutional sector concentrating about 95% of world public funding for space activities. The United States with a budget estimated at about 40.8 billion euros is the incontestable leader and driver of space affairs based on its budget. Europe, when considering consolidated budgets with about 6 billion euros is the second institutional actor in the world followed by Japan with about 1.2 billions euros public funding dedicated to institutional space activities.

When looking at individual countries, the United States is by far the biggest investor in space followed distantly by France, Japan,[108] Russia, Germany and Italy that all spent more than 1 billion U.S. in 2006 on public space activities, or about 770 million euros (Figure 6).

This hierarchy has been very stable in recent years. However, actors like Russia, China, India and South Korea are increasingly investing in space and are expected to rapidly catch-up with European space powers like Italy and Germany. But for those emerging economies (Russia, India and China) relying solely on the absolute volume of institutional investment, it could be misleading to assess their national space efforts because of the significant differences in production costs from country to country as well as in the local standard of living and local purchasing power.

When looking at the top 10 space institutions by budget, unsurprisingly, the list is dominated by a series of U.S. governmental agencies, with 5 out of 10 (Table 4). The DoD is the biggest space agency in the world with an estimated budget of

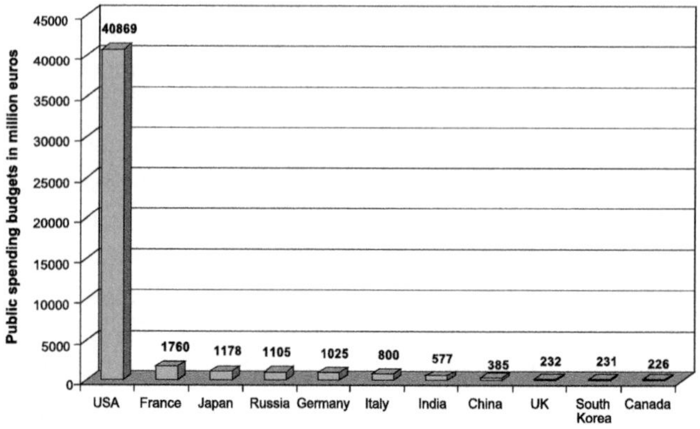


**Fig. 6:** *Estimation of the institutional space budgets in 2006 of the major space powers.*

more than 17 billion euros for 2006 followed by NASA with about 12.7 billion euros. When taken together those two organizations concentrate about 60% of all institutional budgets spent on space in the world. The United States has also two intelligence-related agencies in the top 5, one in charge of developing and operating dedicated space assets, the NRO and another one to exploit the data gathered, the NGA (Table 4). The United States has also another institution is this list, one dedicated to operational meteorology and oceanography, NOAA. ESA is the second-largest civilian space agency in the world after NASA, preceding JAXA, Roskosmos, CNES and ISRO (Table 4).

**Tab. 4:** *Estimation of the top 10 space institutions according to their space budget in 2006.*

| Rank | Agency | Country | Agency type | Value in Billion Euros |
|---|---|---|---|---|
| 1 | DoD | USA | Military | 17.3 |
| 2 | NASA | USA | Civilian | 12.7 |
| 3 | NRO | USA | Military | 7.6 |
| 4 | ESA | Europe | Civilian | 2.9 |
| 5 | NGA | USA | Military | 2.0 |
| 6 | JAXA | Japan | Civilian | 1.2 |
| 7 | Roskosmos | Russia | Civilian | 1.1 |
| 8 | CNES | France | Civilian | 1 |
| 9 | NOAA | USA | Civilian | 0.8 |
| 10 | ISRO | India | Civilian | 0.6 |

## 3.2. Overview of commercial space markets

The overall commercial space sector is estimated at about 86 billion euros in 2006. Satellite-based products and services represent the greatest portion of commercial global revenues, and particularly digital broadcast services (DBS) and ground station and equipment (Table 5).[109]

After few difficult years, the satellite industry has begun to rebound as new applications are driving demand for services such as High Definition Television or online web mapping. Nonetheless, a breakdown of world space industry revenues by sector, i.e., satellite services, launch industry, satellite manufacturing and ground equipment, shows that not every segment experienced growth in 2006 (Figure 7).[110]

Satellite services are becoming increasingly important and are now the major source of revenues for the space sector due mainly to the increasing demand for satellite television and for space-based telecommunication services (including, voice and data). However, the development of new services linked to space-based positioning and new business models in Earth observation with the development of web-based portal using satellites imagery are becoming increasingly important for the space industry in general and for satellite services in particular. In broad terms, the satellite services market is made up of three sectors: Direct Broadcast Services (DBS), the Fixed Satellite Services (FSS), and Mobile Satellite Services (MSS). DBS represented according to the SIA about 78% of the total satellites

**Tab. 5:** *Estimated breakdowns of global space revenues in 2006.*

| Type | Value in Billion Euros |
|---|---|
| Satellite manufacturing (commercial) | 2.3 |
| Launch industry (commercial) | 1.12 |
| Ground Stations and equipment | 22.15 |
| Digital Broadcast Services (DBS) | 37.53 |
| Fixed Satellite Services (FSS) | 9.08 |
| Mobile Satellite Services (MSS) | 1.54 |
| GPS equipment | 11.54 |
| Insurance | 0.65 |
| Orbital tourism | 0.2 |
| **Total** | **86.11** |

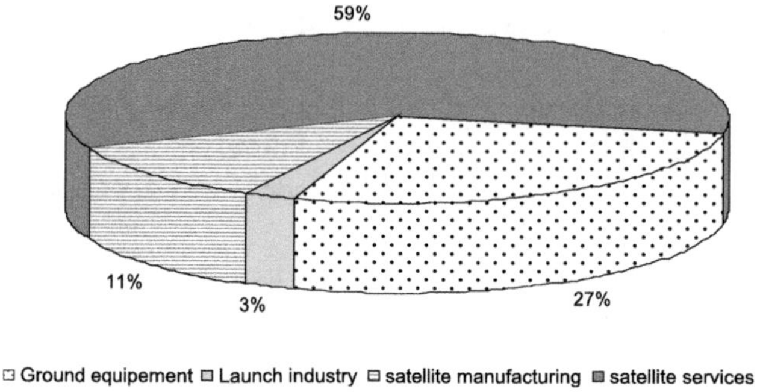


**Fig. 7:** *World satellite industry revenues by sector in 2006 (source Futron/SIA).*

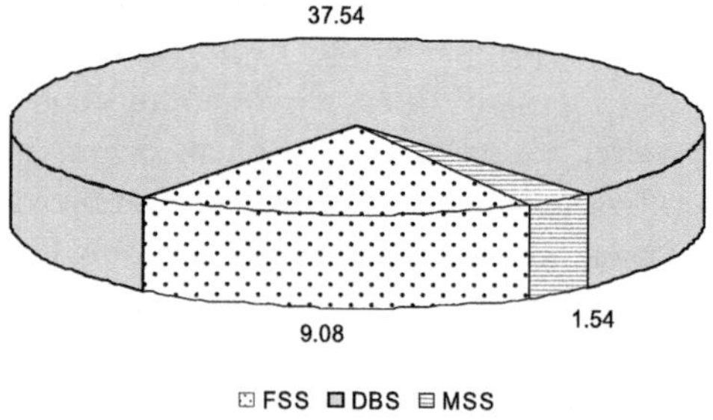


**Fig. 8:** *World satellite services revenues in 2006 (source Futron/SIA).*

services revenues in 2006 up from 63% in 2000. DBS is followed by FSS that represented 19% of world satellite revenues in 2006 up from 31% in 2000. MSS represented in 2006 only 3% (Figure 8).

SIA estimates that in 2006, DBS revenues were about 37.54 billion euros with direct-to-home television services representing the largest portion of satellite services revenues. The other main segment of the DBS sector is satellite radio, and this segment is an increasing domain of revenues growth for the space industry. Satellite radio revenues in 2006 were about 1.22 billion euros, earned by three firms: XM Radio, Sirius Satellite Radio and WorldSpace.[111]

FSS revenues are estimated by the SIA to be at about 9.08 billion euros in 2006 up from 7.53 billion euros in 2005. For the SIA, the FSS market includes telephone, data, and video transponder leasing as well as remote sensing. Revenues from global commercial satellite remote sensing grew by 16% from 2005 to 2006, driven by new and continuing military and intelligence imagery contracts, as well as expanding civil and commercial imagery markets, including online mapping services.

Following the crisis created by the fiasco subsequent to the deployment of the LEO constellations at the end of the last decade, MSS is now a mature market segment enjoying healthy growth and with good prospects with the introduction of the latest generation of mobile satellites and services.[112] MSS revenues increased 18% mainly due to growth in voice traffic to about 2.3 billion euros. The MSS market is dominated by three MSS providers: Iridium Satellite LLC, Globalstar and Inmarsat, as well as several regional operators.

In 2006, total satellite manufacturing revenues of about 9.23 billion euros were up from 6 billion euros in 2005. Government payloads are estimated to have generated 75% of total manufacturing revenues, while commercial payloads generated around 2.3 billion euros.

The SIA estimates the total launch vehicle industry at about 2.1 billion euros in 2006 including commercial and governmental launches that declined by 10% from the previous year. Based on estimates from the United States Federal Aviation Administration (FAA) in 2006, the revenues of the 21 commercial launches are evaluated at about 1.12 billion euros. Europe generated the most revenues for commercial launch services (Figure 9).[113]

The overall revenues of the ground equipment sector grew by around 14% in 2005 to about 22.15 billion euros in 2006 according to SIA data. Ground equipment now accounts for the second-largest share of global space industry revenues. The biggest driver behind this revenue growth is end-user equipment, particularly, for key consumer services such as satellite radio and direct-to-home television. Ground equipment revenues includes infrastructure elements such as mobile terminals, gateways, control stations, as well as end-user equipment such as very small aperture terminals (VSATs) and ultra small aperture terminals (USATs), direct-to-home broadcast dishes, satellite phones and digital audio radio satellite (DARS) equipment. Therefore, this number overstates the revenue associated with ground station infrastructure per se, but it does not include revenues for end-user electronics that

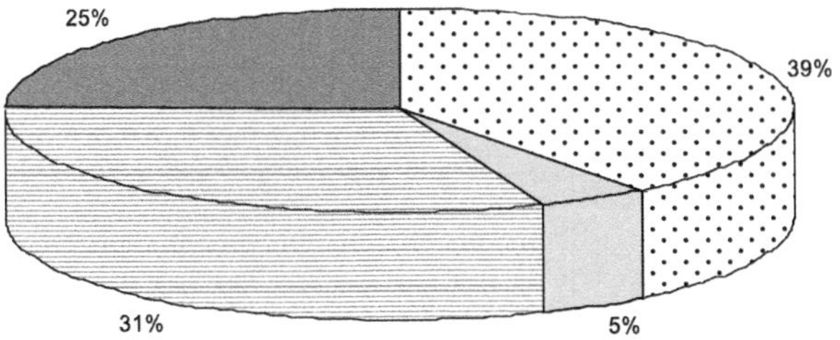


**Fig. 9:** *Commercial launch revenues in 2006 (based on FAA estimation).*

incorporate GPS chip sets such as personal digital assistants (PDAs) or cell phones. However, as the potential new uses for satellite navigation devices and services are constantly expanding, the position, navigation and timing sector is a rapidly growing market. The global market for GPS devices hit 15 billion U.S. dollars in 2006 (or about 11.54 billion euros) according to the GPS Industry Council, a trade group based in Washington. And, this global market is expanding at a rate of 25–30% annually.[114] According to the European manufacturer, TomTom, the European and North American market for personal navigation devices will grow to around 18 million units in 2007, up from over 10 million units in 2006.

Besides the traditional commercial space sectors, two emerging markets: orbital and suborbital spaceflight, have gained momentum in the last years, and 2006 was no exception. In September 2006, Anousheh Ansari became the fourth private astronaut to travel to the ISS onboard a Russian Soyuz followed on April 2007, by Charles Simonyi.[115] The former chief architect of Microsoft, Simonyi, booked his flight, like the others, through the U.S. firm Space Adventures, and is reported to have paid between 20 and 25 million for a 13-day trip into space (between 15.3 and 19.2 billion euros).[116]

## 3.3. Space industry evolution

In the last decade, the space industry has witnessed increasing competition. The most visible and direct effect of this phenomenon is the multiplication of consolidation, mergers and the formation of strategic alliances, and consequently, the shrinking number of prime contractors. A first wave of consolidation and rationalization occurred in the United States in the 1980s, and of the 20 major space companies only 3 "prime" ones were left in the mid-1990s (Boeing, Lockheed Martin and Northrop Grumman). Similar developments took place in Europe in the 1990s. A second wave of consolidation and rationalization has now been observed in recent months in major space-faring countries.

### 3.3.1. Industrial evolution in Europe

Following the acceptance of the merger of Alcatel with U.S.-based Lucent Technologies by the White House in October 2006, and after months of on-and-off-again discussions, Thales made an offer for a merger with Alcatel Space the same month which would mean Thales acquiring Alcatel's 67% stake in the satellite manufacturer Alcatel Alenia Space and a 33% stake in the space services company Telespazio, along with a pair of secure telecom activities. Then, following the review by EC's Competition Directorate of the pending takeover of Alcatel's share in those two space ventures due to a possible market dominance that could result from the deal,[117] on 4 April 2007, the EC announced that under the EU Merger Regulation the proposed acquisition was accepted. This represented the final regulatory approval enabling the creation of Thales Alenia Space.

In 2006, EADS Space changed its name to EADS Astrium, but without any operational management changes. However in the recent months, the European conglomerate has seen a series of changes to the shareholding structure. Following the reduction of the DaimlerChrysler shares in EADS from 30% to 22.5% in April 2006 in a deal worth 2 billion, a group of 15 banks paid DaimlerChrysler 1.5 billion euros in February 2007 to temporarily take over a 7.5% stake in EADS.[118] This transaction meets the combined objective of reducing DaimlerChrysler's holding in EADS to 15%, while retaining the German influence in EADS. DaimlerChrysler will retain voting rights for the full 22.5% stake in order to maintain the Franco-German balance of power in the aerospace group. This agreement may be dissolved as of July 2010 when DaimlerChrysler has the right to buy back the shares or transfer them to the German and French governments, and Lagardère. Russia's second-largest bank VneshTorgBank (VTB) acquired 5.02% of EADS in 2006, and the Gulf state of Qatar announced in spring 2007 its interest in acquiring a stake of up to 10% in EADS.

In the field of space-based telecommunications, on 30 March 2006, the Luxemburg-based SES Global announced that it completed the full acquisition of New Skies Satellites. The integration of New Skies' assets aimed to strengthen SES' industry position, and extend SES' presence in emerging markets like India, the Middle East, Africa and Latin America. In Spring 2007, SES in a deal with General Electric (GE) Capital, sold also its minority stakes in AsiaSat and Star One to GE, plus an underused satellite over the Pacific Ocean, withdrawing GE from the SES shareholder structure.

Finally, on 21 June 2007, Germany's OHB Technology announced that it had purchased German space-component manufacturer Kayser-Threde for 5.95 million euros.[119] This is the second strategic acquisition of OHB in recent years following the purchase of a majority stake in MT Aerospace in mid-2005.

### 3.3.2. Industrial evolution in the United States

On 7 September 2006, Lockheed Martin announced that it was selling its International Launch Services (ILS)[120] stake to Space Transport Inc., a privately-held corporation based in the British Virgin Islands. Following the exit of Lockheed Martin, ILS is now owned 51% by Space Transport Inc. and the remaining 49% is owned by Khrunichev. The main consequence of Lockheed Martin's decision is that ILS no longer offers the same back-up services with the Atlas V rocket as it did in the past, whereby customers were able to be transferred between the Proton and Atlas launchers if the first vehicle was grounded or otherwise unavailable. Following this transaction, Lockheed Martin transferred the marketing operation and future commercial Atlas V vehicles to Lockheed Martin Commercial Launch Services (LMCLS).

On 3 October 2006, the U.S. Federal Trade Commission (FTC)[121] cleared the United Launch Alliance (ULA) deal to go forward with the Boeing and Lockheed Martin merger of their launch manufacturing and service businesses to address the non-commercial U.S. government launch market. Following FTC approval, the joint venture officially began operations on 1 December 2006. Under the ULA, Boeing and Lockheed Martin will jointly sell Lockheed Martin Atlas V and Boeing Delta II and IV rockets to the U.S. government and, particularly, to the DoD within the framework of the Evolved Expendable Launch Vehicle (EELV) programme. On 14 December 2006, the first launch was carried out under ULA auspices when a Delta 2 rocket carried a classified NRO payload into Low Earth Orbit.

On the satellite manufacturing side, Alliant Techsystems (ATK) announced on 3 April 2007 that it intended to acquire Swales Aerospace, a provider of satellite components and subsystems, small spacecraft and engineering services for NASA, the DoD and various commercial satellite customers.

In the field of space-based Earth observation, the Orbimage acquisition of Space Imaging was completed in January 2006, creating a new company called GeoEye that is now the largest commercial remote sensing company in the world. Orbimage bought Space Imaging after the latter was up for sale in early 2005, principally because it was not awarded an important NGA NextView contract. Orbimage, on the other hand, did receive the second NextView contract in September 2004 for the development of a next-generation high-resolution remote sensing satellite now known as GeoEye-1.

In the emerging domain of satellite radio services, on 19 February 2007, XM Radio and Sirius Satellite Radio announced an agreement under which the companies will be combined in a tax-free, all-stock merger of equals. Nevertheless, XM Satellite Radio and Sirius Satellite Radio will have to overcome intense regulatory scrutiny to complete their proposed merger.

In the domain of space-based telecommunications, on 3 July 2006 Intelsat announced the completion of its merger with PanAmSat Holding Corporation creating the world's largest commercial FSS provider. Intelsat acquired all of the outstanding common shares of PanAmSat for approximately 2.4 billion euros. Then, the UK-based private-equity investor BC Partners announced that it would purchase a 76% stake in Intelsat Holdings Ltd[122] in a deal agreed on 19 June 2007. Intelsat's current owners will receive about 3.5 billion euros in cash from BC Partners. The transaction is expected to take between six and nine months to close pending regulatory approvals.

### 3.3.3. Industrial evolution in Russia

For many decades, Russia's space industry vied for first place with the United States in both military and civil space applications. However, since the implosion of the U.S.S.R. in the 1990s funding issues have greatly reduced effectiveness, degrading the space industrial base and infrastructure and thus hampering global competitiveness. In the last decade, the Russian space industry has relied on lower costs of production, low marginal costs, already proven technologies and joint ventures with foreign companies to gain commercial market shares, and thus revenues, particularly in the business of launch services. However, higher costs of production (principally labour and raw materials) and a fragmented industrial base, as well increasing technological obsolescence, particularly, in satellite manufacturing have pushed the Russian government to take action to sustain its domestic space industry. The marked economic recovery of Russia in the past few years due mainly to high world prices for energy in conjunction with the renewed involvement of Russia in major topics of world affairs have led the Russian government under President Putin's leadership to develop an ambitious space policy and generally overhaul its industrial base.

In this context, in October 2005, the Russian government adopted a new Federal Space Programme (2006–2015) that attempts to halt the decline of the country's industrial base and ends years of under-funding.[123] The new research and procurement strategy calls for increased funding in the civil space programme and the consolidation of the Russian space industry into 10 large holdings by 2010, and possible into three or four in 2015. This new strategy entails a wave of capital investment designed to increase Russia's share of the global marketplace for space-related goods and services, and to narrow the technology gap with other space powers. The new research and procurement strategy calls for the civil space programme to receive about 305 billion rubles (about 8.9 billion euros) for space activities in 2006–15, as well as 182 billion rubles in private investments (about 5.3 billion euros).

One of such holdings has already been created. On 9 June 2006, President Putin ordered the merger of a number of Russian space enterprises to form a new state-owned company named the Information Satellite Systems (ISS) that will become the flagship of Russia's satellite manufacturing industry. Pending governmental approval, the Krunichev State Research and Production Space Centre of Moscow will be the centrepiece of another holding that would include two rocket-engine makers and one manufacturer of satellites and launchers. Roskosmos has also set up one such holding centred on the Scientific Research Institute of Space Device Engineering in Moscow, while another holding will be built around the Scientific Production Association of the Machine-Building.

### 3.3.4. Industrial evolution in Japan

Japan's overall space policy is currently in transition, as a new bill is to be submitted shortly to the Diet by the ruling Liberal Democratic Party (LDP) for the establishment of a "Basic Law of Space Activities". The bill proposes to establish a new institutional framework to structure the space policy-making process more coherently in three areas, one being the competitiveness of the Japanese space industry.

Since 1990 and the agreement with the United States to open government procurement for non-R&D satellites to international bidders, the Japanese space industry has been plagued by inefficiencies. In particular, because of the nature of government-funded R&D projects, the Japanese space industry was not concerned with improving its international competitiveness, because it believed that it was sufficient to receive institutional R&D contracts for its survival.[124] Therefore, the Japanese satellite industry focused efforts on R&D and technological development satellites and public funding rather than on the commercial market.[125] However, the proposed Basic Law is urging the government, as well as industry, to use the public budget to strengthen industrial capabilities and the autonomous business base within the scope of Public-Private-Partnership (PPP) mechanisms. These efforts are intended to foster the "industrialization" of Japan and increase Japan's industrial competitiveness and reduce its dependence on public funding.[126] In this context, the Japanese Diet passed legislation (the so-called Fundamental Law on the Promotion of Geospatial Information Activities) on 23 May 2007 that commits the government to fund the development and launch of the initial satellite in a planned GPS augmentation system, plus a period of in-orbit testing. This satellite is foreseen to be part of the future Quazi Zenith Satellite System (QZSS) three-satellite constellation.

### 3.3.5. Industrial evolution in China

Two documents formulated by the Chinese government in 2006 "The Outline of the 11$^{th}$ Five-Year Programme for National Economic and Social Development" and "The National Guideline for Medium-and Long-term Plans for Science and Technology Development (2006–2020)" put space industry in an important position. These documents subsequently led to the development of a new plan for China's space industry, and thus to a new White Paper entitled "China's Space Activities in 2006". The main factors behind this new document were the principles of independence and Chinese-led initiative to meet the needs of the national modernization drive, and therefore have a clear industrial slant.

In the White Paper, the aims of China's space activities are listed, including among other things, meeting "the demands of economic construction, scientific and technological development, national security and social progress; and to raise the scientific quality of the Chinese people, protect China's national interests and rights, and build up the comprehensive national strength".[127] Top priorities involve developing and operating a series of high-level space missions (e.g. high-resolution Earth observation system, new launcher development etc.).[128] These initiatives aim to increase Chinese industrial competitiveness by developing new space assets.

Following the overall political impulse to improve Chinese space industrial competitiveness, 2006 and 2007, were already marked by several commercial successes. On 14 May 2007, China's first satellite export was successfully put into orbit from the Xichang Satellite Launch Centre using a Chinese Long March 3B. Nigcomsat-1 is a 5-ton multi-band telecommunications satellite built for Nigeria based on China's DFH-4 satellite platform developed by China Aerospace Corporation. The satellite and the launch were purchased as a package and insured for about 230 million euros. The Venezuelan government has also ordered a similar spacecraft for launch in 2008.

### 3.3.6. Industrial evolution in India

India has now entered the "global open-commercial market", on 23 April 2007 as for the first time it has launched on board its indigenously developed PSLV a foreign payload won after an international competition (the Italian astronomical satellite Agile). ISRO announced on 18 April 2007 that, following the contract signed with Eutelsat in February 2006,[129] Antrix Corporation, which is the commercial arm of ISRO, it would be building a communications satellite for

Avanti Screen Media, a British company, with a potential scheduled launch in 2009.[130] Like for the Eutelsat W2M satellite, Antrix will be responsible for building the platform while the transponders for the satellite would be sourced from Europe. These industrial achievements demonstrate the increasing competitiveness of the Indian industrial base.

## 3.4. Industrial overview

An in-depth sector analysis is required to assess the competitiveness of the industrial base of major space-faring countries, as well as the latest trends and developments of the main segments of the space sector. Because of the strong relations between the launch sector and satellite industries, neither can flourish without the other and each must take the other's overall level of activities into consideration. The launch sector requires a steady stream of payloads, and both satellite manufacturers and satellites operators need consistent and reliable access to launch vehicles. Thus, to asses the overall state of the space industry, a closer look will be given to these three segments: the launch sector, the satellite manufacturing sector and the commercial telecommunications satellite operators.

### 3.4.1. The launching sector

2006 was a particularly dynamic year for the launch sector.[131] Launch providers from the United States, Russia, Europe, China, Japan, India and the multinational consortium Sea Launch, conducted a total of 66 launches in 2006.[132] Overall, 21 commercial orbital launches occurred worldwide in 2006 representing about 32% of total launches. Whereas non-commercial launches are particularly important for the United States, and the Asian space powers, commercial launches are the core of Europe's and Sea Launch's activities. Four launch "operators", Europe, Russia, the United States and Sea Launch accounted for the 21 commercial launches. Commercial orbital launches represented only 40% of Russia's launches total, but it was nonetheless the leader of the commercial launch market with almost 48% of the total commercial launches. The U.S. as mentioned above relied principally on the non-commercial market, and in 2006, Lockheed Martin's Atlas V conducted the single commercial U.S. launch. This launch was under the operation of ILS prior to the transfer of operation to LMCLS. All of European launches were commercial ones representing about 24% of the global commercial market share, which is as much as Sea Launch.

A look at the actual number of launches shows that Arianespace[133] conducted five launch campaigns in 2006, which is just as many as its rivals ILS and Sea Launch. However, due to its dual-launch capability it launched two times as many missions as its competitors. The three world's principal commercial launch-services providers conducted five launches each, however, all had different strategies for 2007 and different results in the number of launch orders signed in 2006.

In 2006, Arianespace won twelve orders. Of those contracts won, eleven were for telecommunications satellites and one for an Earth observation satellite. These twelve satellites represent the equivalent of seven Ariane 5 launches; two will be in single launch configuration, one for the Helios 2B satellite and the other for the TerreStar 1 telecommunications satellite, while five will be dual launches. This buoyant activity illustrates, in particular, the renewed competitiveness of Ariane 5 in the global commercial market-place. Arianespace also increased its launch rate in 2006 to meet the demands of its customers. For 2007 Arianespace expects to conduct six commercial Ariane 5 ECA launches carrying twelve satellites. Seven Ariane 5 flights are scheduled for 2008 and eight for 2009, by which time Soyuz and Vega will also be operating from the Guiana Space Centre (GSC).[134] With Ariane 5, Soyuz and Vega Europe will be able to undertake all types of launch mission from the CSG, strengthening Arianespace's position of full-fledged operator covering the whole range of payloads.

As Lockheed Martin is no longer ILS's principal shareholder, ILS now focuses exclusively on Proton launch vehicles, and in 2006, it was been reported to have signed four contracts.

In 2006, Sea Launch[135] was reported to have signed four contracts in 2006 for the heavy-lift Sea Launch from its mobile platform and five for Land launch vehicles from Baikonur's Cosmodrome. However, after a historical year for Sea Launch, its first launch of 2007 ended in a failure when the Zenit 3SL launch vehicle carrying SES New Skies satellites was destroyed as it was lifting off the platform Odyssey on 30 January 2007.[136] Consequently, following the launch failure, several customers sought alternative suppliers to assure they were on orbit to meet regulatory or customer requirements, and several clients switched to the competition, particularly, to Arianespace, because slots on Arianespace's 2007 manifest recently opened up with the delay of the launch of the Automated Transfer Vehicle (ATV) to the ISS. Sea Launch expects to be back in business in fall 2007, and is considering launching from Baikonur as early as September 2007. However, Sea Launch operations from its floating platform will take a little longer to return to flight because of the repair and recertification of the platform following the 30 January 2007 on-pad launch failure.

### 3.4.2. The satellite manufacturing sector

Space-based telecommunication is the most mature market of all space applications and therefore constitutes the core business for satellite manufacturers. Consequently, the health of the satellite telecommunications market determines the sustainability of the space industry to a great extent. Looking at the satellite manufacturing market share of the GEO communications satellites is therefore a good proxy for assessing the vitality of a national space industry, as it reflects its competitiveness in the most lucrative segment of the satellite manufacturing sector.

In 2006, a total of 101 satellites were launched, with 23% of those being commercial. The United States was the leader of commercial satellites manufactured and launched in 2006, holding 60% of the market. This domination is due largely to the good performance of Lockheed Martin and Space Systems/Loral (Figure 10). Europe had a market share of 26% of all commercial satellites manufactured with 3 satellites built each by Alcatel-Alenia Space and EADS Astrium. In the segment of non-commercial satellites launched in 2006, Orbital Sciences Corp was highly active (Figure 10).

In 2006, with the emergence of new satellites operators and new commercial ambitions 43 new GEO telecommunication satellites were ordered (Table 6). The U.S. manufacturers won 16 contracts for GEO telecommunication satellites, 12

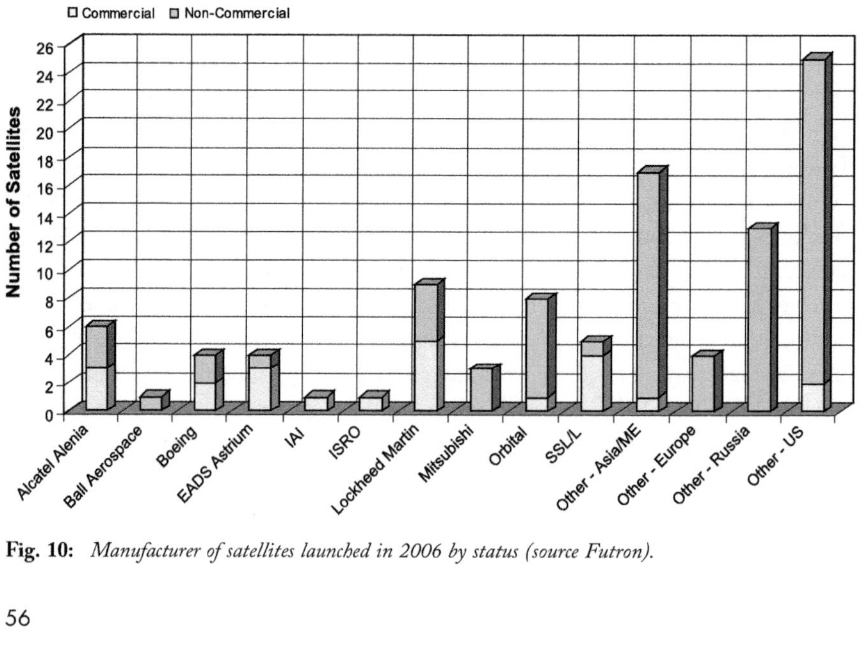


**Fig. 10:** *Manufacturer of satellites launched in 2006 by status (source Futron).*

**Tab. 6:** *Total GEO communications satellite orders in 2006.*

| Primes | Commercial | Non-commercial |
|---|---|---|
| USA | 12 | 4 |
| Europe | 12 | 3 |
| Russia | 3 | 3 |
| China | 1 | 3 |
| India | 0 | 2 |
| **Total** | **28** | **15** |

were commercial and 4 non-commercial. Europe followed with an overall total of 15 satellite contracts won with 12 commercial satellites and an additional 3 non-commercial satellites (Table 6). Russia won 6 contracts split evenly between commercial and non-commercial satellites, followed by China with 4 satellites and India with two (Table 6). China and India were the only two countries where the non-commercial total exceeded the commercial telecommunication orders, and Japan has been reported has not having being awarded a single GEO telecommunications satellites order in 2006.

Twenty-eight GEO commercial telecommunications satellites were ordered in 2006. Space System Loral and EADS Astrium both signed seven commercial GEO telecommunication satellites in 2006, having an aggregated 50% market share, followed by Alcatel-Alenia Space with five satellites orders won. Europe was thus particularly active with 12 new GEO telecommunication satellites, the same amount as the United States. Alcatel Alenia Space and Space System Loral were able to win three orders outside their domestic markets (for European manufacturers, Europe is viewed as a single market) demonstrating the competitiveness of their products and services. Astrium followed with two satellites orders won outside Europe (Figure 11).

However, while the overall telecommunications satellite manufacturing segment is dominated by European and U.S. companies, 2006 was marked by the confirmation of the entry of new satellites manufacturers from the "South" (China and India) in this highly lucrative market. Despite problems with the newly introduced hardware, it seems that China wants to reinforce its effort in telecommunications satellites development.[137] China's signed a contract with the Venezuelan government authorities for a large telecommunications satellite (Venesat-1) based on the new DFH-4 satellite platform. As in the case of its first export sale to Nigeria (Nigcomsat-1), this deal includes a launch aboard a Chinese Launch March rocket.

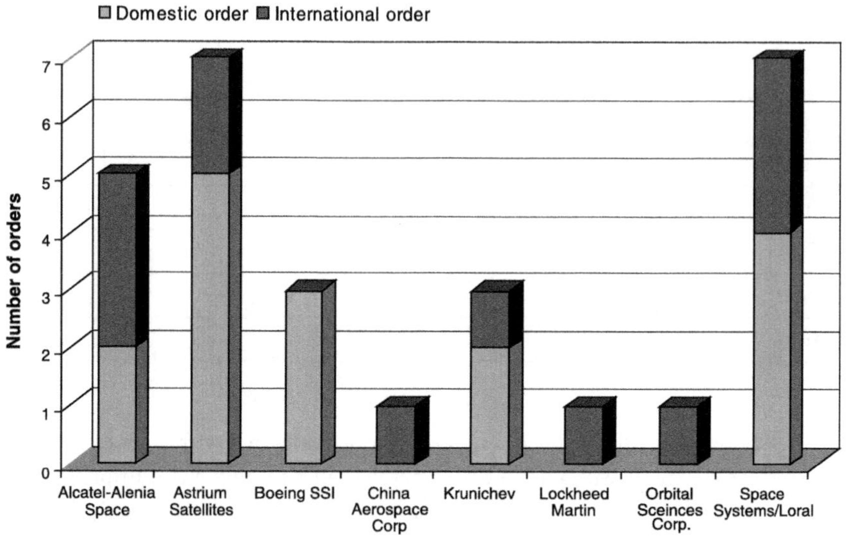


**Fig. 11:** *Commercial GEO satellite orders won in 2006 by manufacturers.*

Meanwhile, India's Antrix Corp, an organization that is part of ISRO, has teamed with Astrium Satellites to offer a small commercial telecommunications satellite product that competes with Orbital Sciences Corp's platform. The Astrium-Antrix joint venture posted two wins in 2006 Eutelsat's W2M, planned for launch in 2008 and for UK's company Avanti Screen Multimedia PLC for HYLAS (Highly Adaptable satellite), which is to deliver broadcasting, multimedia and broadband services. Antrix Corporation signed another order in April 2007. Finally, Russia with Krunichev signed three satellites orders. Albeit the aggregated total of these three players is rather limited, the volume of activity of their respective domestic satellite manufacturer base is expected to increase in the years to come following the evolution of their industrial policies.

Fifteen GEO non-commercial telecommunications satellites were ordered in 2006 (Table 6). The U.S. manufacturers had four orders, three to Boeing Satellite Systems International and one to Lockheed Martin. It was followed by Europe (all for Alcatel Alenia Space), China and Russia, all with three orders each. Finally, India's Antrix Corporation had two satellites orders in 2006. In 2006, no single satellite manufacturer was able to win a non-commercial GEO telecommunications satellites outside its captive domestic market (Figure 12).

When looking at the individual performance of companies, Alcatel Alenia Space was the world leader by the number of contracts won (Figure 13) in 2006,

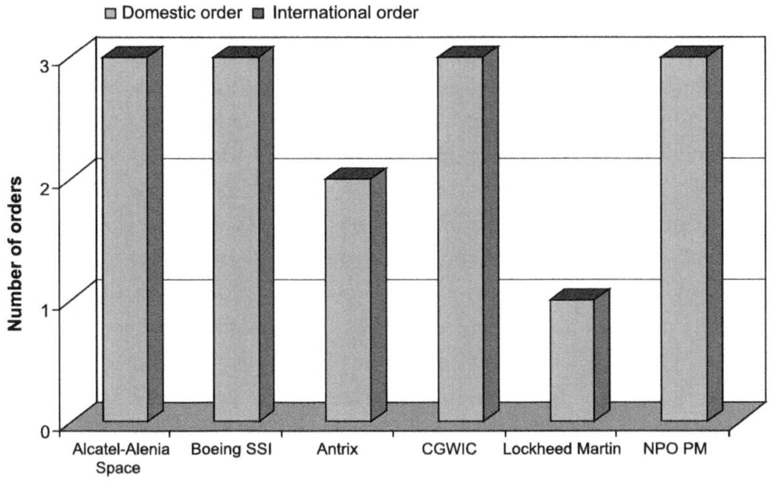


**Fig. 12:** *Non-commercial GEO satellite orders won in 2006 by manufacturers.*

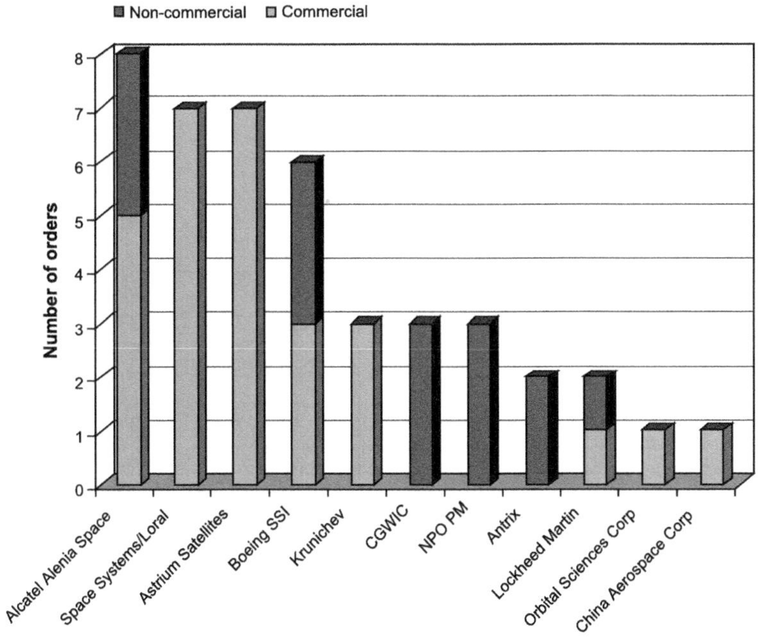


**Fig. 13:** *GEO commercial and non-commercial satellite orders won in 2006 per satellite manufacturer.*

followed by another European company, EADS Astrium, as well as the U.S.-based Space Systems/Loral, and Boeing SSI. This quartet concentrated 65% of contracts won in 2006 (Figure 13).

### 3.4.3. Satellite operators

As already mentioned, space-based telecommunications is the major source of revenues in the value-added services sector, with a major role being played by the FSS segment which is a major driver of the space industry.

In 2006, the consolidation in the sector of satellite operators witnessed in recent years continued: PanAmSat and New Skies Satellites were purchased by Intelsat and SES Global respectively modifying the global hierarchy. Following these acquisitions, Intelsat is now the largest company that leases transponders on satellites it operates in geostationary orbit. Despite the purchase of New Skies Satellites by SES Global, the Luxemburg-based company has dropped to the second rank in 2006. The combined Intelsat-PanAmSat has 51 satellites in orbit and eight satellites on order, compared to 36 satellites in orbit for SES and 23 for Eutelsat, the world's third largest FSS operator. The top three companies were responsible for about 64.5% of the 6 billion euros generated by the telecommunication sub-segment of the FSS industry in 2006, with the top two alone accounting for more than half. After this trio, there is a major quantitative gap, with the other operators having only between eleven and three satellites in orbit (Table 7).

**Tab. 7:** *Top 10 FSS operators in 2006 (Adapted form Space News[138]).*

| Rank | Company | Location | 2006 Revenue in million $ | Satellites in Orbit | Satellites on Order |
|---|---|---|---|---|---|
| 1 | Intelsat Ltd | Bermuda/USA | 2100 | 51 | 8 |
| 2 | SES Global | Luxembourg | 1900 | 36 | 9 |
| 3 | Eutelsat | France | 1050 | 23* | 4 |
| 4 | Telesat Canada | Canada | 411 | 7 | 2** |
| 5 | JSAT Corp | Japan | 326 | 8 | 1 |
| 6 | Star One SA | Brazil | 195.8 | 5 | 2 |
| 7 | SingTel Optus | Australia/ Singapore | 191.8 | 4 | 2 |
| 8 | Loral Skynet | USA | 164 | 5 | 1 |
| 9 | Hispasat | Spain | 159.1 | 3 | 1 |
| 10 | Russian Satellite Communications Co | Russia | 152 | 11 | 4 |

* Including some partially-owned satellites;
** Satellites under construction

# 4. The security dimension

In 2006/2007, security was again pushed to the top of the agenda across the globe, and it is being increasingly recognized that space assets form a system that is key to the requirements of modern armed forces to ensure global security. Apart from the traditional space powers, there are now more and more countries that have committed significant efforts to obtaining dedicated military satellites programmes or "multi-purposes assets", leading to an "internationalization of the militarization of space".[139]

## 4.1. The global space military context

The overall economic value of the global space sector is estimated at about 133 billion euros, including about 26 billion euros spent on military space affairs. By contrast, according to the Stockholm International Peace Research Institute (SIPRI), global military expenditure in 2006 is estimated to have reached 1204 billion U.S. dollars[140] or about 926 billion euros. However, while as already mentioned, traditional as well as new institutional actors are increasingly procuring military space assets, world space military expenditure is like overall military spending, extremely unevenly distributed between countries. Only a limited number of countries devote a substantial amount of funding to security-related space activities. These countries are: Canada and the United States in North America; Belgium, France, Germany, Greece, Italy, Spain, the United Kingdom in Europe; Iran, Israel and Turkey in the Middle East; China, India, Japan, Russia and South Korea in Asia; and Argentina and Brazil in South America. Other countries pursue some military space activities, but of very limited size and scope.

### 4.1.1. Europe

Although Europe invests in defence programmes through several countries, total European spending on military space fluctuates between 650 and 750 millions euros depending on the year, versus about 5 billion euros being spent annually on civil space projects. France has the largest military space programme in Europe both in terms of budget and capacity, but also for its "political will" and policy developments. In 2006, it spent about 72% of the European total space security-related expenditure with an estimated public spending of 469 millions euros devoted to military space activities. However, while France has seen its budget stagnating in recent years, more modest historical contributors like the United Kingdom, Italy, Germany and Spain have seen their investment in security-related space activities increase, as they have been developing or procuring new national capabilities such as COSMO-Skymed in Italy or SAR-Lupe in Germany. Other countries like Belgium or Greece rely on cooperative programmes to reap the benefits of military space assets (i.e. access to Helios data).[141]

Europe due to its limited military space infrastructure depends heavily on civilian and commercial space systems for military and security activities. This European characteristic is due mainly to the aforementioned limited budget capabilities devoted to space capabilities. However, this policy also follows a technological logic, since many space systems have both commercial and security applications, as for instance Earth observation systems. There is a trend in Europe to increasingly develop "multi-use" programmes (projects used for both civilian and military purposes). For instance, the future "Pleiades – COSMO-SkyMed" project will be an explicit dual-use programmes.[142] However, the real new trend emerging in the last few years is that some of the European security-related programmes are managed under PPP-schemes (yet solely for telecommunications systems) such as Skynet 5 (UK) and Spainsat or XTAR-EUR (Spain).[143] In the PPP approach, the system is fully dedicated to the national authorities in times of crisis, but the managing organization can commercialize the capability for the rest of the time. This "European approach" to procure military space assets[144] is unique in the world, but needs to be validated in the long-run; nonetheless an increasing number of European countries are considering such avenues for their future military needs (i.e. France).

Due to the aforementioned tight defence budgets no single European country can afford to develop a wide range of space assets like in the United States or in Russia on its own. As a result, European stakeholders are starting to realize that if they want access to greater variety of space-based military systems they need to pool resources both financially and technically. Current efforts for the coordination, harmonization, and consolidation of the different space activities within Europe are thus taking place to avoid the duplication and minimize unwanted redundancy. As already mentioned, cooperation exists in the form of exchange or leasing of the capacities of national systems. For instance, data collected by SAR-Lupe will be provided to the French government in exchange for data from Helios. In an innovative approach for European military assets in 2006, Italy and France launched a new cooperation approach to develop a dual-use telecommunications satellite called Athena-Fidus dedicated to military telecommunication, as well as civil government and potentially commercial applications.[145] The satellite scheduled for launching in 2011 also spurred the interest of Belgium and the U.K. authorities. Furthermore, because European space capabilities are both modest and very fragmented, the need to rationalise the activities of the different national entities to complement each other's capabilities and avoid the unwanted duplication of capabilities has been recognized. Moreover, following the 2001 BOC[146] initiative in strategic geospatial imagery signed by six European Chiefs of Defence Staff (Belgium, France, Germany, Greece, Italy and Spain) on 13 December 2006, an agreement between the same six aforementioned countries (plus Sweden as an observer) was signed calling for the common development of a future space-based reconnaissance satellite system called MUSIS (MUltinational Space-based Imaging System for surveillance, reconnaissance and observation). This future system will aim to answer the whole range of data collection requirements, from the political decision-making support to the military operation support.

While the argument that Europe should develop integrated defence and security-related space capabilities is not a recent one, it has, however, been given a new impetus in recent years following the development in building up European defence and the increasing willingness of the EU to share the responsibility for global security. Nonetheless, at the difference with European countries, the EU is currently only involved in "soft" security operations such as humanitarian missions rather than military or "hard" security missions. At the EU level, initiatives are being developed to respond to new security requirements, and space-based systems are acknowledged as an answer to emerging security needs and are now recognized as "enablers" that can give the EU the full capability to act independently in conflict prevention and crisis management tasks to support its CSFP, and the European Security and Defence Policy (ESDP) and is an important asset to ensure EU's security[147]. To illustrate this increasing political ambition for developing security space efforts, several important texts and programmes have been approved at the EU-level in the last years. In particular, the newly European space policy also covers several aspects linked to space security. The Resolution adopted on 22 May 2007 deals prominently with security and defence issues, and while recognizing the intrinsic dual nature of space activities, it affirms the need to set up a "structured dialogue" with the competent bodies of the Member States and within the EU Second and Third Pillars including the European Defence Agency for optimizing synergies between defence and civil space technologies and programmes.[148] Along the same lines, the Resolution does not preclude the use of GMES and Galileo by military users, and therefore, recognizes the implicit dual-use nature of the future services proposed by those programmes. In overall terms, the Resolution clearly states the strategic importance of space for Europe in demonstrating its independence and its readiness to assume global responsibilities.

GMES is the principal space programme of the Commission that is seen as having a clear security mandate to support the CFSP and ESDP. Consequently, with the interest in the use of GMES information services for security applications, a series of programmatic activities have been taking place in recent months. Moreover, of the three GMES "Fast Track Services" aiming to develop pre-operational services that will be introduced by the end of 2008, of particular relevance for the security aspects is the emergency response Fast Track service, because emergency cannot be decoupled from an overall "security" context. A GMES service related to security applications is also currently under consideration. In this context, a preliminary workshop "GMES: the security dimension" organized by the EU Institute for Security Studies (EU-ISS) in cooperation with the Council (DG E VIII), the Commission, the EU Military Staff, the EU Satellite Centre and the European Defence Agency was held in Paris on 16 March 2007. Over 100 experts and EU officials attended this seminar whose main purpose was to study the security applications and implications of GMES. Participants looked at building synergies around common interests in GMES across the three pillars, the applications that could be used to enhance European security, and finally, the issues of governance associated with dual use capabilities.

In 2006/2007, ESA has also initiated a series of projects dedicated to security issues following the new five year plan entitled Agenda 2011 released in October 2006 which, among other things, looked at future programmes including the synergies between civil and defence services. In particular, ESA is pursuing a study on space debris, as well as an activity to define the European needs for a Space Situational Awareness Architecture as a first step towards its realization. ESA is also involved in the GMES programme and has initiated a series of GMES projects such as RESPOND, which aims to provide support to humanitarian relief, disaster reduction and reconstruction or MARISS that aims to support European Maritime Security Services through ESA's GMES service element.

### 4.1.2. United States

While the U.S. is the country investing the most in military space,[149] it is also the most conceptually advanced in military space affairs. A series of recent high-level documents have been released in 2006/2007 underlining the strategic nature of space activities for the U.S. The new U.S. National Space Policy asserts that the U.S. assets must be unhindered in carrying out their space duties, and that "freedom of action in space is as important to the United States as air power and sea power".[150] The new policy is designed to ensure that U.S. space capabilities are protected in a time of increasing challenges and threats. While the document is largely similar in the area of space security to its 1996 predecessor,[151] one significant change declares that "the U.S. will oppose the development of new legal regimes or other restrictions that seek to prohibit or limit U.S. access to or use of space; and that proposed arms control agreements or restrictions must not impair the rights of the U.S. to conduct research, development, testing, and operations or other activities in space for U.S. national interests".[152] In this new policy, the Bush Administration also announced its determination to sustain its efforts in military space and states that it "considers space capabilities – including the ground and space segments and supporting links – vital to its national interests".[153] Because, in spite of the fact that the United States has considerably more capabilities than the next most capable actors and mobilizes means with no common measure to those of all the other countries, it is also the actor the most dependent on its space capabilities.[154] In this context, the U.S. military space doctrine has begun to focus on the need for "counterspace operations" to prevent adversaries from accessing space as the policy states explicitly that "the United States will: preserve its rights, capabilities, and freedom of action in space; dissuade or deter others from either impeding those rights or developing capabilities intended to do so; take those actions necessary to protect its space capabilities; respond to interference; and deny, if necessary, its adversaries the use of space capabilities hostile to its national interests".[155] However, while the new policy appears to be more focused on unilateral national security concerns than its predecessor, it does nonetheless identify new areas for international military cooperation, in particular the sharing of intelligence and capacity for improved space situational awareness.

Several major documents were also released by other U.S. Federal Agencies. The National Geospatial-Intelligence Agency (NGA) released on 12 January 2006 a document entitled "National System for Geospatial-Intelligence Statement of Strategic Intent" emphasizing the importance of using geospatial intelligence information to effectively respond to global threats.[156] The DoD unveiled its new Quadrennial Defense Review (QDR)[157] in February 2006,[158] whereby space activities are dealt with in three parts: enhancing existing intelligence, surveillance and reconnaissance (ISR) capabilities; achieving greater "jointness" and network-centricity; and reshaping the defence enterprise as a whole. The QDR acknowledged explicitly the United States' existing superiority in all space capability areas and pledged to maintain this advantage by keeping at least one technology generation ahead of any foreign or commercial space power. The DoD announced also it will continue to develop responsive space capabilities to ensure reliable access to space while increasing the survivability of space assets via improved space situational awareness, protection and "other space control measures".[159] The QDR aimed also to remedy to existing procurement problems in DoD space programmes through a comprehensive reorganization of the acquisition process to ensure stable costs, while maintaining schedule and performance.

Following the Chinese's ASAT test on 11 January 2007, initial high-level talks on potential Sino-American cooperation in civilian space activities that started in late 2006, including joint exploration ventures, have been suspended by the Bush Administration. According to NASA spokesman Jason Sharp, the test of a kinetic anti-satellite weapon undermined an agreement reached between President Bush and Chinese President Hu Jintao during an April 2006 summit.[160] Nonetheless, while the test has been widely criticized in the U.S., standing by the aforementioned new space policy, the United States oppose the development of new legal regimes or other restrictions that seek to prohibit or limit U.S. access to, or use of space. China's test of a ground-based medium-range ballistic missile as anti-satellite weapon will therefore not cause the U.S. to open negotiations on a new treaty that would place limits on what countries can do in space, but will more likely foster more R&D research in the domain.

### 4.1.3. Russia

In the first year of its new Federal Space Programme (2006–2015), Russia increased its space budget by as much as one-third compared to 2005.[161] While this ten-year plan has considerably enlarged the military space budget, other military space funding is also being allocated through the State Armaments (defence and security) Programme for 2007–2015, and the special federal programmes "Global Navigation System" and "Development of Russian Space Centers in 2006–2015"[162]. This overall budgetary boost of military space activities for 2006/2007 is part of an overall effort to upgrade and modernize Russia's military in-orbit infrastructure.

### 4.1.4. Japan

As mentioned earlier, Japan's space policy is in transition, following the changes to the security environment in the post-Cold War era[163] (particularly the missile launch of North Korea over Japan). A new bill was submitted on 20 June 2007 to the Lower House of Diet for the establishment of the "Basic Law for Space Activities" that aims to reconsider the assumption of the "exclusively peaceful purpose" clause in the Diet resolution of 1969. In particular, the second point of the Basic Law focus on the question of security and the flexible interpretation of the excusive peaceful nature of Japanese space activities. This change in the interpretation does not aim to promote an aggressive use of space, but aims to allow

Japan to use space assets for crisis management and disaster monitoring in the Asian region or peacekeeping missions in distant territories.[164]

While the "Basic Law for Space Activities" is not yet been voted on, the Japanese Defence Agency (JDA) became the Ministry of Defence in January 2007 (and is now autonomously responsible for strategic planning) and set up a new "Strategy Planning Office" which includes space as one of the pillars of strategic policy, illustrating the paradigm shift in Japanese space activities. The aforementioned policy debate was rounded off by the successful launch of one IGS satellite in 2006 and two in 2007, thus completing existing assets in orbits and giving Japan an Earth observation constellation dedicated to "security" issues.

### 4.1.5. China

It is difficult to evaluate Chinese military capabilities, as China is very secretive about its military activities and military space is no exception. However, contrasting remarkably with its long-standing pacifist stance in international fora, in 2006/2007, China demonstrated in an unusual fashion a more aggressive position in space. First, it launched a reconnaissance satellite in April 2006 (Yaogan 1) then reportedly used a ground-based laser to illuminate several U.S. spy satellites flying over Chinese territory.[165] These "events" as well as the January 2007 anti-satellite test (ASAT) destroying one of its own ageing weather satellites demonstrate Chinese advances in military space, and in particular, its offensive capabilities. Furthermore, a new report to the U.S. Congress by the U.S.- China Economic and Security Review Commission using open source material highlighted that 30 Chinese ASAT concepts have been formulated by the People's Liberation Army (PLA), and some of these concepts involve covert deployment of anti-satellite weapons system to be used in a surprise manner.[166]

### 4.1.6. India

In 2006/2007, India continued to consider a structural change vis-à-vis the possible establishment of a military Aerospace Command, however with a mandate that remains vague.[167] This overall reflection is part of a wide process that is considering increasing the role of military applications and defence forces in India's space activities. India has not yet launched any explicitly military satellite; however, the Aerospace Command would presumably use space-based assets for military needs, but also dedicated military satellites. In this context, India's armed forces will get their first dedicated military satellite, Cartosat 2A, in summer 2007.

This internal debate vis-à-vis its increasing involvement in military space affairs seems to have gained some extra momentum following the Chinese ASAT test in January 2007.

## 4.2. Space military activities in 2006

In 2006, 18 dedicated military satellites or explicitly recognized "dual-use" satellites were launched into space, representing 15% of all payloads launched into space. Eight countries (China, France, Germany, Japan, Russia, South Korea, Spain and the United States) launched dedicated military space assets in 2006 compared to more than 20 for civilian space (Figure 14). Two countries launched their first military-related assets: Germany and South Korea.[168]

When comparing the levels of activity by country in 2006, Russia was the world's space leader in military space activities according to the number of payloads launched, followed by the United States with four payloads launched. Europe launched three space military assets, while China, Japan and South Korea launched only one satellite (Figure 14). The quantitative domination of Russia and the

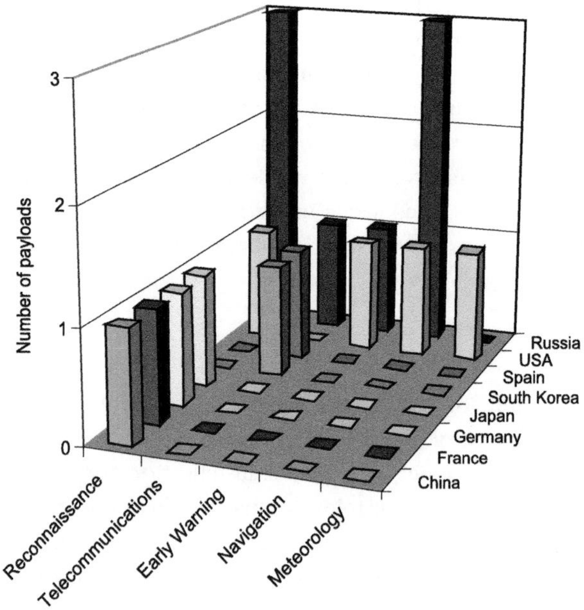


**Fig. 14:** *Military payloads launched in 2006 per country.*

United States is completed by the fact that they are also the ones with the more varied and advanced capabilities, covering both operational and tactical missions. They were, for instance, the only ones to launch early-warning and navigation satellites to refurbish their respective space infrastructures.

In the first half of 2007, a series of new defence and security-related assets were launched or tested.

On 11 January 2007, China successfully tested a direct anti-satellite test (ASAT) system. The target of the test was an old Chinese meteorological spacecraft, Feng Yun-1C (FY-1C) residing in an orbit of 845 by 865 kilometres with an inclination of 98.6°. The spacecraft was hit by a ballistic interceptor launched near Xichang.[169] On 24 April 2007, the U.S. Missile Defense Agency (MDA) successfully launched an experimental satellite NFIRE (Near Field Infrared Experiment) that is designed to collect images of a boosting rocket to improve understanding of missile exhaust "plume" observations and plume-to-rocket body discrimination. NFIRE plans to help validate the MDA's choice of kill vehicles and tracking sensors. The first of UK's Skynet 5 spacecraft (Skynet 5A) was successfully launched on Mars 2007 onboard an Ariane 5. Then, on 7 June 2007, the first of four Italian COSMO-SkyMed X-band radar satellites was placed into orbit. The three other Cosmo-Skymed satellites will be launched in the next two years. Finally, the second SAR-Lupe satellite was launched on 3 July 2007. The remaining three will be launched by 2008.

In the domain of missile defence, talks between Washington, Prague and Warsaw have been underway since January 2007 to host components of the U.S. missile defence system on the territory of the Czech Republic and Poland. The United States would like to station a radar on a former military base in the Czech Republic, while Poland would host ten interceptor missiles. The purpose would be to defend the U.S. mainland and Europe against missiles launched from Iran and elsewhere in the Middle East. The Czech and Polish sites would enlarge the architecture of the Ground-Based Midcourse Defense System (GMD) protecting the United States and currently consisting of a network of sensors and missiles stretching from Japan to Alaska designed to protect U.S. territory. However, the possibility of implementing an U.S. missile shield in Europe has led Russia to take a strong political stance creating fears of a potential new Cold War. As a direct consequence of these plans, U.S.-Russian cooperation on nuclear arms reductions has frozen and Russia is threatening to build new categories of nuclear weapons if the United States goes ahead and deploys missile defences in Europe.[170]

## 4.3. Threats to the space environment

An assessment of the trends and developments to the space security environment over the 2006/2007 period show that a series of events occurred in the domain of space-based telecommunications jamming and orbital debris creation.

In 2006, a series of jamming and piracy events occurred in the commercial satellite sector, notably a persistent jamming of the mobile satellite communication signal provided by Thuraya Satellite Telecommunications from three widely-separated locations inside Libya. The Liberation Tigers of Tamil Eelam (LTTE) have also been using a vacant Ku-band transponder on an Intelsat satellite to broadcast its messages in Sri Lanka and the surrounding region. This later event indicates the increasing capabilities and role played by transnational actors in misusing space assets.

A series of satellite break-ups occurred in 2006/2007 with a total of 14 events creating orbital debris leading to a total of almost 12 000 orbital debris, which are 10 centimetres in diameter or larger, catalogued by the U.S. Space Surveillance Network (USSSN) by July 2007 (Table 8).

In particular, following the aforementioned ASAT-test in January 2007, an estimated 35 000 pieces of debris of at least one centimetre in diameter have been created.[171] According to NASA, this event represents the single worst contamination of LEO in the past 50 years.[172] The extent of the debris cloud created by the destruction of the Feng Yun-1C meteorological satellite is increasing with new

**Tab. 8:** *Orbital debris per major space country as of 4 July 2007 as catalogued by the U.S. Space Surveillance Network (source NASA).*

| Country/ Organization | Payloads | Rocket Bodies and Debris | Total |
|---|---|---|---|
| China | 62 | 2234 | 2296 |
| CIS | 1362 | 2913 | 4281 |
| ESA | 37 | 36 | 73 |
| France | 45 | 316 | 361 |
| India | 33 | 106 | 139 |
| Japan | 101 | 73 | 174 |
| USA | 1069 | 3120 | 4189 |
| Other | 386 | 55 | 441 |
| **Total** | **3095** | **8859** | **11954** |

data gathered.[173] The Feng Yun-1C debris cloud extends from 200 to 4000 kilometres in altitude, with the highest concentration near the break-up altitude of approximately 850 kilometres.[174] The debris orbits are rapidly spreading and will encircle the Earth by the end of 2007.[175] As the majority of debris will remain in orbit for many decades, the debris from Feng Yun-1C are posing greater collision risks for spacecraft in LEO, as the debrisfrequently transit extremely populated orbits, posing new threats to current and future spacecrafts. Several satellites have already executed specific collision avoidance manoeuvres such as NASA's Terra spacecraft on 22 June 2007.[176]

---

[1] "World Economic Outlook database". International Monetary Fund. April 2007. http://www.imf.org/external/pubs/ft/weo/2007/01/pdf/text.pdf.

[2] The external aspects of energy policy remain within the competence of EU Member States' foreign ministries and a matter of national sovereignty.

[3] Since 1984, the Framework Programme has been the EU's main instrument for funding research and development.

[4] Article 9 of the 1946 Constitution of Japan states that "Aspiring sincerely to an international peace based on justice and order, the Japanese people forever renounce war as a sovereign right of the nation and the threat or use of force as means of settling international disputes".

[5] The IPCC was established in 1988 by two United Nations organizations, the World Meteorological Organization (WMO) and the United Nations Environment Programme (UNEP) to periodically evaluate the current understanding of climate change and its relationship to society, based mainly on peer reviewed and published scientific and technical literature.

[6] Jean-Louis Fellous's article "The IPCC Report – In Need of Earth Observations" in the second part of the Yearbook investigates the specific link between Earth observation and climate change.

[7] In the IPCC context, "very likely" means an assessed likelihood of at least 90%, while "likely" means at least 66%.

[8] Despite the general consensus on this first SPM, several experts are criticizing the findings of this report as new developments might have been missed, since a cut-off date was set for submission of scientific paper and other data in December 2005. Moreover, the extent and consequences of the rise of sea level is considered by many has being underestimated in this SPM.

[9] There have been three IPYs over the last 125 years: 1882–1883, 1932–1933 and 1957–1958. The last IPY provided the foundation for much of the polar science knowledge existing today.

[10] H5N1 is a subtype of the Influenza A virus.

[11] So far, all but a handful of cases of human sickness have been caused by direct contact with sick birds, suggesting the virus is unable to move easily among humans. However, it is feared that with continued exposure to people, the virus could mutate further and develop that ability.

[12] The 1918–1919 Spanish Influenza infected an estimated 1 billion people and claimed as many as 50 million lives.

[13] Eurostat. "Research & Development in the EU: preliminary results". Press release 6/2007. 12 Jan. 2007.

[14] "Review of China's National Innovation System". OECD. 27 Aug. 2007. http://www.oecd.org/document/28/0,3343,fr_2649_34273_39033180_1_1_1_1,00.html.

[15] "Science, Technology and Industry Outlook 2006". OECD. 4 Dec. 2006. www.oecd.org/sti/outlook.

[16] Eurostat. "Research & Development in the EU: preliminary results". Press realease 6/2007. 12 Jan. 2007.
[17] Ibid.
[18] GBAORD are all appropriations allocated to R&D in central government or federal budgets and therefore refer to budget provisions, not to actual expenditure.
[19] Eurostat. "Government budget appropriations or outlays on R&D". Press release 17/2006. 20 Oct. 2006.
[20] GBAORD expressed as a percentage of GDP removes the individual weights of the countries and therefore allows a comparison of GBAORD across the various entities/countries.
[21] Eurostat. "Government budget appropriations or outlays on R&D". Press release 17/2006. 20 Oct. 2006.
[22] Ibid.
[23] It should be noted that U.S. data for GBAORD exclude the socio-economic objective "Research financed from general university funds" as there is no federal support via GUF and therefore systematically underestimated (Eurostat).
[24] Patents are an exclusive right issued by authorized national bodies to inventors to allow patent holders to make use of and exploit their inventions for a limited period of time (generally 20 years). Thus, for a certain time and within a certain geographical area they provide protection for innovations. Patents can be granted to firms, individuals or other entities as long as the invention is novel, non-obvious and industrially applicable. In return for the rights, the applicant must disclose information relating to the invention for which protection is sought.
[25] Eurostat. "National patent statistics". Press release 9/2007. 16 Jan. 2007.
[26] Ibid.
[27] The latest year for which triadic patent family data are available is 2000. As indicated by the OECD it can take up to four years between the date of the priority application and the availability of information on patent applications to the EPO and JPO. As a triadic patent family is not counted until the USPTO has granted a patent, the time lag can be even longer due to the U.S. procedure. The data used by the OECD for 2003 are estimations.
[28] Eurostat. "National patent statistics". Press release 9/2007. 16 Jan. 2007.
[29] "Compendium of Patent Statistics 2006". OECD. 19 Jan. 2006. www.oecd.org/sti/ipr-statistics.
[30] In its analysis, the OECD defines space-related patents using a mixture of IPC codes and keywords. The principle IPC class used are B64G "Cosmonautics; vehicles or Equipment thereof", G01S "Radio direction-finding; radio navigation; determining distance or velocity by use of radio waves; locating or presence-detecting by use of the reflection or re-radiation of radio waves"; H01Q "Aerials"; and "Radio transmission systems": H04B7/185, /19 and /195 that represent space-based or airborne stations, Earth-synchronous stations and non-synchronous stations. These were reported to have been chosen only if the title of the patent application contains one or more of the following phrases: "GPS", "global position", "satellite", "remote sensing", "earth observation" and "geographic information system".
[31] Apart from approval of budgetary matters, UNGA resolutions are not binding on the members.
[32] The International Charter "Space and Major Disasters" that was introduced at the UNISPACE III conference held in Vienna, Austria in July 1999, by the European and French space agencies (ESA and CNES) aims at providing a unified system of space data acquisition and delivery to those affected by natural or man-made disasters through Authorized Users. Each member agency has committed resources to support the provisions of the Charter. The Canadian Space Agency (CSA) signed the Charter on 20 October 2000. Then, in September of 2001, the National Oceanic and Atmospheric Administration (NOAA) and the Indian Space Research Organization (ISRO) also became members of the Charter. The Argentine Space Agency (CONAE) joined in July 2003. The Japan Aerospace Exploration Agency (JAXA) became a member in February 2005. The United States Geological Survey (USGS) has also joined the Charter as part of the U.S. team. BNSC/DMC became a member in November 2005. Finally, the China National Space Administration (CNSA) joined in May 2007.

[33] The General Assembly Fourth Committee on Special Political and Decolonization has played a crucial role in advancing space cooperation in the past, but has not been involved in space issues recently.
[34] Specialized Agencies are autonomous organizations working with the United Nations.
[35] ECA, ECE, ECLAC, ESCAP, ESCWA, FAO, IAEA, ICAO, ISDR, IMO, ITU, UN-DESA, UN-DPKO, UN-OCHA, UN-OOSA, UNDP, UNEP, UNESCO, UNHCR, UNIDO, UNITAR, UNODC, World Bank, WHO, WIPO, WMO.
[36] GEOSS is trying to build synergies between individuals national, regional and international systems in an effort of trying to get more value for money for the governments and operators of these systems and to provide comprehensive, coordinated Earth observations (both space borne and *in situ*) to provide vital information for society. It has been established on a voluntary and legally non-binding basis, with voluntary contributions to support its activities.
[37] Peter, Nicolas. "The Changing Geopolitics of Space Activities". Space Policy 22 (2006): 100–109.
[38] Ibid.
[39] APSCO aims to enhance mutual understanding and confidence among Member States through cooperation in space science, applications and technology development to promote regional political stability and security integration, but also social and economic benefits to the Asia-Pacific region. APSCO aims to serve as a platform for the members of the organization to develop multilateral cooperation in space science, technology and applications.
[40] "Disaster Management Support System in the Asia Pacific Region". Sentinel Asia. http: //dmss.tksc.jaxa.jp/sentinel/.
[41] The sixth conference is to be held in Guatemala in 2009.
[42] Peter, Nicolas and Plattard Serge. "The European Space Policy: Europe's New Compass" ESPI Flash Report #1. May 2007. http://www.espi.or.at/images/stories/dokumente/flash_reports/flash-report1-espi-europe_in_space.pdf.
[43] European Commission. European Space Policy COM (2007) 212. Brussels, 26 Apr. 2007.
[44] European Commission. Staff Working Document. European Space Programme – Preliminary elements SEC (2007) 504. Brussels, 26 Apr. 2007.
[45] The HSPG consists of representatives of key government stakeholders of the EU/ESA, Member States, the European Defence Agency (EDA), the EU Satellite Centre and Eumetsat.
[46] John Logsdon's article "The new European Space Policy as seen from across the Atlantic" in the second part of the Yearbook analyses in more details and greater length the content of the European Space Policy, as well as its similarities and differences with the new U.S. national space policy.
[47] European Commission. European Space Policy COM(2007) 212. Brussels, 26 Apr. 2007.
[48] European Commission. Staff Working Document. European Space Programme – Preliminary elements SEC(2007) 504. Brussels, 26 Apr. 2007.
[49] EU Council. Resolution on the European Space Policy DS 471/07. Brussels, 16 May 2007.
[50] It also takes note of the European Space Programme.
[51] ESA. Agenda 2011 – A Document by the Director General and Directors. Paris, Oct. 2006.
[52] The Long-Term Plan is a tool for planning and managing ESA's financial resources allocated by ESA Member States, the EC and other organisations over a ten-year period.
[53] ESA. Agenda 2011 – A Document by the Director General and Directors. Paris, Oct. 2006.
[54] Ibid.
[55] The ECS concept was developed as a means to facilitating accession to the ESA Convention.
[56] The Presidency of the Council of the European Union refers to the responsibility of presiding over all aspects of the Council of the European Union, when exercised collectively by a government.
[57] Council's Operational Programme for 2006 submitted by the incoming Austrian and Finnish Presidencies; 22 December 2005.16065/05: "The Presidencies will continue to work on different aspects relating to the European Satellite Radio Navigation System GALILEO, paying particular attention to security, safety and financial issues as well as services, the international cooperation and the negotiation of the concession contract relating to GALILEO. Consequently, the Presidencies will make every effort to reach final agreement on the proposed regulation on deployment and operation".

[58] Following this initiative a series of meetings have been held by European regions in 2006/2007. Those meetings aim to develop and structure a "European Network of Regions using space technologies". A formal institutionalisation of this network is expected on 18 December 2007 during a Founding Conference to be held in Toulouse, France.
[59] Even with its two flagship programmes (Galileo and GMES), this is the case.
[60] However, other resources can be funnelled through other thematic priorities, e.g. security, to space-related activities.
[61] The amount of EC funds directly or indirectly dedicated to space varying year to year as a function of the number of calls for proposal issued, the quality of projects presented and their costs is estimated that the EC spent around 80 million euros on space activities through various mechanisms in 2006.
[62] This budget allocation might increase dramatically in the upcoming months following the problems of the Galileo programme.
[63] The European GNSS Supervisory Authority (GSA) was established by Council Regulation (EC) 1321/2004 of 12 July 2004 (and amended by Council Regulation (EC) No 1942/2006).
[64] Eumetsat will soon have 21 Member States with the full membership of Slovenia.
[65] Serbia Montenegro has recently expressed its willingness to join the organization.
[66] In the article "The Cabal Report of the French Parliament on space policy – A blue print for European space ambition or another cry in the wilderness?" published in the second part of the Yearbook, Kevin Madders provides a view on the "Cabal-Revol" Report and its impact on the European space policy debate.
[67] Taverna, Michael. "Germany Ups Space Spending". Aviation Week & Space Technology (5 Aug. 2006): 29.
[68] Ibid.
[69] Ibid.
[70] Nativi, Andy and Taverna, Michael. "Italy Plans Space Spending Hike". Aviation Week & Space Technology (27 Mar. 2006): 42.
[71] Ibid.
[72] Ibid.
[73] "Parliament Initiates U.K. Space Priorities Study". Space News Briefs 10 Aug. 2006. http://www.space.com/spacenews/archive06/Briefs_080706.html
[74] In his article "The new UK approach" published in the second part of the Yearbook, Klaus Becher provides a more in-depth analysis of the current space context in the United Kingdom.
[75] The BNSC was set up in 1985 to co-ordinate civil space activities across a number of Government Departments and Research Councils that have interests in space. It is a voluntary partnership of 11 Government departments and Research Councils: the Department of Trade and Industry (DTI); the Office of Science and Innovation (OSI); the Department for Education and Skills (DfES); the Department for Transport (DfT); the Ministry of Defence (MoD); the Foreign and Commonwealth Office (FCO); the Department for Environment, Food and Rural Affairs (Defra); the Council for the Central Laboratory of the Research Councils (CCLRC); the Natural Environment Research Council (NERC); the Particle Physics and Astronomy Research Council (PPARC); and the Met Office. It also acts as the UK point of contact with ESA, the EC, and space-faring countries.
[76] President George W. Bush signed off the new space policy on 31 August 2006. This document supersedes the September 1996 version of the directive.
[77] United States. The White House. US National Space Policy. 31 Aug. 2006. www.ostp.gov/html/US%20National%20Space%20Policy.pdf.
[78] In the 110th Congress a yearlong spending resolution, also known as the Byrd/Obey plan (from Senator Robert Byrd (D-W.Va.) and Representative David Obey (D-Wis.) the new appropriations committee chairmen), will force all agencies besides the DoD and Department of Homeland Security (DHS) to stay within their 2006 spending levels in 2007.
[79] United States. Congressional Research Service (CRS). The National Aeronautics and Space Administration's FY2006 Budget Request: Description, Analysis and Issues for Congress. 24 Jan. 2006.

80 Singer, Jeremy. "Air Force Budget, New Development Approach Reflect Congressional Scrutiny of Cost Growth". Space News (13 Feb. 2006): 13.
81 Singer, Jeremy. "US National Shield Dominates MDA Budget". Space News (13 Feb. 2006): 14.
82 Depending of the definition, the MDA can be considered as a space agency rather than an agency involved indirectly in space: As such its budget of 9.3 billion U.S. dollars (about 7.15 billion euros) can also be considered as space budget.
83 Permjakov, Viktor. "In 2007, the Russian space budget will amount to more than 50 billion rubles". Annanews (18 Sept. 2006). http://www.annews.ru/news/detail.php?ID=26397.
84 For Russia, as well as for emerging powers like China and India assessing the domestic commitment to space activities only using budgets figures is misleading due principally to major purchasing power differences with Western space powers.
85 Zaitsev, Yuri. "State Council Meets to Discuss Space Matters". RIA Novosti (30 Mar. 2007). http://en.rian.ru/analysis/20070330/62869340.html.
86 In his article "Basic law for space activities: a new space policy of Japan for the 21st century" published in the second part of the Yearbook, Kazuto Suzuki provides an insider view on the Japanese decision-making process as well as a more in-depth background on the recent evolution of the Japanese space context.
87 Currently, Japan's use of space is limited to non-military purposes under a strict interpretation of the 1967 Outer Space Treaty as required by the Japanese constitution and a subsequent parliamentary resolution incorporating the treaty into its domestic law.
88 The Japanese fiscal year starts in April.
89 The first White Paper on "China's Space Activities" was released in 2000. This document introduced China's situation in space technology, space applications and space science. It also provided guiding principles and core policies for China's space activities, as well as sets of priorities for international cooperation.
90 It is largely synonymous with the Central People's Government and is the chief administrative authority of the People's Republic of China. It is chaired by the Premier and includes the heads of each governmental department and agency.
91 The Information Office of the State Council is an administrative office under the State Council, the chief administrative body of the People's Republic of China.
92 As indicated in the White Paper, China value highly international cooperation and in the past five years it has signed 16 international space cooperation agreements and memoranda with 13 countries, space agencies and international organizations.
93 "Defence" and "security" are mentioned only three times throughout the entire document, and there is no mention of the China National Space Administration (CNSA) partnership with the People's Liberation Army (PLA), which is responsible for the operation of China's launch sites and its spacecraft, as well as the Human spaceflight programme.
94 While China's military establishment may participate in producing policies and initiatives, they do not independently publish doctrinal documents as in the United States. for instance.
95 China. China National Space Administration. "Full text: China's Space Activities in 2006". 12 Oct. 2006 www.cnsa.gov.cn/n615709/n620682/n639462/79381.html.
96 Ibid.
97 China. Commission of Science Technology and Industry for National Defense of China (COSTIND). "Eleventh-Five-Year-Plan of the Science Space Program". 19 Mar. 2007 www.cnsa.gov.cn/n615709/n620682/n639462/94761.html.
98 Ibid.
99 Ibid.
100 Morring, Frank. "In Orbit Chinese Space Lags U.S., Russia By 15 Years, Manager Says". Aviation Week & Space Technology (4 Oct. 2006): 15.
101 Jayaraman, K.S. "India To Develop Regional Navigation System". Space News (22 May 2006): 14.
102 Jayaraman, K.S. "Indian Space Budget Funds Astronaut Capsule". Space News (5 Mar. 2007): 17.

[103] The decades-old rivalry with China for regional supremacy seems also to be a major driver in such initiative.
[104] He-suk, Choi. "Science Ministry vows to turn Korea into a global space leader by 2015". Korea Herald (15 Jan. 2007). http://www.koreaherald.com/
[105] Kompsat-2, known in South Korea as Arirang-2, carries an optical imager with a 1-meter ground-resolution in panchromatic mode. It is also equipped with a multispectral instrument with a 4-meter resolution.
[106] Ankara TDN-Defense Desk. "Bidders vie for Turkey's Military Satellite". Turkish Daily News (19 Aug. 2006). http://gbulten.ssm.gov.tr/arsiv/2006/08/19/01.htm.
[107] The objective in producing this chapter is to establish a consistent and solid baseline of figures that are reliable and to the extent possible to provide an easily identifiable estimated measure of the size of the global space sector. Governmental space expenditures are not always easy to obtain as not every country and space agency publishes detailed annual expenditure on space activities. Moreover, given the opaque nature of defence budgets, tracking military space budget is extremely difficult since space is not identified as a separate line item in most national defence budgets. Sizing the commercial space sector is also difficult due to the discretion surrounding commercial contracts etc. Consequently, the overall size of the space sector can simply be only approximated, and estimates will vary from one study to another.
[108] The data for Japan comprise only the JAXA budgets.
[109] DBS revenues include revenues that are often not considered by many analysts as being space related as they provide "just" content. Nonetheless, DBS use satellite assets and drive both the satellite manufacturing sector as well as the demand of launch services. Furthermore, one of the fast-growing markets that is often excluded from many analysis, the localisation/navigation market for products has been included in those results.
[110] There are many estimates of revenues available in the literature, but in order to ensure consistency and to eliminate double counting, the Satellite Industry Association's (SIA) Satellite Industry Survey data are being used despite several approximations; nonetheless, this data set has been completed with other information. The results of the study "State of the Satellite Industry Report" released in June 2007 sponsored by the SIA and prepared by Futron Corporation covers four satellite industry segments (satellites services, launch industry, satellite manufacturing and ground equipment). This study is based on questionnaire targeting large companies operating on the satellite industry. The data is augmented with publicly available data and other industry research to calculate total industry revenues. All launch industry and satellite manufacturing revenues are recognized in the year of launch, not the year the contract is awarded. And, all revenues are in then-year dollars (not adjusted for inflation).
[111] XM Radio and Sirius Satellite Radio operate mainly in North America, while WorldSpace is a global satellite radio.
[112] Among the 2006's highlights was the return to the stock market of Globalstar and Orbcomm as both raised sufficient capital on the U.S. Nasdaq market to permit them to move forward on second generation satellite constellations. In the mean time, Inmarsat acquired ACeS and launched its new service Inmarsat BGAN.
[113] The low level of revenues of Russian launch services providers compared to their high level of activity results from the fact that they are particularly active in the LEO market which has a lower price tag than launches to deliver satellites to GTO, the most lucrative market. On the contrary, Arianespace's good results with a similar number of launches, for instance, Sea Launch, can be explained by the fact that it is launching two payloads at the same time, and has therefore higher revenues per launch.
[114] Kramer, Andrew E. "GPS alternatives take flight". International Herald Tribune (3 Apr. 2007). http://www.iht.com/articles/2007/04/03/business/gps.php
[115] Both Ansari and Simonyi followed the Americans Dennis Tito (2001) and Greg Olsen (2005), and the South African Mark Shuttleworth (2002).
[116] In the their article entitled "Space Entrepreneurship – Status and Prospects" published in the second part of the Yearbook, Joeg Kreisel and Burton Lee provide a greater inside view of the current activities of space entrepreneurs in Europe and the United States.

[117] The EC was concerned that Thales could gain an unfair advantage by combining Thales Traveling Wave Tubes (TWTs) with related Alcatel components and subsystems.
[118] The consortium is composed of seven private institutions that are buying 60% of the stake and are Allianz, Commerzbank, Credit Suisse, Deutsche Bank, Goldman Sachs, Morgan Stanley and Sal Oppenheim. The eight public banks and regional states that bought the remaining 40% of DaimlerChrysler's share are led by KfW, the state development bank.
[119] "OHB Technology AG acquires Kayser-Threde GmbH, Munich". OHB Press Release (21 June 2007).
[120] ILS was a joint-venture formed in 1995 between Lockheed Martin, and Khrunichev State Research and Production Space Center, and RSC Energia for the purpose of co-marketing their respective rocket launch services, the Russian Proton and the Lockheed Martin Atlas V.
[121] The Federal Trade Commission is an independent agency of the U.S. government and its principal mission is the promotion of consumer protection, and the elimination and prevention of anticompetitive business practices.
[122] Intelsat Holdings, Ltd. is an entity formed at the direction of funds advised by or associated with Apax Partners Worldwide LLP, Apax Partners Inc., Apollo Management V, L.P., MDP Global Investors Limited and Permira Advisors, LLC.
[123] The Russian cabinet endorsed also the "2015 Strategy on Development of the Rocket and Space Industry" on 06 July 2006.
[124] Suzuki, Kazuto. "Transforming Japan's space policy-making". Space Policy 23 (2007): 73–80.
[125] Ibid.
[126] Ibid.
[127] China. China National Space Administration. "Full text: China's Space Activities in 2006". 12 Oct. 2006. http://www.cnsa.gov.cn/n615709/n620682/n639462/79381.html.
[128] Ibid.
[129] Antrix Corporation in cooperation with EADS Astrium will manufacture the W2M satellite to Eutelsat. The transponders for the satellite will be built by EADS Astrium, while the satellite platform will be built by Antrix Corporation.
[130] "ISRO builds communication satellites for European clients". The Hindu News. 19 Apr. 2007.
[131] Several differences can be observed when comparing the following results with other studies due to methodological discrepancies. The following definitions apply to the launch sector analysis. A commercial orbital launch is defined as a primary payload for which the contract was internationally competed (the launch opportunity was available in principle to any capable launch service provider) and/or the launch is privately financed without government support. A primary payload is defined as the payload with the greatest mass for the concerned launch. Finally, launches are attributed to the country in which the main vehicle manufacturer is based, except on the case of Sea Launch which is designed as multinational. However, when considering, for instance, ILS, when a launch is done using an Atlas V, the launch is viewed as a U.S. launch, and when a Proton is used, the launch is considered Russian. Finally, no distinction has been made between the Ukrainian and Russian launch systems as major shareholders in most Ukrainian launch providers, as well as launch manufacturer are Russian.
[132] Of the 66 worldwide orbital launches, there were four launch failures including one commercial (Proton) and three non-commercial (Falcon1, Dnepr and GSLV).
[133] Arianespace has 23 shareholders from ten European countries including the main shareholders French space agency CNES (32.53%), EADS (28, 59%), and all European companies participating in the construction of Ariane launchers.
[134] Arianespace will act as launch service operator of the Vega launcher for five consecutive launches following the qualification flight within the framework of the Vega Research and Technology Accompaniment (VERTA) programme decided at the ESA Council Meeting at the Ministerial level in 2005. These launches will mainly carry payloads built by ESA.

[135] Boeing is the majority shareholder (40%) of Sea Launch. Other partners include S. P. Korolev Rocket and Space Corporation Energia of Russia (25%), Aker ASA of Norway (20%), and SDO Yuzhnoye/NPO Yuzhmash of Ukraine (15%).
[136] A Russian-Ukrainian government commission concluded that the failure was caused when a foreign object found its way into the first-stage engine and caused it to shut down.
[137] The first model of its new platform DFH-4 (Sinosat-2) was launched on 29 October 2006, but experienced severe in-orbit problems soon after.
[138] De Selding, Peter. "Top Fixed Satellite Service Operators". Space News (25 June 2007): 10–13.
[139] The terms "military" and "security" are used interchangeably in the text when referring to space issues, as beyond the minor semantic difference, the use of space assets for military or security purposes overlap considerably.
[140] SIPRI. SIPRI Yearbook. Oxford: Oxford University Press, 2007.
[141] European countries can join national programmes by paying their share to the programme's leader, and receiving a proportional part of the services offered. The pooling of financial resources and the exchange of data or even sometimes satellite-tasking time is thus a well-integrated and common procedure in Europe.
[142] Pleiades will be a French two-satellite optical military reconnaissance system and COSMO-SkyMed will be an Italian constellation of X-band SAR satellites.
[143] In the case of Skynet programme, Paradigm Secure Communications (a subsidiary of Astrium Services) signed a contract in 2003 (and then revised it in December 2005) with the British Defence Ministry that could be worth up to 3.66 billion British pounds (about 5.4 billion Euros) through May 2021 provided a number of options are exercised. In addition to the purchase and operation of the Skynet 5s (three guaranteed Skynet 5 spacecraft), the contract covers the purchase, management and running of the Skynet 4 fleet and the provision of the ground segment. This PPP is the first involving full outsourcing of military satellite telecommunications to the private sector.
[144] Europe is trying to replicate the Skynet 5 PPP model with the Galileo Programme, but is having trouble in doing so.
[145] The Franco-Italian Athena-Fidus agreement was signed on 22 June 2006 by the Heads of the French and Italian space agencies.
[146] "Besoins Operationnels Communs" or "Common Operational Requirements for Global European Earth Observation System by Satellites" in English.
[147] Ibid.
[148] EU Council. Resolution on the European Space Policy DS 471/07. Brussels, 16 May 2007.
[149] Military space budgets are highly concentrated in the United States who spends more than 95% of the (known) world public funding.
[150] United States. The White House. "US National Space Policy". http://www.ostp.gov/html/US%20National%20Space%20Policy.pdf.
[151] United States. Presidential Decision Directive/NSC-49/NSTC-8. National Space Policy. 14 Sep. 1996.
[152] Ibid.
[153] Ibid.
[154] Xavier Pasco in his article published in the second part of the Yearbook entitled "Controlling the freedom of using space: the White House space policy dilemma" provides relevant background on the evolution of U.S. space policies over the years, and in particular vis-à-vis the increasing security dimension of domestic policies in the United States.
[155] United States. Presidential Decision Directive/NSC-49/NSTC-8. National Space Policy. 14 Sep. 1996.
[156] United States. NGA. The National System for Geospatial-Intelligence Statement of Strategic Intent. 15 Mar. 2007. http://www.nga.mil.
[157] The QDR analyses strategic objectives and potential military threats. It is the main public document describing the United States' military doctrine.

[158] United States. US Department of Defence. Quadrennial Defence Review Report. 6 Feb. 2006 http://www.defenselink.mil/qdr/report/Report20060203.pdf.
[159] Ibid.
[160] Gertz, Bill. "US halts China space ventures". The Washington Times (2 Feb. 2007). http://findarticles.com/p/articles/mi_hb5244/is_200702/ai_n20944434
[161] Popovkin, Vladimir. "Russia's Space Defenses Stage a Revival". RIA Novosti (4 Oct. 2006). http://en.rian.ru/analysis/20061004/54509604.html.
[162] Ibid.
[163] Currently, Japan's use of space is limited to non-military purposes under a strict interpretation of the 1967 Outer Space Treaty as required by the Japanese constitution and a subsequent parliamentary resolution incorporating the treaty into its domestic law.
[164] Suzuki, Kazuto. "Transforming Japan's Space Policy-making". Space Policy 23 (2007): 73–80.
[165] These incidents were reported publicly by the director of the National Reconnai ssance Office, Donald Kerr in October 2006.
[166] Pillsbury, Michael P. "An Assessment of China's Anti-Satellite and Space Warfare Programs, Policies and Doctrines". Commissioned Research Study (Jan. 2007). http://www.uscc.gov/researchpapers/2007/FINAL_REPORT_1-19-2007_REVISED_BY_MPP.pdf.
[167] "India Begins Work on Space Weapons Command". Space Daily. 12 Apr. 2006.
[168] South Korea asset is considered a military spacecraft due to its explicit "dual-use" nature.
[169] Götz Neuneck, in his article entitled "The China's ASAT Test – A warning shot or the beginning of an Arms race in space?" published in the second part of the Yearbook, provides a more detailed technical analysis of China's FY-1C destruction.
[170] In his article "The US missile defence shield in Europe" published in the second part of the Yearbook Thomas Valasek provides a view on the recent developments in the installation of component of the U.S. missile defence system in Poland and the Czech Republic and its potential impact on both Russia and Europe.
[171] United States. Nasa Orbital Debris Program Office. Orbital Debris Quarterly News 11.2 (2007) http://orbitaldebris.jsc.nasa.gov/newsletter/newsletter.html.
[172] Ibid.
[173] United States. Nasa Orbital Debris Program Office. Orbital Debris Quarterly News. 11.3 (2007) http://orbitaldebris.jsc.nasa.gov/newsletter/newsletter.html.
[174] Ibid.
[175] Ibid.
[176] Ibid.

# Developments in space policies programmes and technologies throughout the world and in Europe

Nicolas Peter

## 1. Space policies and programmes

A series of major policy and programmatic milestones as well as reinforcing trends were the highlights of the year 2006 and the first half of 2007. These included the increasing internationalization of access to space and space applications as well as the new race to the moon.

### 1.1. Highlights in activities and programmes

In the domain of space transportation, independent and reliable access to space is still limited to the space powers particularly for human access to space. However, emerging space powers are improving the reliability of their launch vehicles and newcomers in the space sector are moving forward to develop their own space transportation infrastructure independently or in cooperation (i.e. South Korea's Korea Space Launch Vehicle 1 or KSLV-1).

In the domain of space science and exploration following the long hiatus after the loss of Columbia in February 2003, the assembly of the International Space Station (ISS) resumed in September 2006. While human spaceflight activities were dynamic and successful for Europe in 2006, in China, delays in the testing of an indigenously developed spacesuit have postponed the third Shenzhou mission that had been planned for early 2007 to 2008. Fuelled by an increasing regional rivalry with China, the Indian space agency, ISRO is eager to start a human spaceflight programme and is testing some preliminary technologies. An increasing number of space agencies have also planned lunar orbiter and lander missions leading to an unprecedented race to the moon. However, while Earth's natural satellite is a target of choice for major space-faring countries, the Red Planet has also shifted to the centre of attention for exploration plans.

In 2006/2007 also marked the confirmation of the trend witnessed in recent years of China and India increasingly becoming competitive exporters of space technology, particularly, in the domain of space-based telecommunications confirming their emerging status as major space powers. In the field of position, navigation and timing (PNT), while a continuously growing number of users and applications benefit from the U.S. Global Positioning System (GPS), the new initiatives and alternatives are gaining momentum. Europe, Russia, Japan, China and India have taken further steps towards developing their own systems. In 2006/2007, a successful Earth observation mission came to an end and new assets were launched, with Europe being particularly active in this domain. Progress was achieved, above all, in making the Global Monitoring for Environment and Security (GMES) programme and its Fast Tracks Services operational.

A series of new technologies and design were tested in the last few months, specifically in the field of propulsion, information technology, spacecraft operations, thus paving the way for future breakthroughs. Progress was also achieved in the development of sub-orbital commercial activities in the United States and in Europe.

## 1.2. Highlights in partnerships

Over the years, international cooperation has been a central element of the strategy of most countries involved in space activities. However, because partnerships are not static but highly dynamic,[177] there were new developments in 2006/2007.

Trans-Atlantic relations have historically been a major part of Europe's space cooperative activities, and have been led primarily by European countries and ESA with a strong focus on Earth observation and space sciences. However, a new actor on the European side, the European Union (EU), is gaining weight in this axis of cooperation following the GPS-Galileo negotiations as well as the June 2005 EU-US Summit.[178] During the bilateral Summit, a EU-US Agreement was signed that provides for an annual dialogue meeting on space, focusing mainly on space applications under the leadership of the European Commission (EC) for Europe and of the State Department for the United States. The last EU-US Summit took place in Washington D.C. in the first half of 2007. Another axis of trans-Atlantic cooperation is the Eumetsat-National Oceanic and Atmospheric Administration (NOAA) axis on space-based meteorological activities. Russia is also a priority for Europe and the importance of space cooperation with Russia has been reaffirmed as a top priority for Europe on many occasions. In particular, Russia, the EU and ESA signed a cooperation agreement on space technologies and activities on 10 March

2006. The areas of specific interest are space applications (satellite navigation, Earth observation and satellite communications), access to space (launches and future space transportation systems), space exploration and the ISS, and space technology development.[179] The main axis of European-Russian cooperation, however, remains the one on launch vehicles. Russia participates with ESA in the Advanced Crew Transportation System (ACTS) project to explore crew-transport vehicle designs. Furthermore, Europe and Russia are also cooperating on hosting the Russian mid-class Soyuz at the Guiana Space Centre (CSG).

In the United States, a major contemporary axis of cooperation focuses on space applications like the GPS-Galileo, or the aforementioned Eumetsat-NOAA cooperation as well as space exploration-related activities. On 6 December 2006 NASA rolled out its strategy and rationale for robotic and human exploration of the moon centred around a lunar polar outpost to achieve a sustained, human presence on the moon, NASA also detailed its plans for a "global exploration strategy" guiding the future international coordination and collaboration efforts. This proposal evolved into a "Global Exploration Strategy – The Framework for Cooperation" agreed by fourteen space agencies on 31 May 2007 at a meeting in Spineto, Italy.[180] NASA also signed an agreement on 18 June 2007 to launch the James Webb Space Telescope (JWST) aboard an Ariane 5 rocket. The launch of the successor of the Hubble Space Telescope is tentatively scheduled for 2013. ESA provides the launch as part of its planned contribution to the mission and the two agencies will also split the instrument development for the telescope. However, one of the partnership highlight of 2006/2007 is the fact that the United States entered into high-level talks on civilian space activities with China. Following a meeting in Washington in April 2006, U.S. President George W. Bush and Chinese President Hu Jintao agreed on a visit of Mike Griffin to China. Griffin was the first NASA Administrator to make an official visit to China in September 2006. However, following the Chinese's anti-satellite test on 11 January 2007 that destroyed the Feng Yun 1C weather satellite, the initial high-level talks on potential Sino-American cooperation in civilian space activities have been suspended by the Bush Administration.

The renewed space interest at the highest-political level in Russia combined with the budgetary increase devoted to space activities has led to the reinforcement of several cooperation projects and partnerships. In 2006/2007, Russia continued to cooperate with Europe and the United States on human spaceflight and space transportation activities. For instance, in April 2007, Roskosmos signed a 719 million U.S. dollars (about 553 million euros) modification to the current ISS contract with NASA relating to crew and cargo logistics services through 2011. However, Russia is also looking to increase its cooperative activities with emerging space powers. It is looking to intensify its partnership with China in lunar exploration beyond the traditional sales of Russian equipment to the Chinese government. A dedicated cooperation agreement was signed in March 2007 whereby a Chinese small satellite will be launched along the Russian sample return mission "Phobos Explorer". China and Russia are also cooperating in space astronomy. China will participate in the "Radioastron" programme scheduled to be launched in 2007/2008 and it will also launch the Russian ultraviolet observatory in 2010. While Russia is cooperating with China, it is also collaborating with India on updating Russia's Global Navigation Satellite System (Glonass) as part of a broad space cooperation plan that also encompasses space sciences. In particular an Indian instrument for solar physics and solar-terrestrial sciences will be flying on board the Russian Coronas-Photon satellite.[181] Russia's revitalization of its space programme has led also to an increasing number of space cooperation agreements with non-space-faring countries as a tool of foreign diplomacy with agreements and negotiations with South Africa, Venezuela, Iran, Thailand, Vietnam, Chile, and the United Arab Emirates, as well as direct joint-project with South Korea.

In 2006/2007, Japan has continued to focus its cooperation efforts on Research and Development (R&D) activities as well as on scientific activities with established space powers like the United States and Europe. However, following the creation of the Asia-Pacific Space Cooperation Organization (APSCO), a regional space organization under Chinese leadership, JAXA is now trying to re-launch the Asia-Pacific Regional Space Agency Forum (APR-SAF) that started in 1993. In particular, JAXA is aiming at supporting Asian countries in various applications programmes and, in particular, in Earth observation and education programmes as a tool of foreign relations in the region.

Continuing the recent trends observed, in 2006/2007 China has been more open to participation in international fora and has been more engaged in international cooperation projects following its technological achievements and its nascent ambitions in the area of space sciences and exploration. In this context, China has attended various international exploration workshops and meetings in 2006/2007. As already mentioned, it also signed a dedicated agreement with Russia in March 2007 to contribute a small satellite to be launched along the Russian sample return mission "Phobos Explorer" to the Martian moon Phobos in October 2009. However, an extraordinary development was that China entered in direct bilateral cooperation talks with the United States on potential civilian space cooperation. However, those talks were suspended by the United States following the January 2007 anti-satellite test. While China is increasingly cooperating with major space-faring countries, it is also reinforcing its partnership with countries from the Southern hemisphere by facilitating access of those countries to space technologies as a foreign policy tool. An example is Nigeria when China's export-credit agency granted a 200 million U.S. dollars (about 153 million euros) loan to the Nigeria government in January 2006 to complete the financing package of Nigeria's Nigcomsat 1 telecommunications satellite that was subsequently built and launched by Chinese companies. China decided to donate data-receiving stations for its weather satellites to seven nations around the Pacific Rim and Indian Ocean (Bangladesh, Indonesia, Iran, Mongolia, Pakistan, Peru and Thailand[182]) that are all members of the Asia-Pacific Space Cooperation Organization (APSCO).

Despite the difficulties of Indo-European cooperation within the framework of the Galileo programme, India is developing its cooperation with Europe, and besides Europe's participation in Chandrayaan-1, it has agreed to include an atmospheric sounder called Rosa from the Italian Space Agency (ASI) on board its Oceansat-2 satellite. The Indo-French joint space mission, Megha-Tropiques, to study the atmospheric water circulation in the tropical belt has also made further progress. Moreover, plans for a mini-satellite altimeter to be launched in 2009/2010 in cooperation with France are moving forward. India is also co-operating with Russia on updating Russia's Global Navigation Satellite System (Glonass) as part of a broad space cooperation plan. However, India is undergoing a major paradigm shift and is evolving beyond its traditional domain of space activities of that are Earth observation, space-based telecommunications and unmanned access to space and moving into space exploration and space science activities. In this context, ISRO after it successfully tested recently technology, is working on its first unmanned mission, Chandrayaan-1, to be launched to orbit the moon in early 2008 with the payloads from ESA and from the United States. Furthermore, an Indian instrument will be flying on board the Russian Coronas-Photon satellite. In addition to this cooperative venture with established space powers, "South-South" cooperation is still a major axis for India. In this context, during the November 2006 visit to New Dehli by Chinese President Hu Jinto, India and China, despite their regional rivalry, indicated their desire to cooperate in the use of space technologies for peaceful purposes. Potential areas of cooperation include remote sensing, communications, meteorology, distance learning, disaster management and launch services. Nonetheless, China's ASAT test in January 2007 has slowed down any further endeavours. However, another example of "South-South" cooperation is the one between India and Brazil, following the visit on 16 May 2006 of an ISRO delegation to Brazil to meet their counterpart from the Brazilian Space Agency (AEB). At this meeting, ISRO signed an agreement with its Brazilian counterpart on 4 June 2007 under which ISRO will equip a Brazilian Earth station with instruments that will enable it to receive and process data from India's remote sensing satellites.

# 2. Space transportation

Independent access to space is still the privilege of the space powers, particularly, human space flight. However, emerging space actors are improving the reliability of their domestic launch vehicles, and newcomers in the space sector are also developing space transportation infrastructure.

## 2.1. Europe

The past few months have been particularly important for Europe. Following two successful flights in 2005, the generic qualification of the ECA version of Ariane 5 was confirmed in 2006, thus providing conclusive evidence that Ariane 5 ECA had acquired fully operational status. Starsem[183] performed two launches in 2006 with the Soyuz 2-1a and Soyuz 2-1b that successfully placed the MetOp-A and Corot satellites, respectively, into orbit validating the two new Soyuz versions to be flown from the Guiana Space Centre (CSG). On 26 February 2007, the construction site of the Soyuz launch base in French Guiana was officially opened. However, the first launch of a Soyuz rocket from Kourou is now slated for early 2009 mainly because of problems in building a giant flame deflector under the new launch pad.

In 2006/2007, the Vega light launcher programme passed major milestones with the first ground test firings of the P-80, Zephiro 23 and Zephiro 9 solid rocket stages. Construction of the mobile gantry at the GSC is also underway. The first commercial Vega launch from the GSC is planned for 2009. The Vega small launcher, alongside the Ariane 5 heavy-lift launcher and Russian mid-class Soyuz at the GSC will complete the range of launchers operated by Arianespace.

Europe's space tug, known as the ATV (Automatic Transfer Vehicle) is also getting closer to its inaugural mission to ISS after an extensive three-year test campaign. The first launch of the ATV, named Jules Verne, has been pushed back to January 2008 at the earliest. It will now be launched after the Columbus orbital laboratory. The ATV will be a vital ISS add-on, re-supplying the station with dry and liquid cargo, periodically re-boosting it into a higher orbit, and removing waste on departure. But, the ATV is also a key part of barter services to be provided by ESA in return for power, air, water and other Columbus utilities furnished by the United States.

ESA also agreed on 22 June 2006 to participate in an 18-month long effort with Russia to explore crew-transport vehicle designs for missions to ISS, the moon and beyond. An ESA manned space programme board meeting on 12 July 2006 approved an 18–22-millions euros funding envelope for a detailed study project to determine a possible European role in cooperation with Russia. The Russian proposal formally called Clipper has evolved to an effort called Advanced Crew Transportation System (ACTS) an initiative that aims to replace the aging Soyuz spaceship with a new system able to serve space exploration as well as ISS. Japan and Italy have also indicated interest in joining ACTS. Initial plans suggest the use of both the ESA launch site in Kourou and the Baikonur Cosmodrome in Kazakhstan, both of which could be compatible with Soyuz-derived launchers. By 2008, a decision should be taken on whether to participate in full-scale development. At a meeting in Moscow on 14 February 2006, French and Russian government officials renewed their commitment to the five-year joint programme (Oural) whose ultimate goal is the development of a commonly designed heavy-lift rocket for commercial and government payloads that would be ready for service around 2020.

## 2.2. United States

Design review and early developments have started in recent months on NASA's next-generation launch infrastructure: the Crew Launch Vehicle (CLV) now named Ares-1, and the heavy-lift Cargo Launch Vehicle (CaLV), now named Ares-5 planned for moon and Mars exploration missions. On 31 August 2006, NASA made the long-awaited announcement that Lockheed Martin[184] won a 3.9 billion U.S. dollars contract (about 3 billion euros) through 7 September 2013 against the bidder team of Northrop Grumman and Boeing to help NASA design and build the Crew Exploration Vehicle (CEV), now named Orion.[185] This contract covers design, development, construction, testing and evaluation of the first two spacecrafts. The contract could be worth as much as 8.15 billion U.S. dollars (about 6.26 billion euros) through 7 September 2019, depending on how many spacecraft are ordered by NASA. NASA wants this vehicle in service no later than 2014 to ferry Astronauts and cargo to ISS and later transport up to four Astronauts to the moon.[186] The number of vehicles NASA will ultimately buy depends on a variety of factors, including how many flights NASA can get out of each vehicle and how many missions it needs to conduct. However, due to the gap between the end of the shuttle programme in 2010 and the introduction of Orion crew exploration vehicle in 2014, U.S. President George W. Bush signed on 22 November 2005 the U.S. Senate bill S.1713, opening the possibility for NASA to purchase services related to the use of the Soyuz and Progress vehicles from Russia needed to support the ISS.

Subsequently, before the end of the agreement by which Russia had been supplying services from these vehicles terminated in April 2006, Russia and the U.S. agreed on 6 January 2006 on a price of 21.8 million U.S. dollars (about 16.7 million euros) for NASA Astronauts to travel to and from the ISS on Soyuz rockets until 2011.

On 18 August 2006, NASA announced that it selected Space Exploration Technologies (SpaceX) and Rocketplane Kistler (RpK) to share 500 million U.S. dollars (about 380 million euros) in funding under the Commercial Orbital Transportation Services (COTS) demonstration programme to help complete the development of new launchers and spacecraft, and demonstrate by 2010 that they can safely deliver cargo to the ISS.[187] Under the terms of the COTS Space Act Agreement, SpaceX and RpK must show NASA that they are making technical, programmatic and financial progress in order to continue receiving financial assistance. Neither company, however, is guaranteed a NASA re-supply service contract even if they successfully complete their demonstration, as NASA intends to hold another competition at the end of the decade for the actual service contract.

On 1 February 2007, NASA announced that it had agreed to cooperate with PlanetSpace Inc and Transformational Space Corp (t/Space) to facilitate the development of capabilities to deliver goods and people to orbital destinations by signing non-reimbursable Space Act Agreements. Those agreements involve no agency funding, but establish milestones and objective criteria by which the companies can gauge their own progress and aim to provide those companies with up-to-date technical requirements and specifications for crew and cargo flights to the ISS.[188] Furthermore, in April 2007, NASA announced that it signed a 719 million U.S. dollars (about 553 million euros) modification to the current ISS on contract with Russia's Federal Space Agency, Roskosmos for crew and cargo logistics services through 2011,[189] NASA but that it still planned to rely on its COTS programme to provide space station logistics starting in 2010. In this context, NASA announced on 18 June 2007 that it had signed three more non-reimbursable Space Act Agreements with three others private firms, Constellation Services International (CSI), SpaceDev and Spacehab, that like with PlanetSpace Inc and t/Space aim to field new space transportation system that could carry cargo and/or Astronauts to the ISS.

The year 2006 also witnessed the first flight of a new U.S. launcher, the privately developed Falcon 1 by SpaceX that is part of the Falcon series of launch vehicles (Falcon 1, 5 and 9) for the commercial market, as well as for government clients. The first attempt to launch Falcon 1 was in March 2006. The unsuccessful launch, which took place at the SpaceX private launch complex in the Pacific Ocean on the Omelek Island in the Kwajalein Atoll, carried the FalconSat-2 satellite for the U.S.

Air Force Academy. The failure of the Falcon 1 rocket on its maiden flight was due to a busted nut that led to a fuel leak according to the findings of a government review board charged with the investigation of the 24 March 2006 failure. Nonetheless, Falcon 1 reached space on its second test flight on 20 March 2007, but suffered a second stage malfunction that prevented it from reaching the planned orbit.

Finally, the joint Boeing-Lockheed Martin company entitled United Launch Alliance (ULA) was cleared in October 2006 by the U.S. government to address the non-commercial U.S. government launch market. On 14 December 2006, the first launch carried out under ULA auspices took place when a Delta 2 rocket carried a classified NRO payload into Low Earth Orbit (LEO).

## 2.3. Russia

In 2006/2007, a series of failures in the ex-U.S.S.R launch vehicle family was particularly dramatic for the Russian industry.[190] In this context, in line with the revived Russian space ambitions presented in the new Federal Space Programme (2006–2015), which is an attempt to improve the competitiveness of the country's industrial base, the efforts in the domain of space transportation infrastructure were increased in the past few months. The development of the replacement of the Soyuz space capsule used to support ISS and its associated infrastructure, as well as the maintenance and development of the Baikonur Cosmodrome facilities and the development of internationally competitive rocket technology are the major items of this new Federal Space Programme. In the new space-spending plan, 27.32 billion of rubles (about 797.06 million euros) would be assigned to the development of the various Russian spaceports in a two-step approach.[191] The first phase (2006–2010), will focus on the development of new launch pads to launch military payload on Soyouz-2 from Plesetsk as well as the testing of the new heavy launch vehicle Angara. About 9.53 billion of rubles (about 278 million euros) were allocated to this first phase.[192] The second phase (2011–2015), will consist of the development of the infrastructure needed by the Angara launch vehicle in Plesetsk and the transfer of all Russian military launches on Russian territory. The second phase is allocated a budget of 17.79 billion of rubles (about 519 million euros).[193]

Russia's revitalization of its space programme has led to an increasing number of space cooperation agreements including agreements on space transportation with non-space-faring countries as a tool of foreign diplomacy. In particular, Russia will help Brazil on its VLS-1 launch vehicle (Veículo Lançador de Satélite) and South Korea on its Korea Space Launch Vehicle 1 (KSLV-1). Russia is also participating with Europe in several space transportation programmes. It is involved with ESA in a programme to explore crew-transport vehicle designs for missions to ISS, the moon and elsewhere. Furthermore, as already mentioned, the French and Russian government officials recommitted themselves in February 2006 to the five-year joint programme (Oural) whose ultimate goal is the development of heavy-lift rocket for commercial and government payloads.

## 2.4. India

Despite the launch failure on 10 July 2006 of India's Geosynchronous Satellite Launch Vehicle (GSLV) from ISRO's new launch pad at the Satish Dhawan Space Centre on Sriharikota Island that destroyed the domestically built Insat 4-C television broadcasting satellite, Indian leadership including President Abdul Kalam, one of India's space pioneers, has been supportive of its space programme and has even increased its financial support to space activities. In particular, ISRO's budget for the 2006/2007 fiscal year, funds have been allocated to the development of a new vehicle (GSLV Mk3) that will be able to lift the planned 4-ton satellites to geostationary transfer orbit (GTO) and 10-ton into LEO. Furthermore, following the efforts to improve the reliability and performance of its launch vehicle fleet, India is now entering the "open-commercial market", as for the first time, it launched a foreign payload in April 2007 after winning an international competition (an Italian scientific satellite) on board its indigenously developed Polar Satellite Launch Vehicle (PSLV).

## 2.5. Emerging actors

Besides the traditional space powers, other space actors have been working on developing autonomous access to space and the associated infrastructure in 2006/2007.

Israel, following the failure of its fourth Shavit-1 launch in September 2004 due to a problem during the separation between the second and third stage successfully launched its latest reconnaissance satellite in June 2007 in the maiden flight of the Shavit-2 launch vehicle. Shavit-2 is now the workhorse of Israel's space activities.

In its quest to become a space power, the South Korean government is overseeing launch-related projects in addition to satellites development and its nascent human spaceflight programme. One such project is the construction of the Oinarodo Space Centre. The construction project started in 2002 and is scheduled for completion in late 2007. The space centre will serve for launching the KSLV-1 currently under development, which has been developed with technical cooperation from Russia and is scheduled for its first launch in 2008. It is also envisioned that Korea will have developed an indigenous space launch vehicle entirely from local technology by 2015.[194]

In Brazil, following the pre-launch accident that took place during the third launch campaign of the small satellite launcher VLS-1 in August 2003, the development of sounding rockets as well as of the VLS-1 have continued. Moreover, as mentioned above, Brazil is cooperating with Russia to improve the reliability of the VLS-1 launch vehicle. Furthermore, Brazil has worked to refurbish and modernize the Alcantara Launch Centre (CLA) in 2006/2007.

## 2.6. International comparison

In 2006, 66 launches distributed over seven countries/entities took place. However, it is clear that the number and capacity of the launch vehicles varies widely among the space actors. In 2006, when comparing the levels of activity country by country, Russia was the world's space leader according to the launch rate criterion followed distantly by the United States. Twenty five Russian vehicles were launched using ten different launchers (Figure 15). Ten of those launches were commercial ones and 15 were non-commercial, with five dedicated to re-supplying the ISS, four other to civil governmental launches and six to military launches. U.S. vehicles carried out a total of 18 launches in 2006 on seven different launchers,

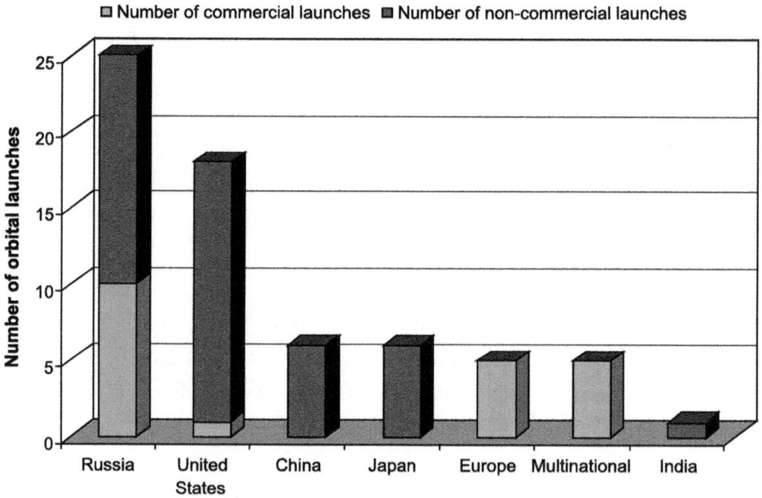


**Fig. 15:** *Worldwide commercial and non-commercial orbital launches per country/entity in 2006.*

17 of these launches were non-commercial and only one was commercial (Figure 15). Japan conducted six launches using its two launchers M-5 and H2A, and all launches were non-commercial (Figure 15). China also conducted six launches in 2006, up from five in 2005, but less than in 2004 with a record of eight. China used three types of rockets (Long March 2, 3 and 4). Europe conducted five launches in 2006 with its Ariane 5, all commercial ones, like Sea Launch and its Zenit 3SL.[195] Finally, India performed only one launch in 2006, but it failed to reach orbit. The versatility of the launch vehicle fleet therefore reflects the national capabilities of a country and the importance it gives to independent access to space.

The launches were distributed over 25 different launch systems (Figure 16). Soyuz was the most widely used launch system with 11 launches, followed by Delta 2 and Proton, with six launches each, and Ariane and Zenit 3SL with five orbital launches (Figure 16).

While the number of launches by each country indicates the number of vehicles produced by a country, the degree to which space bases are used also directly reflects the hierarchy of the states in their capacity as major space actors and indicates to which extent a country is maintaining and improving its infrastructure.

When looking at the countries involved in the launch sector, the ex-U.S.S.R. was the most active in 2006 with 36% of all launches taking place from Russia or a former member of the U.S.S.R (i.e. Kazakhstan). The United States launch sites

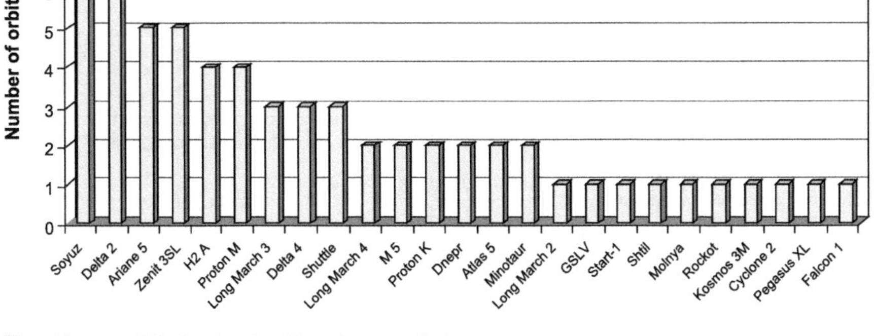


**Fig. 16:** *2006 Worldwide orbital launches per vehicle in 2006.*

saw 26% of all launches performed in 2006, followed by China, Japan and Europe with a similar volume of activity. The analysis of the level of activities of launch sites therefore, also reveals certain persistent historical factors, such as the domination of Russia and the United States.

However, while it is acknowledged that space-faring countries must have a complementary fleet of launch vehicles to address the various commercial markets or national needs, what is interesting to note is that major space actors rely, for historical or strategically reasons, on several launch bases. Russia used five different launch sites in 2006, including a mobile platform (a submarine in the Barents Sea), while the United States used four different bases (Figure 17). In 2006, China used three distinct launch sites to launch its rockets, three from the Xichang launch site, two from Tai Yuan and one from Jiquan (Figure 17). Japan used both of its launch sites in 2006 while India and Europe used only one base. Finally, in 2006 Russia's Sea Launch project probably relied solely on its mobile platform "Odyssey" for the last time to launch rockets, as it plans to operate launches from Baikonur as of the fall 2007.

In 2006, 17 launch sites performed at least one launch including two mobile platforms one form the Pacific Ocean, and one from a Russian submarine in the Barents Sea. Mobile platforms therefore represented 9% of the volume of launch site activity, while 91% of all launches were performed form fixed launch sites. Baikonur in Kazakhstan (operated by Russia) was the busiest space base in 2006 with 17 launches followed by Cape Canaveral in the U.S. with 10 launches (Figure 17).

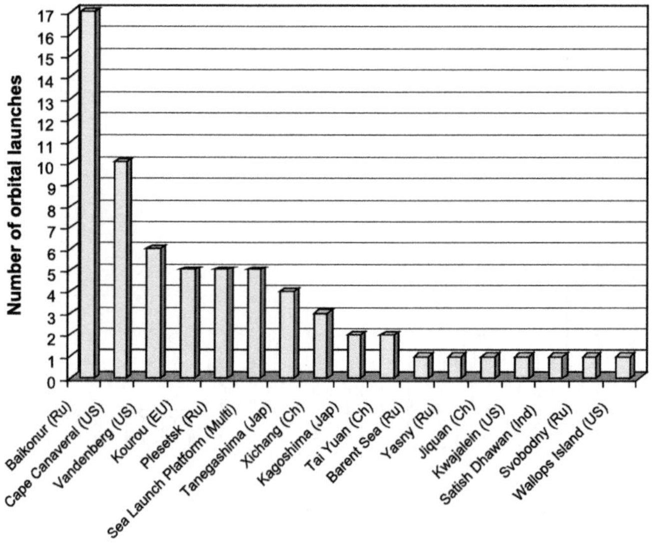


**Fig. 17:** *2006 launches by launch site.*

Furthermore, in 2006/2007 some bases were reactivated after being dormant for quite some time, and new ones were inaugurated (i.e. Yasny launch base also known as the Dombarovskiy missile base). In early 2007, China has also been reported to have chosen the site for its new satellite centre. The Wenchang Satellite Launching Centre will be the fourth in China and is likely to be the launch site for the next generation of rockets to be put in use at the beginning of the next decade.[196] Furthermore, Russia has an agreement with Kazakhstan to lease the Baikonur spaceport until 2050 and is also considering developing another spaceport for its manned space mission. However, other countries that did not launch rockets in 2006 are also improving or building launching complexes. For instance, South Korea is building its first launching pad the "Oinarodo Space Centre" scheduled to be operational by 2008, while Brazil is refurbishing the Alcantara Launch Centre.

## 2.7. New developments and trends

While the prospects of the launch sector are better than in the past few years, there are a number of trends that might have far reaching consequences for the launching sector in the near future.

Firstly, 2006 saw the confirmation of the general trends of higher prices in the launching sector, partially due to the higher cost of raw materials and production (particularly in Russia and Ukraine).

Secondly, following the Sea Launch failure in January 2007 and the frenzy to find a launching alternative, on 18 June 2007, SES purchased ten satellite launch slots from Arianespace and ILS, to give availability guarantees for satellites to be placed into orbit between 2008 and 2013. Each of the ten satellites (five for each company) are part of a bulk launcher procurement contract that features one primary launcher and a back-up guaranteed by the other company. In addition, each company agrees to make the launch slots available for each satellite, to avoid a situation in which satellite delays and a subsequent launcher delay grounds one of them. SES will also be able to switch from one launch-services provider to the other as late as three or four months before the launch without penalty. This marks the first bulk purchase of launch services by a private entity to have a guaranteed access to launch capabilities and thus a reduced price for access to space.

Thirdly, following the recent trend whereby each commercial launch services provider has a full manifest for the years ahead, there is the prospect of increased competition of China and India in the commercial market. The consequences of the entry of those two actors in the launch services market could be similar to the entry of the Russian launch vehicles in the commercial market in the 1990s that led to an overcapacity in the market. For instance, while India has launched foreign payloads as piggybacks in the past, the launch of Italy's Agile spacecraft as primary payload on board a PSLV on 23 April 2007 is a stepping stone for the Indian space programme, as the contract to launch this satellite was won in an open competition. However, while Chinese and Indian launchers are improving their reliability (particularly the Chinese launchers), U.S. launch services providers are also increasingly looking at the commercial market to expand their revenues stream. Lockheed Martin Commercial Launch Services (LMCLS) signed its first contract in February 2007 to launch Inmarsat 4 F3 mobile communications satellites onboard an Atlas V by early 2008, marking the return of Lockheed Martin to the commercial market. While Boeing withdrew the Delta 4 from the commercial market in mid-2003, citing weak launch demand, it has been reported considering returning the Delta 4 launch vehicle to the commercial market to help to make up for a shortfall in government launches later this decade under the request of the U.S. Air Force. Moreover, the U.S. launch services provider, Orbital Sciences Corp., which has built its success on providing small satellites and small launch vehicles to both commercial and government customers, is considering an expansion of its rocket business to be able to carry much larger satellites into orbit than its current aircraft-launched Pegasus and the ground-launched Taurus and Minotaur. Finally, with the partial success of the second Falcon 1 vehicle launch in March 2007, SpaceX announced that it will start commercial launches to LEO in the fall of 2007. While Falcon 1 is not a competitor for the lucrative GEO market, SpaceX is, however, developing other launchers: Falcon 5 and Falcon 9 that in the latter case would be a potentially serious competitor in the commercial market if development stays on track. In this context, SpaceX was granted a five-year licence by the U.S. Air Force on 26 April 2007 to conduct launches from the Cape Canaveral Air Station for both its future Falcon 5 and Falcon 9 launch vehicles. SpaceX has been operating up-to-now from a private launch complex in the Pacific Ocean's Kwajalein Atoll, but moving to Cape Canaveral is expected to be more attractive to future customers.

Therefore, while the prospects of the launch sector are better than in the last few years, fierce competition is expected in the years ahead.

## 3. Space sciences and exploration

Space sciences and exploration are the emblematic domains of space activities in which primarily the space powers are active. However, newcomers are showing a growing interest in becoming involved in these activities to gain international prestige.

## 3.1. Human spaceflight activities

The year 2006, was an important one for human spaceflight activities in Europe. ESA astronaut Thomas Reiter from Germany became the first European to undertake a long-duration mission onboard the ISS following his dispatch on the Shuttle mission (STS-121).[197] Thomas Reiter was also the first non-U.S., non-Russian astronaut to become a permanent crew member. With the mission dubbed "Astrolab", ESA inaugurated the long-term presence of European Astronauts onboard the ISS and was also the first long-duration human spaceflight to the ISS to draw on the support of a European control centre. In December 2006, ESA astronaut Christer Fuglesang became the first Swedish and the first Nordic astronaut in space as a Mission Specialist on flight STS-116. He met with fellow ESA astronaut Reiter, who has been a permanent crew member since 6 July 2006. It was the first time that two ESA Astronauts flew together on the same ISS mission. Both returned safely to Earth on 22 December 2006.

After 10 years of development, and a long delay following the Columbia shuttle accident, the 10.3-ton Columbus orbital module was shipped to the Kennedy Space Centre for launch to the ISS. The Columbus laboratory is ESA's biggest contribution to the ISS. It will be permanently attached to the space station and will provide internal payload accommodation for research experiments into material sciences, fluids physics, life science as well as applications in the field of space science, Earth observation and technology.

The overall ISS programme situation has stabilised following the long hiatus after the loss of the Shuttle Columbia in February 2003. The assembly of the station resumed after the confirmation of the shuttle's flight status and safety. Three shuttle missions were carried out in 2006. The first one was the STS-121 with the Discovery orbiter in April 2006, followed by the STS-115 with the Atlantis orbiter in September 2006. On 9 December 2006, the STS-116 mission with the Discovery shuttle performed the first night launch in four years. In the first half of 2007, there was only one shuttle launch, the STS-117 with the Atlantis, propelled into orbit on June 2007. These missions allowed the assembly of the ISS to continue, and in particular, to add new solar panels (S3/S4) providing an additional 20-Kilowatts of power to the station in June 2007.

On 2 March 2006, space agency leaders announced a new schedule for completing the assembly of the ISS. The United States agreed to operate 16

Shuttle flights to launch the major elements of the station before retiring the Shuttle fleet in 2010. However, on 17 April 2007 NASA announced a revised launch schedule for the upcoming Space Shuttle missions following a review of repairs to the insulation on the Shuttle's external fuel tank, which was damaged during a hail storm over NASA's Florida launch site in February 2007. The third Space Shuttle flight of 2007, the STS-120 mission, with ESA astronaut Paolo Nespoli, which will also carry the Italian-built Node 2 connecting module into orbit, is now scheduled for October 2007. The flight STS-122, which will see the launch of the Columbus laboratory as well as ESA Astronauts Hans Schlegel and Leopold Eyharts, is now due for launch from NASA's Kennedy Space Centre in December 2007.

In China, the third Shenzhou mission that had been planned for late 2007 and that will carry three Astronauts,[198] one of whom is to do an extravehicular activity (EVA), has been delayed by about six months. The launch is now projected to take place in early 2008 to complete testing of an indigenously developed spacesuit.

While India has well-developed space capabilities in the field of Earth observation and telecommunications and has long since demonstrated its ability to build sophisticated launch vehicles and satellites for national development needs, its space agency ISRO is eager to start a human spaceflight programme, and to autonomously launch its first manned flight by 2014–15 and land an Indian astronaut on the moon by 2020. A manned spaceflight programme marks a very big step for India. ISRO estimated that the project leading to a first manned flight will cost from 2.5 to 3 billion U.S. dollars a year[199] (or about 1.9 to 2.3 billion euros). The proposal of developing an Indian human spaceflight programme was presented to Indian Prime Minister Mamohan Singh on 17 October 2006, and on the latter's advice, the proposal was presented by ISRO Chairman, Gopalan Madhavan Nair, in a brainstorming session to a cross-section of the scientific community that met in Bangalore on 7 November 2006.[200] Initial funding began on 1 April 2007. The decision of the development of a man-rated GSLV has been taken and actions initiated.[201] ISRO already validated its re-entry technology in January 2007 with the successful recovery of its space capsule in the Bay of Bengal. The Space-capsule Recovery Experiment (SRE-1) was launched on a PSLV on 9 January 2007 and stayed in orbit for 12 days. SRE-1 carried a metallurgical and a biomaterial experiment, and aimed to provide experience in spacecraft recovery, thermal protection, guidance, navigation, as well as control and recovery.

Besides the traditional space powers, a variety of new actors have increased their involvement in space in the last few months, mostly in cooperation with Russia. In particular, the astronaut Marcus Pontes became the first Brazilian in space and went onboard the Soyuz TMA-8/12S mission in March 2006 for a 9-day trip to ISS. A visit to the ISS by the first Malaysian astronaut was approved by Malaysia and Russia on 19 May 2006 as part of a contract offset for Malaysia's 1 billion U.S. dollars (or about 770 million euros) order for 18 Russian Sukhoi Su-30 MKM fighter aircraft. Dr. Sheikh Muszaphar Shukor Al Masrie bin Sheikh Mustapha will be the first Malaysian astronaut or "Angkasawan" in October 2007, and will not fly as a tourist but as a full member of the ISS science and research programme, conducting experiments for the Malaysian Ministry of Science. Furthermore, on 25 December 2006, the final two runners in the race to become Korea's first astronaut were chosen by the Ministry of Science and Technology and the aeronautics and space agency of South Korea, the Korea Aerospace Research Institute (KARI). The final candidate that will travel to ISS aboard a Russian Soyuz in April 2008 will be selected later this year. Russia has also been reported to be considering the possibility of cooperation with Venezuela, in particular, in the training of a Venezuelan astronaut for a possible trip to space at the earliest in fall 2008. Iran would like also to send its first astronaut in space with the help of Russia, as well as Thailand, but no dates have been proposed for these two cooperation projects.

## 3.2. Lunar exploration

Several major space agencies are working on lunar orbiting and landing missions in the context of preparations for future space exploration, and this is creating an unprecedented race to the moon.

SMART-1 mission, the first ESA Small Mission for Advanced Research in Technology, ended with a controlled impact into the moon's Lake of Excellence on 3 September 2006 as planned after an almost three-year mission that demonstrated several advanced technologies and produced valuable scientific results. Launched on 27 September 2003, SMART-1 used solar-electric propulsion that gradually raised its apogee until it was captured by lunar gravity and entered into orbit around the moon in mid-November 2004. It was the first European spacecraft to travel to and orbit around the moon. Its lunar science investigations included studies of the chemical composition of the moon and of the geophysical processes of Earth's satellite for comparative planetology, but also studies in preparation for future lunar exploration. It produced the first comprehensive inventory of the key chemical elements in the lunar surface.

In Europe, the main lunar exploration activities are undertaken under ESA's leadership. However, several countries are gearing up their own national initiatives. The German Aerospace Centre, DLR, issued a contract in summer 2007 to OHB-System AG for Mona Lisa, a 10-month study of a future lunar-lander exploration project. The British National Space Centre (BNSC) and NASA signed an agreement on 19 April 2007 to jointly study how the two organizations might work together on future lunar and planetary exploration, illustrating the new ambitions of the UK in space.[202]

In the United States, the Lunar Reconnaissance Orbiter (LRO) is being developed by NASA to prepare a new mission for future moon exploration that should be launched at the end 2008 and will survey measurements of the lunar surface. A current proposal is for the LRO to be launched together with the Lunar Crater Observation and Sensing Satellite (LCROSS) that will target the Shackleton Crater, the preliminary location of US human moon landings and study the lunar regolith and characterizing potential polar ice deposits.

With increasing funding, Russia is also reenergizing its lunar and planetary programmes.[203] No Soviet lunar mission was launched after 1976 and no planetary missions were flown after the Mars 96 launch failure in 1996. However, Russia, which pioneered and then abandoned robotic exploration of the moon after the loss of the Space Race and the collapse of the U.S.S.R., is starting the development of its first lunar mission in more than 30 years.[204] On 1 June 2006, it announced an ambitious programme for resuming its robotic exploration of the moon. Russia is planning to launch an ambitious lunar penetrator mission in the 2009–2012 timeframe that could be followed by a lunar rover in 2015–2016.

The current efforts of JAXA focus on reaching out to public constituencies in the country to increase awareness about space and consequently, support for space activities. Nonetheless, following the release of a new plan in 2005, the so-called JAXA Vision, JAXA has defined as a goal to erect a manned lunar base by 2030. Astronauts would be sent to the moon by around 2020 to start construction of the base to be completed by 2030. In this context, JAXA is interested in participating in the future Advanced Crew Transportation System with Europe and Russia, but no dedicated funds have yet been earmarked. Still, the current JAXA President, Keiji Tachikawa, has been using its personal funds as initial seed money to start internal studies. Furthermore, the JAXA orbiter mission SELENE (SELenological and ENgineering Explorer) planned to be launched in mid-2007 by an H-IIA rocket will complete the international space race to the moon. It will carry more than 10 instruments (including spectrometers for X-rays, gamma rays and charged particles, a multi-band imager and a laser altimeter) and the mission includes a number of sub-satellites that will contribute to modelling the global lunar gravity field with high precision and resolution.

In line with its development in human spaceflight China is planning to launch its first lunar mission, Chang'e-1, in 2007, in the context of a three-step approach before the potential human exploration of the moon. Subsequent developments in the Chang'e programme include remote controlled lunar rovers (Chang'e-2) to be launched in 2012[205] and an unmanned sample return mission (Chang'e-3).

India is also working on its first unmanned mission, Chandrayaan-1, to be launched to orbit the moon in early 2008 with the objectives of studying lunar origin and evolution, producing a 3-D atlas and carrying out chemical mapping. The payload includes international instruments, from ESA and from the United States.

## 3.3. Mars exploration

While Earth's natural satellite is a target of choice for major space-faring countries, the Red Planet has also become the centre of attention for exploration plans.

The ESA Mars Express mission, launched on 2 June 2003, that arrived at Mars on 25 December 2003 continues to deliver new scientific results related to the water ice in polar caps, the presence of sulphates, the evidence for recent volcanic and glacial activity, methane detection in the atmosphere and signatures of atmospheric escape. Mars Express has completed its nominal mission lifetime and has been in orbit around the Red Planet for one Martian year or 687 earth days. In September 2005, ESA took the decision to extend the mission by an additional Martian year, starting in December 2005.

On 11 June 2007, an ExoMars programme board meeting gave the go-ahead for the so-called "Enhanced Baseline Mission" concept using a heavy lift launcher.[206] ExoMars that aims to put a rover on the Martian surface in 2013 is the first of ESA's Aurora flagship missions to be assessed. The estimated cost of the mission has now risen to 250 million euros above the original approved 650 million euros, but without including the initially foreseen Mars orbiter to act as a dedicated transmission relay. Furthermore, preparations of the MARS-500 isolation study in conjunction with Russia's Institute of Biomedical Problems is underway, and a "Call for Candidates" has been issued during the Paris Air Show on 19 June 2007.

In the United States, NASA's Mars Global Surveyor, which has been in orbit since September 1997, stopped its operation on 2 November 2006. Over its operational lifetime it provided impressive images of the surface of Mars, and in particular, of the clear evidence for a "sapping" origin of many channels, probably from melting of subsurface ice, suggesting the possible existence of liquid water in the recent past on Mars.

NASA's Mars Odyssey, the spacecraft mainly devoted to the mapping of chemical elements and minerals on the Martian surface that arrived in the vicinity of the planet on 24 October 2001 is still orbiting the Red Planet. In its extended mission phase, collecting scientific data and relaying communications from NASA's two Mars rovers to Earth. Among others, Odyssey's accomplishments include the measurement of radiation, a prerequisite for future human exploration. The two identical NASA Mars Exploration Rovers (Spirit and Opportunity) that were launched by NASA on 10 June and 7 July 2003, respectively arriving on 4 and 25 January 2004 have by far exceeded their initial 90-day warranties on Mars and still collects precious information. Major results to date include evidence of evaporated minerals in exposed rock layers, and other signs strongly suggesting the presence of standing bodies of water on Mars when the rocks were formed.

NASA's Mars Reconnaissance Orbiter (MRO), launched in August 2005, arrived on Mars on 10 March 2006 and was successfully inserted into an elliptical orbit, and was then aerobraked into its final operating orbit in September 2006. MRO seeks to characterize the surface, subsurface and atmosphere of Mars, and to identify potential landing sites for future missions. It will be completed by NASA's Phoenix mission that will use a lander that was intended for use by 2001's Mars Surveyor lander prior to its cancellation. It will carry a complex suite of instruments that are improved variations of those that flew on the lost Mars Polar Lander. Phoenix will land on the icy northern pole of Mars, and during the course of the 150 Martian day mission, it will deploy its robotic arm and dig trenches up to half a meter into the layers of water ice and analyse the soil samples collected.

Following the release in December 2006 of NASA's strategy and rationale for robotic and human exploration of the moon, the NASA Administrator has also tasked NASA's Exploration Systems Mission Directorate (ESMD) to begin a formal assessment of potential approaches for sending humans to Mars in 2007. However, it is not expected to have the first human Mars expedition begin until the late 2020s.

In July 2006, Russia also announced the advanced development of a robotic sample return mission to the Mars moon Phobos. Planned for launch in 2009 from the Baikonur Cosmodrome, the Phobos spacecraft would land, drill one meter into the Mars moon's surface, collect soil and rock samples, and return it to Russia in 2012. The lander will be designed to remain on Phobos' surface for a year. Other goals for the mission include mapping the surface of Phobos, measuring its mass and dimensions, and analyzing the plasma and dust around the Martian moon. The Russian Space Research Institute, IKI, is developing the sensors for the Phobos spacecraft, with participation of ESA, the French space agency CNES and U.S. researchers. Russia is also looking at an intensifying its partnership with

China in exploration to evolve beyond the traditional sales of Russian equipment to China. A dedicated cooperation agreement whereby a Chinese small satellite to be launched along the Russian sample return mission "Phobos Explorer" to the Martian moon Phobos was signed during a three-day visit to Russia by the Chinese President Hu Juntao in March 2007. The agreement signed by the CNSA Roskomos follows recent pledges by Moscow to work closely with Beijing on the exploration of both Mars and the moon, and consequently, a more substantive Russian-Chinese cooperation in exploration may be expected in the future.

As already mentioned, while China signed a dedicated agreement with Russia in March 2007 to contribute a small satellite to be launched along the Russian sample return mission "Phobos Explorer" to the Martian moon Phobos in October 2009, India is also beginning to define its first Mars orbiter for launch as early as 2013.

## 3.4. Saturn exploration

While the ESA Huygens probe successfully landed on Titan on 14 January 2005, which is Saturn's largest moon, becoming the first spacecraft to land in a world in the outer solar system, the NASA Cassini spacecraft, launched jointly with Huygens, is still in orbit around Saturn making an extensive survey of the ringed planet and its moons. The multispectral images acquired are revealing new details of the cloud morphology and atmospheric circulation patterns of Saturn. Furthermore, during recent a flyby of Enceladus, a small icy moon of Saturn, Cassini discovered a water vapour cloud around the planet, indicating geyser-type water volcanism. The nominal 4-year Cassini-Huygens mission ends in June 2008, but a 2-year Cassini mission extension (mid-2008 to mid-2010) is under consideration by NASA.

## 3.5. Venus exploration

Venus Express, a follow-up from ESA's Mars Express mission was launched on 9 November 2005 by the Russian Souyuz-Fregat launcher from the Baikonur Cosmodrome. After a 153-day cruise, the spacecraft was inserted into a 24-hour period polar orbit around Venus and began observations in spring 2006. Venus Express seeks to investigate the Venusian complex dynamics and chemistry, and the interactions between the atmosphere and the surface, as well as the interactions between the atmosphere and the interplanetary environment to better understand the evolution of the planet. In 2006/2007, the Venus Express continued to perform routine operations, with the spacecraft and the active instruments all performing nominally and providing a continuous stream of data.

## 3.6. Solar observation

A fleet of space weather detection satellites and a space-based observatory are now operational providing more accurate alerts for the arrival time of Earth-directed solar ejections.

The mission of ESA's Solar and Heliospheric Observatory (SOHO) spacecraft launched on 2 December 1995 was extended on 1 June 2006 from April 2007 to December 2009. SOHO is designed to study the internal structure of the sun, its extensive outer atmosphere and the origin of the solar winds. The scientific data gathered by its 12 instruments in the last 11 years at the Lagrangian point L1 have contributed to understanding the interactions between the sun and the Earth's environment.

The four satellites and instruments of the ESA Cluster II mission are operating nominally. This mission is investigating the small-structure of the Earth's plasma environment and the interaction between the charged particles of the solar wind and Earth's atmosphere.

The ESA-NASA Ulysses deep-space mission continues to operate successfully almost 17 years after its launch. In April 2007, Ulysses completed its third passage over the sun's south polar cap and will be over the northern polar cap at the end of 2007.

STEREO (Solar TErrestrial RElations Observatory) was launched on 25 October 2006 aboard a Delta-2 launch vehicle from Cape Canaveral. STEREO will provide a unique view of the solar-Earth system with two nearly identical observatories – one ahead of Earth in its orbit, the other trailing behind – that will trace the flow of energy and matter from the sun to Earth and will capture 3-dimensional views of the sun. This mission will enable the exploration, the origin, evolution, and interplanetary consequences of solar coronal mass ejections.

JAXA's Hinode (formerly Solar-B) was successfully launched from Uchinoura Space Centre on 23 September 2006 onboard a M-5 launch vehicle. Hinode will, for the first time, provide quantitative measurements of the full vector magnetic field on small enough scales to resolve elemental flux tubes. Its three telescopes will enable the exploration of the origins of the outer solar atmosphere, the corona, and the coupling between the fine magnetic structure at the photosphere and the dynamic processes occurring in the corona. It is expected that Hinode will drastically advance the understanding of the relationship between various energetic processes and solar magnetic fine structure.

The Chinese Double Star satellites and their instruments that follow the footsteps of ESA's Cluster II mission of studying the effects of the sun on the Earth's environment are operating normally. Double Star is the first mission launched by China to explore Earth's magnetosphere, but also the first programmatic collaboration with ESA. It involves two satellites called TC-1 and TC-2 launched respectively on 29 December 2003 and 25 July 2004. The Double Star observations are synchronized with ESA's four Cluster satellites so that all six spacecraft are studying the same region of near-Earth space at the same time.

## 3.7. Outer solar system space probes

On 24 August 2006, the International Astronomical Union's (IAU) General Assembly in Prague (Czech Republic) adopted a statement declaring that only the eight solar system planets from Mercury to Neptune are true planets, and that Pluto and other small round objects in the solar system are to be called "dwarf planets." That decision, however, will continue to be debated by astronomers for at least several years.

Integral, the international mission under ESA's leadership is the first space observatory that can simultaneously observe objects in gamma rays, X-rays and visible light looking at gamma-ray bursts and regions in the Universe thought to contain black holes has been operating successfully since 2002. In particular, Integral has recently discovered radioactive iron-60 in the interstellar space of our galaxy.

ESA's Rosetta spacecraft is travelling to meet Comet 67P/Churyumov-Gerasimenko in 2014. After entering orbit around the comet, the spacecraft will release a small lander onto the icy nucleus, then spend the next two years orbiting the comet as it heads towards the sun.

COROT (COnvection, ROtation and planetary Transits), a mission led by CNES with ESA participation, was launched successfully on 27 December 2006 by a Soyuz launcher and placed in a polar circular orbit around Earth at an altitude of 896 kilometres allowing the continuous observation of two regions in the sky for more than 150 days each. COROT began its scientific operations on 2 February 2007. It is the first satellite dedicated to the search for, and the study of, planets orbiting other stars. Its two scientific objectives are the search for extra-solar planets and the probing of stellar interiors. COROT will monitor about 120 000 stars with its 30-centimetre telescope in search for planets around other stars. It promises to find undiscovered exoplanets during its two-and-a-half-year mission, and to expand the knowledge toward ever smaller planets. It will also be the first mission capable of detecting rocky planets several times larger than Earth (or even smaller) around nearby stars.

On 18 June 2007, ESA and NASA signed a cooperation accord on ESA's LISA Pathfinder technology-demonstration mission, set for launch in the first quarter of 2010. The satellite is designed to prove technologies for a later spacecraft, called LISA (Laser Interferometer Space Antenna) that will test the theory of general relativity and look for gravity waves in space. ESA is in charge of the design manufacturing and launch, while NASA will supply the Disturbance Reduction System Package for the spacecraft.[207]

On 22 March 2006, a joint mission between China and France (SMall Explorer for Solar Eruptions (SMESE)) was launched to observe the solar flares and Coronal Mass Ejections (CMEs) for the next Solar Maximum around 2011.

After 30 years of operation, the twin Voyager 1 and 2 spacecraft continue to explore our solar system and are approaching the heliopause and the interstellar space.

After a seven-year round-trip voyage to comet 81P/Wild 2, NASA's Stardust sample return mission returned safely to Earth. The capsule, carrying cometary and interstellar particles successfully touched down in Utah (U.S.) on 15 January 2006. In addition to the dust collected during a close encounter with Comet Wild-2 near Jupiter in January 2004, Stardust's samples included what are believed to be pristine remnants of material that formed the solar system.

NASA's long-planned New Frontiers mission entitled "New Horizons" to Pluto, its moon Charon, the two newly discovered satellites of Pluto, and several objects in the far-distant Kuiper Belt was successfully launched in January 2006. The spacecraft will reach the Pluto-Charon system in 2015 and is scheduled to spend five months studying Pluto and its satellites. New Horizons' primary goals are to characterize the geology, topography and chemical compositions of the surfaces of Pluto and Charon, and to study Pluto's atmosphere, as well as many lower-priority objectives including a search for additional satellites and rings.

The launch of the NASA Dawn mission was postponed to 26 September 2007 due to a series of technical problems on the spacecraft. Dawn is designed to orbit and study two asteroids, Vesta and Ceres. It will use solar electric propulsion to rendezvous with both of them and will observe both with the same set of instruments. Dawn will be able to specify the properties of each asteroid and probe conditions that existed in the early solar system.

On 31 October 2006, NASA Administrator Mike Griffin announced that space shuttle Astronauts would be sent to refurbish the Hubble Space Telescope as soon as May 2008. The decision followed a lengthy risk analysis that concluded that Hubble's life can be extended to 2013 and later without posing undue danger to the lives of the crew that will perform the repairs and upgrades. NASA and ESA signed a Memorandum of Understanding on 18 June 2007 at the Paris Air Show to launch the James Webb Space Telescope (JWST), the successor of the Hubble Space Telescope, aboard an Ariane 5 rocket in 2013. ESA is providing the launch as part of its planned contribution to the mission. It has already made a down payment of about 2.5 million euros to Arianespace for an Ariane 5 ECA. Nonetheless it does not need to convert the agreement into a formal launch contract until three years before the launch.[208] The two agencies will also split instrument development for the telescope, with NASA supplying the Near-Infrared Camera and ESA the new Near-Infrared Spectrograph.

The joint U.S.-German SOFIA (Stratospheric Observatory or Infrared Astronomy) airborne astronomy mission which consists of a Boeing 747SP modified by NASA to carry a 17-ton 2.5 meter infrared telescope provided by the German space agency, DLR, was officially welcomed on 27 June 2007 to NASA's Dryden Flight Research Facility.

With the renewed ambitions of the Russian space programme, Russia is eying to new endeavour beyond Earth orbits in cooperation principally with China and India. In addition to their Mars exploration plans, Russia and China are also cooperating in space astronomy. China will participate in the "Radioastron" programme scheduled to be launched in 2007/2008 and it will also launch the Russian ultraviolet observatory in 2010. Furthermore, Russo-Indian space cooperation has been extended beyond space-based navigation to space sciences with an Indian instrument set to fly in 2007 onboard the Russian Coronas-Photon satellite that aims to study the sun and solar-terrestrial connections physics.

The JAXA Hayabusa (formerly MUSES-C) mission was launched in May 2003 to investigate a near-Earth asteroid and to return a sample of its surface to Earth. Using solar electric propulsion, Hayabusa arrived at the asteroid 25143 Itokawa in September 2005 and studied the asteroid throughout November 2005; however its miniature rover (Minerva) failed to reach the asteroid. Nonetheless, the spacecraft successfully landed on the surface and collected samples. After some technical difficulties with the spacecraft, the collected samples are planned to be returned to Earth in June 2010.

The JAXA Akari (formerly ASTRO-F) mission was launched aboard an M-5 rocket from the Uchinoura Space Centre on 22 February 2006. It is Japan's first infra-red astronomy telescope. Akari's mission is to study protogalaxies and to study star formation. ESA collaborates with JAXA on the mission by providing ground and data-processing support.

## 3.8. International space exploration cooperation

In 2006/2007, remarkable progress was made in the development of the strategic framework for space exploration. On 6 December 2006, NASA rolled out its strategy and rationale for robotic and human exploration of the moon centred around a lunar polar outpost to achieve a sustained, human presence on the moon.[209] The base would be built in incremental steps, starting with four-person crews making several seven-day visits. The first mission would begin by 2020, with the base growing over time, beefed up with more power, mobility rovers and living quarters. The moon base would eventually support 180-day lunar stays, a stretch of time seen as the best avenue to establish a permanent presence there, as well as prepare for future human exploration of Mars. NASA's lunar plan also encourages participation by other nations as well as non-governmental organizations and commercial groups. NASA detailed also its plans of a "global exploration strategy" guiding future international coordination and collaboration efforts that will be refined in 2007.

As a result of the work between representatives of fourteen space agencies, which have met four times since August 2006, on 31 May 2007, at the third ESA/ASI workshop on "International Cooperation for Sustainable Space Exploration", a 25-page report "Global Exploration Strategy – The Framework for Cooperation" was released as the first product of an international coordination process among those agencies.[210] The document develops the case for a globally coordinated space exploration and discusses the rationale for society to explore space and proposes, among other things, a framework for the future coordination of global space exploration. This strategy is designed to introduce minimum standards of interoperability to facilitate cooperation, while permitting individual countries to pursue their own national strategies. The Global Space Exploration Strategy also proposes a vision for globally coordinated space exploration. The societal rationale to explore space is based around five major themes: new knowledge in science and technology, sustained presence – extending human frontiers, economic expansion, a global partnership, and inspiration and education. From this document, the international definition of space exploration may be read as "extend human access and sustainable presence in the Earth-Moon-Mars space" with the five explorations goals being: human missions to near Earth orbits, robotic and human exploration of the moon, human missions to liberation points of the Earth-moon and Earth-sun systems; robotic (and human) exploration of near-Earth objects (NEOs); robotic and human exploration of Mars. Following the adoption of those basic principles, the fourteen signatory agencies set as the next step the creation of a "Coordination Mechanism" in the form of a semi-permanent body is to the coordinate further steps in harmonizing the exploration effort. The formal establishment of an "International Space Exploration Coordination Group" for steering the further implementation of the international coordination process, with the terms of reference having been agreed at the level of the directors of participating space agencies is envisaged before the end of 2007. This group aims to facilitate the exchange of information on space exploration plans. Nonetheless, the overall planning process is clearly led by the United States which has identified space exploration as a core strategy for NASA announced and published a guiding presidential vision for space exploration in February 2004. However, Europe is currently defining its own rationale for space exploration and the first version of the "European long-term Strategy for Space Exploration – The Stakeholders View" will be published in the context of the International Space Exploration Conference from 8 to 9 November 2007 in Berlin.

## 4. Satellite applications

Space technology applications are the most-widely implemented space activities throughout the world.

### 4.1. Space-based telecommunications

2006/2007 was marked by the confirmation of the trend witnessed in recent years of new assets being procured, purchased and launched by new actors both private and institutional particularly from the "South".[211] Moreover, China and India are increasingly becoming competitive exporters of space technology, confirming their increasing status as major space powers.

In May 2007, China launched its first Chinese satellite sold for export, NigComSat-1. In January 2006, China's export-credit agency granted a 200 million U.S. dollars (about 153 million euros) loan to the Nigerian government to complete the financing package of Nigeria's telecommunications satellite that was subsequently built and launched by Chinese companies, thus illustrating the packaging possibilities of Chinese authorities to deploy its high-tech capabilities for foreign policy purposes. However, 2006 was also marked by the breakdown of China's newly launched Sinosat 2 direct broadcast satellite (the largest, most complex spacecraft ever developed by China).[212] Nonetheless, despite problems with the newly introduced hardware, China wants to reinforce its effort in telecommunications satellite development. China's signed a contract with the Venezuelan government authorities for a large telecommunications satellite

(Venesat-1) based on the new DFH-4 satellite platform. And, as in the case of its first export sale to Nigeria, this deal includes a launch aboard a Chinese Long March rocket.

Meanwhile, India's Antrix Corp, the commercial division of ISRO, has teamed up with Astrium Satellites to offer a small commercial telecommunications satellite product. The Astrium-Antrix joint venture won two contracts in 2006 (Eutelsat's W2M, and Avanti Screen Multimedia PLC's HYLAS) and another one in early 2007.

Another example for the growing interest in space-based telecommunications assets of the "South", and particularly of Asian countries, was the announcement on 21 June 2006 by Vietnam that its first communications satellite, Vinasat-1, was being built by LMCSS carrying both C-band and Ku-band transponders and was scheduled for launch by an Ariane-5 rocket. Kazakhstan's first commercial satellite, KazSat-1 for providing television and communications services in Kazakhstan, central Russia, and surrounding central Asia areas, was also launched from the Baikonur Cosmodrome on 18 June 2006 by a Russian Proton-K rocket. Kazakhstan's new national satellite operator KazSat assumed control of the spacecraft on 18 October 2006.

In Europe, following the approval of the "ARTES II" (Advanced Research in Telecommunications Systems) programme at the ESA Council in December 2005, ESA officially announced the signing of a 100 million euros framework contract with the German company OHB on 28 March 2007 to develop a general purpose European Small Geostationary Satellite platform for telecommunication missions. Divided into two parts, the ARTES-II programme initially involves the development and manufacture of the first flight model of a generic bus. The platform will accommodate a payload mass of up to 300 kilogrammes with power consumption of up to three kilowatts over a 15 year mission lifetime. The programme subsequently involves the development, manufacture and launch of a first satellite mission by the end of 2010 with a mission payload to be selected to provide flight heritage and in-orbit demonstration for the platform.

On 15 March 2006, a cooperation agreement was signed between ESA and CNES for the development of Alphabus, Europe's next generation of multi-purpose platform for the high-power payload communications satellite market. The signing of this cooperation agreement followed the signing of a contract in June 2005 by ESA, CNES, EADS Astrium and Alcatel Alenia Space (which has since become Thales Alenia Space) stating the joint commitment to the Alphabus development programme and to the production of the first flight model. The new agreement establishes the arrangements for cooperation between ESA and CNES with respect to the development and qualification of a generic line of large platforms for geostationary telecommunication satellites, the provision of a flight model from this generic line and its validation in orbit. CNES will manage the development of this new platform line, with ESA co-financing, for its part, the development of the selected European equipment making of ESA the first client for the Alphabus line. Alphabus aims to offer Europe reliable solutions matching world demand for very high power satellites and will be commercialised jointly by EADS Astrium and Thales Alenia Space. The Alphabus platform is designed for telecommunication satellites having a payload power consumption of between 12 and 18 kW, and a lift-off mass of between six and eight ton and will be optimised for the European launcher Ariane 5 ECA. Finally, on 20 June 2007 at the Paris Air Show, ESA and Inmarsat announced the signature of a Memorandum of Understanding (MoU), which is a step towards confirming Inmarsat as the first customer for the Alphabus platform with a launch of the Alphasat satellite targeted for 2011.

In the last few months, Europe was also active in the field of military telecommunications. The United Kingdom, which was the first European country to acquire a domestic satellite communications system for defence purposes in the 1970s, upgraded its satellites communications system, and procured new services through an innovative model of Private Financing Initiative (PFI) [213] for the three Skynet 5 satellites. The first Skynet 5 spacecraft (Skynet 5A) was successfully launched in Mars 2007 onboard an Ariane 5. Spain has also demonstrated recently new ambitions in military space with the launch of its own dedicated telecommunication military satellite Spainsat in March 2006, which provides X-band capacity primarily to the Spanish Ministry of Defence under a leasing contract.

## 4.2. Space-based positioning, navigation and timing systems

In 2006/2007, position, navigation and timing (PNT) was a domain of great activity. While a continuously growing number of users and applications benefited from the U.S. Global Positioning System (GPS) new initiatives and alternatives are gaining momentum and Russia, Japan, China, India and Europe have undertaken further steps toward developing their own PNT systems.

As high oil and other natural resources prices have made it possible to balance and increase the Russian institutional budget, and due to the Russian government's intent on revitalizing its national space programme, Russia has undertaken the upgrading of its navigation system, Glonass within the scope of its new Federal Space Programme (2006–2015). Three Glonass M class satellites were launched on 25 December 2006.[214] By the end of 2007, the Russian space agency plans to launch two Proton rockets, each carrying three Glonass spacecraft to replace satellites expected to be retired by then and thus maintain a total of 18 fully operational satellites to ensure that its domestic system can provide continuous coverage over Russia. By late 2009, Russia projects a full constellation of 24 operational satellites. The renewed interest in space at the highest-political level in Russia has also led to the reinforcement of several cooperation and partnerships. In particular, Russia will collaborate with India on updating Russia's Global Navigation Satellite System (Glonass) as part of a broad space cooperation plan.

On 14 April 2006, Japanese government agencies signed an agreement with Advanced Space Business Corporation, an industry consortium, clearing the path for the development and launch of the first satellite of the three-satellite constellation Quasi Zenith Satellite System (QZSS) planned for launch in the first quarter of 2009 aiming to demonstrate an enhanced positioning-location service over Japan. Then, the Japanese Diet passed legislation (the so-called Fundamental Law on the Promotion of Geospatial Information Activities) on 23 May 2007 that commits the government to fund the development and launch of the aforementioned initial QZSS satellite, plus a period of in-orbit testing.[215]

In 2006, China entered into delicate negotiations with Europe regarding the Galileo programme. China has invested about 5 million euros in cash in the Galileo Joint Undertaking (GJU)[216] to become a shareholder in the system on top of contributing to in-kind hardware estimated at about 60–65 million euros for the Search-And-Rescue system plus the investment in China to prepare Galileo ground installations and to promote a domestic industrial base for navigation products and services. However, that shareholding expired at the end of 2006 with the end of the operations of the GJU and the transfer the responsibility of the Galileo programme to the GNSS Supervisory Authority (GSA).[217] The GSA, which is a Community agency of the EU, will be managed only by Europeans, as it will manage mainly Galileo's encrypted, European-government only Public Regulated Service (PRS) signal.[218] The Chinese government is thus moving ahead to build its own global navigation satellite system named Compass/Beidou.[219] In particular, it has raised the potentiality to place its signal atop frequencies reserved for Europe's encrypted Galileo navigation service, and even perhaps over GPS military signal as well.[220] The Compass/Beidou system is expected to provide services to customers throughout China and neighbouring countries by 2008 before being expanded into a global network of navigation and positioning.[221] In this context, on 3 February 2007, China has successfully launched its fourth navigation satellite since 2000 as part of the aforementioned effort to build a domestic positioning system.[222] This Compass satellite was unable to deploy its solar panels correctly after the launch and it took 60 days to fix this glitch,[223] but the Compass satellite has been reported to be working smoothly ever since. On April 2007, China launched its fifth navigation satellite onboard a Long March 3-A rocket from the Xichang launching centre.

India approved the development of an independent Indian Regional Navigation Satellite System (INRSS) on 16 May 2006. The seven-satellite constellation (four geostationary-orbit spacecraft and three in highly elliptical orbits) is planned to be deployed using ISRO's PSLV from the beginning of 2008 and the entire infrastructure is expected to be completes by 2012. Furthermore, like China following the difficulties of the GJU/GSA transition, and the fact that India does not yet have a seat on the GSA board nor access the European-only PRS signal, it has been seeking participation in the joint restoration of the Russian Glonass system.[224] And, in the past months, India and Russia have made further progress in their cooperation regarding the development and use of space-based satellite navigation system as part of a broad space cooperation plan. For Russia, the Indian in-kind contribution would help the refurbishment of the Glonass constellation. On the Indian side, this cooperation would allow it to acquire new technologies and gain precious insight and experience with global navigation systems. It will also allow it to participate in the next-generation navigation satellite system development. The deals based on a no-exchange of fund basis were signed by ISRO Chairman Gopalan Madhavan Nair and Russian Federal Space Agency (Roskosmos) Director General Anatoly Perminov on 17 March 2006 in New Delhi.[225]

The European Geostationary Navigation Overlay Service (EGNOS), Galileo's precursor, is now in operation.[226] On 12 July 2006, ESA announced the transition of EGNOS from the agency's EGNOS System Test Bed (ESTB) to the "production" EGNOS system for the provision of GPS augmentation services over Europe. It is now providing signals to the European user community for PNT, with an improved utilization of the GPS system.

The period 2006/2007 was also marked by progress, albeit limited, on Galileo itself. Nonetheless, 2006 was a symbolic year for Galileo, as the successful operation of the first experimental Galileo test satellite, Giove-A launched on 28 December 2005 on-board a Soyuz rocket from the Baikonur Cosmodrome, permitted European governments to meet the radio frequency reservation deadlines before mid-2006 and to secure the frequency for the system. This successful launch also permitted testing of several key Galileo technologies including rubidium clocks and the simultaneous use of two signal-transmission channels. Giove-A, which was build by Surrey Satellite Technology Ltd. (SSTL) is functioning perfectly and should operate successfully until early 2008. However, the second test satellite, Giove-B, which had been scheduled for launch in 2005, and now December 2007, is experiencing major technical problems as a result of an onboard computer glitch found during testing by its manufacturer, the European Satellite Navigation Industries (ESNI)[227] (formerly Galileo Industries). Consequently, to mitigate risks related to the Giove-B launch, and further programme slips, in March 2007, ESA announced that it intended to issue a new contract for a third Giove test satellite, Giove A2 to SSTL. This new spacecraft is set to be launched by the second-half 2008 if the Giove-B fails at a launch to test key technologies, particularly atomic clocks, and to meet in-orbit operation requirements set out by the International Telecommunication Union (ITU).[228]

The Galileo project has moved from the preparation and specification phase into the development phase, and ESA is managing this development phase on behalf of the EC. In this context, ESA and ESNI signed a one billion euro contract for the In-Orbit Validation (IOV) phase that will be paid out equally by ESA and the EC to ESNI to construct four initial satellites and produce the ground network.[229] However, the four IOV satellites that will serve as a basis for the overall system design have been delayed, as they have encountered numerous technical and management issues that may require a wholesale design review. On a parallel track coordinated with the EC, ESA agreed 15 March 2007 to intervene more heavily in the operation of the industry consortium that is building the first four Galileo test satellites for the IOV phase (ESNI), as this assembly of European space-hardware companies has some trouble working together.

A Green Paper on Satellite Navigation Applications was also presented by the European Commission in December 2006. The aim of this document was to launch a discussion on what the public sector can do to create an appropriate policy and legal framework for supporting the development of satellite navigation applications beyond the financial support for research and the creation of infrastructure. The Green Paper launched a consultation process that addressed the industry, public authorities, consumer groups and individuals in order to identify Galileo's potential commercial and civil applications through a set of questions on applications development, privacy and ethical issues as well as the regulatory environment and others issues. The answers, which were received in spring 2007, will be analysed by the European Commission and used as basis for recommendations to the European Council and the European Parliament.

Besides the technical difficulties, the Galileo programme with its complex governance and funding structure was not able to avoid political disputes.[230] On

12 December 2006, European transport ministers failed to agree where the government body overseeing the Galileo satellite navigation project, the GSA, will be headquartered. Eleven EU Member States are vying to host the organization, and none has been willing to stand down to enable a compromise up to now.[231] Moreover, while on 1 January 2007, the GSA officially took over the tasks previously assigned to the GJU, which was dismantled at the end of 2006,[232] the negotiations between the GSA with the eight-member consortium (Aena, Alcatel, EADS, Finmeccanica, Hispasat, Inmarsat, Thales and TeleOp) were difficult and no agreement on Galileo operations acceptable to both parties has been reached.[233] The particular issues of concern focused on which of the European governments would be responsible for the various types of programme risks under the 20-year Galileo operating license. Out of the nine blocks of risks identified, the main differences of opinion concerned the sharing of the risks associated with the system design and with commercial revenues and market development. However, apart from the divergent opinions between the commercial consortium and the GSA regarding Galileo's risks, another main element blocking the negotiations was primarily the fact that the eight companies have been unable to agree on conditions for incorporating the consortium, determining a workable governing structure, or naming a chief executive officer (CEO).

Following these negotiations problems, in a letter dated 14 March 2007 and addressed to the German Minister Tiefensee as Germany held the Presidency of the EU Council, Transport Commissioner Jacques Barrot expressed serious concerns regarding the success of Galileo, and indicated that the delay thus far and the absence of any signs of progress in the negotiations of the concession contract must be considered a risk for the delivery of the project within the timeline and budget. In his letter, Barrot, among others, gave the consortia until 10 May 2007 to incorporate the Galileo Operating Company (GOC) and appoint a CEO of that company. Consequently, on 22 March 2007 the EU Transport, Telecommunications and Energy (TTE) Council gave the Galileo consortium partners a strict deadline by which the eight companies are expected to solve their internal problems linked to the distribution of responsibilities, organisational structure, risk-sharing and pending financial aspects.[234] The TTE Council indicated that it would consider alternative scenarios, including a possible new call for tenders at its next meeting scheduled in June 2007 if negotiations were not resumed by then. In the mean time, the Council mandated the EC to look into alternative solutions.

In response to the pressure applied by the EC and the TTE Council, the concession group incorporated the GOC in Toulouse, France on 26 March 2007 and elected a CEO. However, on 16 May 2007, the Commission presented six scenarios on the future of Galileo saying that the most realistic and economic option was for the public sector to put the initial infrastructure into place.[235] At the subsequent TTE Council meeting on 6 to 8 June 2007, EU transport ministers recognized the failure of the current concession negotiations and the Commission was asked to go into more details concerning the options for the project's completion it laid out in May, and present those findings in September 2007. The Council, in its Resolution of 8 June 2007, while it re-affirmed the value of Galileo, agreed in principle to a re-profiling of the Galileo programme and recognised the need for additional public funding. In the meantime, on 20 June 2007, the Members of the European Parliament adopted a joint resolution on the financing of Galileo considering that it should be financed in full from the Union's budget, and that the EU budget should be increased accordingly.[236] A majority of the EU-27 is currently backing this option.[237]

## 4.3. Space-based earth observation

On 4 January 2006, the U.S.-French Topex/Poseidon ocean surveillance mission was terminated after 13 years of operations due to a stalled pitch reaction wheel that could not be restarted. Only Jason-1 now provides information on sea surface heights for determining ocean circulation and sea level rise. However, the follow-on Jason-2 satellite altimetry mission is scheduled to be launched in June 2008.

The U.S.-Canadian Cloudsat and the French-U.S. Calipso (Cloud-Aerosol Lidar and Infrared Pathfinder Satellite Observation) satellites were jointly launched on 28 April 2006 by a Boeing Delta-2 rocket. The spacecrafts were placed in the same orbit, allowing the two spacecraft to study the same column of atmosphere with different instruments. The NASA/Canadian Space Agency (CSA) CloudSat's primary goal was to provide the atmospheric scientific data needed to evaluate and improve the way clouds are modelled in global models and help to improve the understanding of their role in influencing climate change. Calipso is a joint NASA/CNES satellite mission designed to provide new information on the role that clouds and atmospheric aerosols play in the Earth's weather, climate, and air quality.

In July 2007, the DLR CHAMP mission (CHAllenging Minisatellite Payload) that was launched on 15 July 2000 from Plesetsk using a Cosmos launcher that was decommissioned. CHAMP was aimed at investigating the structure and dynamics from the solid core to mantle and crust, as well as the interaction between the oceans and the atmosphere in order to create a model of the Earth's gravity. In particular, the data from CHAMP made it possible to calculate a new model of the gravitational field: Eigen-2 and paved the way for the twin NASA/DLR spacecraft Gravity Recovery and Climate Experiment (GRACE) launched in March 2002 that continued to accurately map variations in the Earth's gravity field in 2006/ 2007 by making accurate measurements of the distance between the two satellites.

On 24 January 2006, an H-2A rocket launched Japan's Advanced Land Observing Satellite (ALOS) from the Yoshinobu Launch Complex at the Tanegashima Space Centre into orbit. The primary mission of the renamed Daichi is to improve the mapping of Japan's territory and to support global disaster-monitoring programmes and resource surveying. JAXA completed the initial functional verification and calibration tests of ALOS and began operations on 24 October 2006.

In 2006, South Korea became the newest member of the club of countries with their own high-resolution optical reconnaissance satellite with the successful launch on 28 July 2006 of Kompsat-2 satellite (Korea Multi-Purpose Satellite-2). Also known as Arirang-2, the satellite carries an optical imager with 1-m black-and white resolution and a multispectral imager with 4-meter resolution. South Korea has also purchased a high-resolution radar sensor from Italy for its Kompsat-5 (Arirang-5), satellite to be launched around 2009, with an expected ground resolution of between 1 and 3 meters. Officials announced on 1 December 2006 that South Korea plans to launch a civil-military satellite (Arirang-3A) carrying high-resolution optical and infrared sensors in 2012.

ImageSat International N.V. is a Netherlands Antilles company with offices in Limassol, Cyprus and Tel Aviv, Israel that markets the satellite imagery collected by its Earth Remote Observation Satellite (EROS) satellites. Following EROS 1A launched in December 2000, on 25 April 2006, ImageSat successfully launched its second satellite, EROS 1B using a Start-1 launch vehicle as well.

Following the visit on 16 May 2006 of an ISRO delegation to Brazil to meet their counterpart from the Brazilian Space Agency (AEB), ISRO signed an agreement with its Brazilian counterpart on 4 June 2007. Under the framework of this agreement, ISRO will upgrade a Brazilian Earth station with equipments that will enable it to receive and process data from India's remote sensing satellite series. India is therefore expanding its reach in South America and fostering "South–South" cooperation. In another example of "South–South" cooperation, in spring 2006, China has decided to donate data-receiving stations for its weather satellites to seven nations around the Pacific Rim and Indian Ocean (Bangladesh, Indonesia, Iran, Mongolia, Pakistan, Peru and Thailand[238]) that are all members of APSCO.

In Europe, the Global Monitoring for Environment and Security (GMES),[239] EU's second "flagship" made some progress in 2006/2007 towards becoming operational, and particularly, in the preparation of three first "Fast Track Services" dealing with Emergency Response (ERCS – Emergency Response Core Service, Land Monitoring (LMCS – Land Monitoring Core Service) and Marine Services (MCS – Marine Core Service).[240] Pre-operational validation of these three "Fast Track" services is planned for 2008. They aim to provide on a sustained basis, reliable and timely information in support of public policy makers' needs. A 4th GMES Pilot service is now being set-up, focusing on air quality-monitoring and climatology and a fifth GMES service related to security applications is also currently under consideration.

Progress was also achieved on the GMES architecture following the orientation of the third Space Council held in November 2005 whereby it was underlined that the implementation of a phased operational GMES calls for consolidation of the overall GMES architecture, including the relationship between the functional components and identification of the appropriate governance schemes. The functional GMES architecture is planned to comprise three main layers: an information infrastructure layer with two components representing the "data collection basis" of GMES (both space-based and *in-situ* infrastructure), two levels of GMES Services representing the main "outcome" of GMES (Core Services and Downstream Services) and finally an information management and dissemination layer.

Regarding GMES programme management, in a decision dated 8 March 2006, the Commission announced its willingness to complement the existing GMES management structure (GMES Advisory Council and the GMES Programme Office) by the establishment of the GMES Bureau within the scope of its specialized core team services. Since 1 June 2006, this GMES Bureau has been operating in the Directorate General Enterprise and Industry (DG ENTR) for a period of three years to become the focal point of the Commission's GMES-related activities and strengthen the management of the programme. The Bureau is tasked to prepare, in close coordination with the relevant stakeholders, a proposition for GMES, especially GMES governance and the long-term sustainability of GMES.

ESA was also active in the field of space-based Earth sciences with the adoption in September 2006 of a new science strategy for the future direction of the Living Planet Programme which addresses the continuing need to further the understanding of the Earth system and the impact human activity is having it. New missions are currently being developed and are scheduled for launch in early 2008 (i.e. GOCE – Gravity Field and Steady-State Ocean Circulation Explorer – and SMOS – Soil Moisture and Ocean Salinity). ESA's Envisat launched in 2002 with an expected lifespan of five years was extended in April 2007 for another five to seven years. Furthermore, the CryoSat-2 satellite which will replace CryoSat, which was lost as a result of a launch failure in October 2005, was adopted in 2006 and is scheduled for launch in 2009.

ESA has also geared up its efforts in the development of the space component for GMES. On 18 April 2007, it announced that it has selected Thales Alenia Space as a prime contractor for a C-band Synthetic Aperture Radar (SAR) Earth observation satellite called Sentinnel-1 to be launched in 2011 under a manufacturing contract valued at 229 million euros.[241] The contract was signed on 18 June 2007 at the Paris Air Show.

Following the successful launch of the first European polar-orbiting satellite (MetOp-A) in October 2006, Eumetsat's core mission was expanded to provide operational meteorological observation from LEO. MetOp-A was declared operational on 15 May 2007 after 6 months of commissioning. The full data flow from its 11 instruments is now available to users on an operational basis. This marks the first step outside of Eumetsat's initial perimeter as a data provider from a geostationary Earth orbit satellite fleet. MetoOp-A is the first of three satellites developed under a joint programme being carried out by ESA and Eumetsat which are designed to provide meteorological operational data from polar orbit until 2020. The other two spacecraft in the constellation, MetOp-B and MetOp-C, will fly in 2010 and 2014 respectively. MetOp data will significantly improve weather forecasting by offering unprecedented accuracy and resolution of variables such as temperature and humidity, wind speed, ozone and measurements of trace gases such as carbon dioxide, nitrous oxide and methane.

On 15 June 2007, Germany launched the TerraSAR-X satellite which carries an X-band radar. This satellite is financed by both the public and the private sector. DLR is responsible for the management of the entire project and pays 80% of the satellite's cost. The satellite was built by EADS SPACE, which provided the remaining 20% of financing. DLR approved two new Earth observation missions on 8 March 2006. The X-band synthetic aperture radar, TanDem-X satellite, planned for launch in early 2009, will operate with the TerraSAR-X satellite and will also carry an X-band radar with the same performance features. The second new mission is the EnMAP hyperspectral imager that is planned for launch in 2009 or 2010 to study environmental phenomena worldwide.

In Europe, several new Earth observation dual-use reconnaissance satellites have also been launched in recent months. On 7 June 2007, the first of four Italian COSMO-SkyMed[242] X-band radar satellites was placed into orbit. In the same vein, Germany's dependence on foreign military systems for imagery has led the government to acquire domestic satellite reconnaissance capability and the first SAR-Lupe satellite was on 19 December 2006, and the second, on

3 July 2007. Those satellites are the initial two in a planned constellation of five satellites.

## 5. Technology developments

There were major developments in technology in 2006/2007, particularly, in the field of propulsion, information technology, spacecraft design and operations, as well as suborbital activities.

### 5.1. Propulsion

The small satellite launcher Vega, earmarked for small scientific and Earth observation satellites completed a series of engine-firings in the last few months. The Vega launcher is a single body launcher composed of three solid propellant stages and a liquid upper module. It is designed to lift single or multiple payloads to orbit up to 1500 km in altitude, with a baseline capability of about 1.5 ton to a circular 700-km high sun-synchronous orbit. The Zefiro (West Wind) 9 third-stage solid-propellant rocket engine was successfully test-fired on 23 December 2005. The test included successful operation of the motor's thrust-vector-control system. It was followed by the successful first static firing test of the Zefiro 23 second stage solid rocket motor on 26 June 2006. The first firing on 30 November 2006 of the vehicle's P-80 first-stage solid rocket motor was also successful. In addition by being a critical Vega technology, the P-80 is also a technology demonstrator for future updates of the Ariane 5 and other advanced European launch systems. However, while those tests were all successful, the Zefiro 9 second test failed on 28 March 2007. It is believed that failure was due to a production defect that allowed gases to infiltrate the internal structure of the nozzle. The Vega launcher is nonetheless still being considered for a planned maiden launch in 2008.

Several achievements were also made in the domain of air breathing supersonic combustion ramjet (scramjet) propulsion. Scramjet technology is expected to be able to dramatically reduce point-to-point travel. On 10 January 2006, ISRO announced the successful completion of ground testing on an indigenously designed and built scramjet reaching Mach 6 for nearly 7 seconds. Moreover, a series of test launches in the framework of the research project of the University of Queensland's HyShot supersonic combustion ramjet (scramjet) programme were conducted at the Woomera test range in Australia in 2006/2007. On 25 March 2006, HyShot III was launched and achieved supersonic combustion, and on 30

March 2006, HyShot IV was successfully launched. Finally, HyCAUSE was launched on 15 June 2007 and reached speeds of up to March 10.

## 5.2. Information technology

Following the first satellite bi-directional laser communication on 9 December 2005 between Japan's KIRARI satellite (also known as OICETS or Optical Inter-orbit Communications Engineering Test Satellite) operating in LEO and ESA's Artemis (Agency Advanced Relay and Technology Mission Satellite) spacecraft in its geostationary orbit, on December 2006, Artemis relayed optical laser links from an aircraft. These airborne laser links, established over a distance of 40 000 km during two flights at altitudes of 6000 and 10 000 metres represent a world first. Furthermore, in the framework of preparation for testing laser-optical inter-satellite links, the Tesat company, commissioned by DLR, developed two laser communication terminals (LCTs) with a transmission rate of 5.6 Gbit/s. These two LCTs are now flying as experimental payloads on the German TerraSAR-X radar satellite launched on 15 June 2007 and within the framework of a German cooperation, on the American test satellite NFIRE launched on 24 April 2007.

## 5.3. Spacecraft operations

Two Defense Advanced Research Projects Agency (DARPA) Orbital Express spacecraft were launched on 8 March 2007 from Cape Canaveral aboard an Atlas V. Orbital Express consisted of a next generation serviceable "client" satellite (NextSat) and a prototype servicing spacecraft (Autonomous Space Transport Robotic Orbiter or ASTRO). Both were deployed together into LEO and aimed at demonstrating the technical feasibility of on-orbit servicing. In particular, it achieved for the first time, a fully autonomous rendezvous and soft capture of client spacecraft, satellite-to-satellite refuelling, and replacement of battery and flight-computer orbital replacement units as well as reconfiguration of satellites. Despite some technical problems in the course of the 3-month mission, the ASTRO spacecraft successfully demonstrated the ability to autonomously rendezvous with, refuel and replace batteries aboard NextSat. The technologies developed by DARPA's Orbital Express programme are intended to support a broad range of future U.S. national security, civil and commercial space activities. For instance, refuelling commercial satellites would extend their service life without incurring the construction and launch costs for replacement assets. Furthermore, satellites could be upgraded with technologies that become available after launch, or components that fail could be replaced.

## 5.4. Spacecraft design

Bigelow Aerospace, the Las Vegas-based company developing inflatable orbital habitats, released new details about its business plan at the National Space Symposium in April 2007. Bigelow announced it was planning to launch a series of inflatable modules starting around 2010 that would be capable of hosting between three to six people at a time. Several modules would be linked together to form a single space station, with multiple stations being planned. It aims to have two distinct types of customers. One group of customers, called "sovereign clients", would be Astronauts from national space agencies that would pay just under 15 million U.S. dollars (about 11.5 million euros) for a four-week stay, transportation included. A second set of customers, called "prime clients", would be large companies interested in leasing module space for research.[243] Bigelow Aerospace successfully launched its first subscale test module, Genesis 1, on 12 July 2006. It allowed it to test new technologies, as well as attitude control mechanisms. A NASA Ames Research Center "Genebox" biology experiment was also carried aboard the module and was successfully activated on 25 July 2006. Bigelow's second test module, Genesis 2, was launched on 28 June 2007 also onboard a Dnper 1 rocket and has been operating successfully ever since. Genesis 2 is the second pathfinder space module designed to test and confirm systems for future manned commercial space modules. Bigelow Aerospace is now planning to launch the Sundancer module in the next few years that will be the company's first spacecraft capable of supporting a human crew. During the 2007 National Space Symposium, it announced an agreement with Lockheed Martin to study the use of the Atlas V as a potential launch vehicle for its space station modules and for the prospective space tourist who may want to visit them.

## 5.5. Sub-orbital activities

In the field of suborbital flights, work on the SpaceShipTwo prototype is moving forward, as is the fabrication of the White Knight 2 mothership, and at this point, spaceline operator Virgin Galactic is eyeing early 2010 as the beginning of commercial flights with paying customers. Now, the price of a ticket is now 200 000 U.S. dollars (about 154 000 euros) which covers pre-training, the suborbital trip to an altitude of 109.4 kilometers and post-landing activities. According to its president Will Whitehorn, Virgin Galactic is said to have 20 million on deposit (about 15 million euros) and more than 200 customers who have actually made a financial commitment.[244] The full-size model of the interior of SpaceShipTwo was revealed during a technology show in New York on 28 September 2006. The space vehicle is being built to seat six passengers and two pilots. The commercial operation of Virgin Galactic is expected to take place at New Mexico's Spaceport America.[245] On 26 March 2007, Virgin Galactic signed a Memorandum of Agreement (MOA) with the New Mexico Spaceport Authority setting out the terms under which Virgin Galactic would lease a dedicated terminal and hangar as well as state-of-the-art medical facilities, mission control, and office accommodations for Virgin Galactic personnel at Spaceport America for 27.5 million U.S. dollars for over 20 years[246] (about 21.5 million euros). In April 2007, the residents of the Dona Ana County in southern New Mexico voted positively on a referendum to introduce a tax that would help pay for the aforementioned commercial spaceport, renamed Spaceport America.[247]

On 13 June 2007, EADS Astrium disclosed the basic design of the space plane it proposes to build for a suborbital space tourism venture. The European conglomerate intends to build a four-passenger rocket-equipped jets designed to take off from a normal runway (liquid methane and liquid oxygen engine). EADS Astrium plans to attract as many as 45 000 paying customers per year by 2020 at about 200 000 euros per ticket. The company gives itself until the end of the year to round up

**Tab. 9:** *2006 and 2007 FAA-Permitted Flight Events (source FAA).*

| Flight Date | Operator Vehicle | Vehicle |
|---|---|---|
| 19 October 2006 | Armadillo Aerospace | Pixel |
| 20 October 2006 | Armadillo Aerospace | Pixel |
| 21 October 2006 | Armadillo Aerospace | Pixel |
| 21 October 2006 | Armadillo Aerospace | Pixel |
| 21 October 2006 | Armadillo Aerospace | Pixel |
| 13 November 2006 | Blue Origin | PM1 |
| 22 March 2007 | Blue Origin | PM1 |
| 19 April 2007 | Blue Origin | PM1 |
| 02 June 2007 | Armadillo Aerospace | Pixel |
| 02 June 2007 | Armadillo Aerospace | Pixel |

financial partners for the project that is estimated to cost about 1 billion euros to complete the plane's design and development, and flight-qualifying the proposed vehicle.

In June 2006, the sixth license was issued to a non-federal commercial launch site operator in the United States to the Oklahoma Space Industry Development Authority. 2006 was also the first year of development of reusable suborbital rockets for permitted flights.[248] Six such flights were conducted in 2006 and four in the first half of 2007 (Table 9). Armadillo Aerospace carried out seven of these permitted flights, with five of them using one of its vertical takeoff pads during the Lunar Lander Challenge at the 2006 Wirefly X Prize Cup. Blue Origin conducted one flight in 2006 and two in 2007, which are tests of its first developmental vertical-takeoff, vertical-landing rocket, named Goddard, as part of the New Shepard programme to create a manned suborbital vehicle (Table 9).

On 6 April 2007, the FAA released new guidelines for obtaining an one-year experimental launch permits for reusable spacecraft that will give developers the opportunity to fly and test their vehicles before applying for an FAA launch license.[249] However, none of the flights covered by an experimental permit, like the earlier permit for the developmental reusable suborbital rockets, can be flown for profit. Each permit will cover multiple vehicles of a particular design and will allow an unlimited number of launches when conducted in an area large enough to contain its trajectory that is not close to any densely populated areas, and those permits will be renewable following FAA review.[250]

## 5.6. Innovation policy

Drawing on the experience of the Ansari X-Prize that offered a 10 million U.S. dollars cash prize to the first private team to build and launch a spacecraft capable of carrying three people to an altitude of 100 kilometres twice within two weeks, NASA is increasingly proposing cash prizes to spur the development of advanced technology. However, while previous Centennial Challenges initiatives had focused primarily on robotics technology, in early February 2006, NASA announced a major expansion of its Centennial Challenges prize programme to encompass technologies specifically oriented toward space exploration. In this context, on 18 April 2006, NASA announced the opening registration for five Centennial Challenges competitions: the Astronaut Glove Challenge, the Beam Power Challenge, the Lunar Regolith Excavation Challenge, Moon Regolith Oxygen Extraction (MoonROx), and tether challenge. Then, on 5 May 2006, NASA also joined with the X-Prize Foundation in offering a 2 million U.S. dollars (about 1.5 million euros) prize for the development of vehicles designed to go to the moon. Almost all of these challenges have since held a first competition to win the prize. In particular, the first Lunar Lander Analog Challenge was held on 22–23 October 2006 during the Wirefly X-Prize Cup Expo in Las Cruces, New Mexico. The teams participating in this challenge were rated on demonstrations of their vehicles' abilities to launch vertically, hover in mid-air, land on a target about 100 m away, and then repeat the manoeuvre.[251] However, only the Astronaut Glove Challenge has been won thus far. On 2–3 May 2007, Peter Homer won the 200 000 U.S. dollars (about 153 000 euros) for the first place in this competition.

---

[177] Peter, Nicolas. "The Changing Geopolitics of Space Activities". Space Policy 22 (2006): 100–109.

[178] The EU-US Summit was held on 20 June 2005 in Washington D.C., USA.

[179] Peter, Nicolas. "The EU's Emergent Space Diplomacy". Space Policy 23 (2007): 97–107.

[180] The 14 agency signatories are the national space agencies of Australia,, Canada, China, Germany, France, India, Italy, Japan, Russia, South Korea, the United States, Ukraine, the United Kingdom, and the 17-country ESA.

[181] Russia will also cooperate with India on the development of cryogenic upper stages for Indian launchers.

[182] Morring, Frank. "In Orbit China Donating Weather-Satellite Ground Stations". Aviation Week & Space Technology (4 Mar. 2006): 15.

[183] Shareholders in Starsem are Arianespace, EADS (50% together), the Russian Federal Space Agency and the Samara Space Center (50% together).

[184] Lockheed Martin's Orion industrial team include: United Space Alliance, Orbital Sciences Corp, Honeywell Defense and Space Electronics Systems and Hamilton Sundstrand.

[185] It is a six-person ballistic re-entry capsule meant to replace the space shuttle as the agency's primary manned spacecraft that will become the centrepiece of the U.S. human spaceflight programme.

[186] A preliminary date for the first manned flight into the Earth's orbit has been fixed for 9/11/2014.

[187] Four other companies were COTS finalists. Andrews Space, SpaceDev, Spacehab and Transformational Space Corp (t/Space).

[188] NASA will monitor the progress achieved by the firms in developing their systems via quarterly meetings.

[189] "NASA Extends Contract with Russia's Federal Space Agency". NASA News Release. 9 Apr. 2007.

[190] On 28 February 2006, a Proton-M/Breeze-M vehicle launched from the Baikonur Cosmodrome by International Launch Services (ILS) failed to deliver its Arabsat-4A payload into a proper orbit due to early shutdown of the vehicle's Breeze-M upper stage during its second firing. On 26 July 2006, the first stage of a Dnepr rocket launched from the Baikonur Cosmodrome failed 74 seconds after launch, destroying Ukraine's first Earth surveillance satellite and 17 other small payloads, which included Belarus's first satellite, BelKA. Root cause of the Dnepr failure was subsequently identified by the launch service company ISC Kosmotras (Russia) as a malfunctioning hydraulic drive in the thrust-vectoring system on the first-stage combustion chamber, caused by the failure of defective thermal insulation resulting from a manufacturing error.

[191] "Les Troupes militaires spatiales russes quitteront le cosmodrome de Baïkonour à la fin de 2007 (Roskosmos)". Ria Novosti. 3 Mai 2005. http://fr.rian.ru/science/20060503/47176716.html.

[192] Ibid.

[193] Ibid.

[194] He-suk, Choi. "Science Ministry vows to turn Korea into a global space leader by 2015". Korea Herald (15 Jan. 2007). http://www.koreaherald.com/.
[195] Sea Launch is referred in the upcoming figures as multinational.
[196] "China Picks Site for Satellite Center". AP Press release. 8 Feb. 2007. www.space.com/spacenews/asia/APChinanlaunchsiteweb_020807.html.
[197] Reiter's 6-month mission has been conducted under an agreement between ESA and Roskosmos with Reiter taking a position originally earmarked for a Russian.
[198] Two Chinese astronauts flew during the second Shenzhou manned mission in 2005 and one was onboard the first manned flight in 2003.
[199] Jayaraman, K.S. "ISRO Seeks Government Approval For Manned Spaceflight Program". Space News (13 Nov. 2006): 13.
[200] Ibid.
[201] Discussion with the ISRO representative in Europe in Dec. 2006.
[202] The BNSC has commissioned studies of a lunar lander, a lunar penetrator and high bandwidth telecommunications technologies that could be used from the lunar surface, but no decision has been taken yet to pursue them into hardware development.
[203] Lunar missions are not a major item in the new Federal Space Programme, but are nevertheless a major element of the current Russian overall space strategy.
[204] Major Soviet achievements included the first lunar flyby in 1959; the first lunar far-side photos in 1960; the first semi-soft lander to return images from the surface in 1966; a series of successful lunar orbiters starting in 1966; three robotic sample returns in 1970, 1972 and 1976; and two Lunokhod rovers in 1970 and 1973.
[205] According to press reports, this rover could be potentially nuclear powered.
[206] ExoMars aims to put a rover on the Martian surface in the framework of the Aurora programme.
[207] De Selding, Peter. "NASA Seals Deal to Launch Webb Telescope on an Ariane 5". Space News (27 June 2007). http://www.astronomyandspace.com/spacenews/archive07/webb_0625.html.
[208] Ibid.
[209] The preliminary location is on the rim of the Shackleton Crater on the South Pole.
[210] The 14 agency signatories are the national space agencies of Australia, Britain, China, Canada, France, Germany, India, Italy, Japan, Russia, South Korea, the United States, Ukraine and the 17-country ESA.
[211] The term "South" refers to all developing countries, as well as all Least Developed Countries (LDCs). It rests on the fact that the entire world's industrially developed countries (with the exception of Australia and New Zealand) lie to the north of those developing countries. However, the diversity of countries in the South must be borne in mind. Some countries, such as Argentina, Brazil, China, India, Mexico, South Africa and South Korea, have enviable records of scientific achievement compared to the others. Yet, many countries from the South have not seen any significant development for some time.
[212] The first model of its new platform DFH-4 (Sinosat-2) was launched on 29 October 2006 but experienced severe in-orbit problems soon after as both its solar arrays and its large antennas failed to deploy.
[213] A Private Finance Initiative (PFI) is a method to provide financial support for "Public-Private Partnerships" (PPPs) between the public and private sectors. These projects aim to deliver hardware and services for the public sector. In return, the private sector receives payment above the price that the public sector could have achieved linked to its performance in meeting the agreed standards of provision.
[214] Russia announced that all precision restrictions on Glonass use would be lifted in 2007 to enable more accurate and unlimited civilian and commercial use of the navigation system.
[215] Kallender-Umezu, Paul. "Japanese Government Commits to Funding 1st of the Three QZSS Satellites". Space News (28 May 2007): 5.
[216] GJU was set up in May 2002 by the European Community and ESA to manage the development phase of the Galileo programme.

[217] The role of the GSA is to manage the public interests connected with the Galileo programme and to be the regulatory authority for this programme.
[218] Non-European authorities will not be permitted access to the PRS.
[219] Compass known as Beidou in Chinese would feature 35 satellites with five in geosynchronous orbits and the rest in medium Earth orbits.
[220] Overlaying one signal on another does not necessarily violate ITU protocols as signals can operate without interfering each other, but it precludes the jamming of one signal as it will lead to jamming all signals if all are in the same frequency.
[221] "China launches "Compass" navigation satellite". Xinhua News. 14 Apr. 2007.
[222] The previous three were launched in 2000 and 2003.
[223] "China fixes new navigation satellite". Xinhua News. 11 Apr. 2007.
[224] India signed an agreement to participate in Galileo but has not ratified it.
[225] Jayaraman, K.S. "India to Launch Russian Glonass Satellite". Space News (27 Mar. 2006). http://www.space.com/spacenews/archive06/India_032706.html.
[226] EGNOS, the first European satellite navigation system, is a joint project of ESA, the European Commission and Eurocontrol.
[227] ESNI is owned by the following five companies: EADS Astrium NV, Alcatel Space, Alenia Spazio, Thales, Galileo Systemas y Servicios.
[228] The ITU mandates that rights to orbital slots expire if the slot remains vacant for two years.
[229] The total cost of the development phase, approximately 1.5 billion euros is shared equally between the EU and ESA.
[230] In his article "What's the problem with Europe's flagships Galileo and GMES" published in the second part of the Yearbook Serge Plattard provides a view on the problems of the Galileo programme, as well as some lessons learned.
[231] European heads of State in 2003 agreed that new EU agencies should, where possible, be located in one of the new EU Member States and not in the EU-15, nevertheless several EU-15 states presented a candidate to host the GSA.
[232] The European GNSS Supervisory Authority (GSA) was established by Council Regulation (EC) 1321/2004 of 12 July 2004 (and amended by Council Regulation (EC) No 1942/2006). The task of the GJU set up by Council Regulation of 21 May 2002 was to manage the programme's development phase of the Galileo project and carry out the procedure to select the future concessionaire. The tasks have now been passed on to the GSA. This Community Agency will sign the concession contracts and will be the licensing authority vis-à-vis the private concession holder responsible for implementing and managing the Galileo deployment and operating phases and the European Geostationary Navigation Overlay Service (EGNOS). The role of the Supervisory Authority is therefore to manage the public interests connected with the European GNSS programmes and to be the regulatory authority for these.
[233] The GJU launched a tender for the deployment and exploitation phase on 15 October 2003 to which four different initial bids were submitted in December 2003. One bid was eliminated during the pre-selection phase, and during the summer 2004, one of the remaining bids was withdrawn. In September 2004, the remaining two bidders submitted detailed offers, and in October 2004, the GJU requested additional information from the two consortia. Then, on 10 May 2005, the two consortia proposed to merge their bids. This demarche was accepted on 4 July 2005, by the GJU under three conditions requested by the Commission that were subsequently accepted by a letter from the eight companies involved in the merged bid on 21 October 2005.
[234] EU Council. 2791st Council Meeting, Transport, Telecommunications and Energy. Council Conclusions on the status of the concession negotiations in respect of the Global Navigation Satellite System (GALILEO). Brussels, 22 Mar. 2007.
[235] European Commission. Galileo at a cross-road: the implementation of the European GNSS programmes COM (2007) 261 final. Brussels, 16 May 2007.

[236] The resolution drafted by the Parliament Budget Committee and adopted in plenary (P6_TA-PROV(2007)0272) is not legally binding but needs to be taken into account by the other institutions as the Parliament holds budgetary co-decision powers.
[237] Several countries would like to see individual states contribute the extra funding funds to ESA, which could finance and manage the project.
[238] Morring, Frank. "In Orbit China Donating Weather-Satellite Ground Stations". Aviation Week & Space Technology (4 Mar. 2006): 15.
[239] GMES is an EU-led initiative, in which ESA implements the space component and the Commission manages activities for identifying and developing services relaying both on in-situ and space-borne remote sensing data. These data will be coordinated, analyzed and prepared for end-users. GMES will be built up gradually, it will start off with a pilot phase, which targets the availability of a first set of operational GMES services by 2008, followed by the development of an extended range of services to meet user requirements.
[240] Serge Plattard, in his article "What's the problem with Europe's flagships Galileo and GMES" published in the second part of the Yearbook provides a different assessment of the progresses of the GMES programme.
[241] Sentinel-1, a follow-on for the radar instruments aboard ERS-2 and Envisat. Sentinel-2, carrying a super-spectral land monitoring sensor that will complement data from the U.S. Landsat, is planned for launch in 2011 – 2012. Sentinel-3, also expected to launch in 2011 – 2012, will carry an ocean measuring altimeter and optical and infrared radiometers that will serve as follow-ons to the Medium Resolution Imaging Spectrometer Instrument (MERIS) on Envisat.
[242] Constellation of Small Satellites for Mediterranean Basin Observation.
[243] Bigelow expects to require up to 30 launches a year by the middle of the next decade to transport customers to and from the stations.
[244] David, Leonard. "Virgin Galactic Spaceliner Steps Forward". Space News (26 Feb. 2007): 16.
[245] In December 2005, New Mexico Governor Bill Richardson and Sir Richard Branson, chairman of the Virgin Companies, announced that Virgin Galactic would locate its world headquarters and mission control operations in New Mexico.
[246] Over the next few months, the MOA will be developed into a legal lease agreement.
[247] Virgin Galactic is also looking at launch site outside the United States. On 26 January 2007, during the official inauguration of Spaceport Sweden an agreement was signed with Virgin Galactic and the spaceport authorities. The agreement stated that the two would be working together towards an operational agreement whereby Spaceport Sweden would be the first spaceport outside the United States which Virgin Galactic can use for flight campaigns.
[248] Under the direction and delegation of the Commercial Space Launch Amendments Act of 2004 (CSLAA), enacted on December 23, 2004, the FAA has established an experimental permit regime for developmental reusable suborbital rockets. This allows for more flexibility in vehicle development and test flights prior to or instead of the issuance of a commercial launch license. In contrast to licensed flights, permitted flights cannot carry property or people for compensation or hire, and any damages that may occur under permitted flights are not eligible for indemnification.
[249] Morris, Jefferson. "FAA issues new rules for experimental spacecraft permits". Aerospace Daily & Defense Report (10 Apr. 2007): 2.
[250] Ibid.
[251] The Lunar Lander Challenge is divided into two levels. Level 1 requires a rocket to take off from a designated launch area, fly to an altitude of 50 meters, then hover for 90 seconds while landing precisely on a pad 100 meters away. The flight must then be repeated in reverse, and both flights-along with all of the necessary preparation for each – must take place within a two-and-a-half-hour period. Level 2 requires a team's rocket to hover for twice as long before landing precisely on a simulated lunar surface. The hover times are calculated so that the Level 2 mission closely simulates a true lunar landing scenario.

# PART 2

# VIEWS AND INSIGHTS

# 1. The Cabal Report of the French Parliament on space policy – A blueprint for European space ambitions or another cry in the wilderness?

Kevin Madders

## 1.1. Background

In February 2007, Christian Cabal, member of the Assemblée Nationale, and Senator Henri Revol submitted as rapporteurs of the French parliament's science and technology (S&T) oversight panel, a report to their respective chambers entitled "Space Policy: Daring or Decline", subtitled "How to make Europe world leader in the space domain".[252] The report provides an analysis and policy recommendations over the next ten years, geared towards the position France and Europe should hope to achieve by the 2020s. In accordance with constitutional practice, the report is referred to under the short name, "Cabal Report", based on the name of Assembly rapporteur.

Cabal and Revol, like others on the panel – the Parliamentary Office for Scientific and Technological Assessment of the Assembly and Senate (known by its French acronym OPECST) – were senior parliamentarians, with Cabal highly experienced in science and technology policy and research and development (R&D) financing, and Revol in regional and international policy as well as S&T. Revol, OPECST's president, had already led the committee's focus on space since the late 1990s and had reviewed French policy in 2001.[253] Cabal chaired the Assembly's aerospace study group. They both belonged to the governing UMP (Union pour un Mouvement Populaire), the party behind the presidential candidate, Nicolas Sarkozy in the final months of the government headed by Dominique de Villepin during which the report appeared.

Thus, while in France policymaking is dominated by the executive branch and any opinion on such a report by parliament would normally be of only limited significance, the Cabal Report enjoyed greater interest in these circumstances as a possible indicator of Sarkozy's position on space. And, apart from the boldness of its proposals, there is a distinct bite in the report that distances it from the approach of the President Chirac era. The President Chirac period of the last several years was characterized in the report as one of a drift in French and European space policy (ESP). Part of Sarkozy's electoral achievement lay in creating the same sort of distance on the national stage.

Sarkozy was indeed to triumph at the polls in May 2007 (Cabal, though, lost his seat in the elections to the Assembly). But the report came at a pivotal juncture in the establishment of a first European space policy, adopted in April/May, and took ESP's likely shape into account, while setting out its own vision for European policy.

## 1.2. Scope and main features of the report

The report spotlights the recently heightened emphasis on space in other nations and regions, some of which, especially in the United States, is connected with post-9/11 developments. This sets the scene for backing a firm political initiative and plan for French and European space policy until 2020. We shall concentrate on these fundamentals here.[254]

### 1.2.1. A shift in the political outlook

What is striking in the Cabal Report is a bullishness of the stance on space unseen since the late 1980s.

While Senator Revol is a steadfast advocate of an "energetic" French and European space programme, in 2005, Cabal was still tempering ambition with warnings about the need for budgetary realism and consolidation in the sector, notwithstanding job losses, and admonishing critics to give credit for positive achievements[255] – the "glass is half-full" rebuttal[256] which completed an official litany of restraint that has held sway across Europe following the 1992 Maastricht Treaty and the end of the Cold War.

In stark contrast, the Cabal Report calls for new space missions, better governance and new money. It follows with an equally forceful disavowal of the utilitarian, "market"-based justification for space activities that has held sway in recent times. The report's rationale for the turnaround – to demand that the glass be filled, urgently – is the arrival of "radical changes" and a "sharp upward turn" for space activities worldwide, which bring the strategic role and interests of the state into the centre of the picture.

### 1.2.2. Back to the future

This is good, old-fashioned stuff. For the changes referred to may be new, but the argument itself – that European nations must respond to them – is an old one, indeed, from the early years of the Space Age. Moreover, the report returns to the preoccupations of the 1960s. The missions evoked are grandiose: – today the moon, Mars and solar system exploration, but joined by military missions. Once more, maintaining scientific and technical excellence and avoiding a brain drain are at stake as others advance. And the principal tools needed are again basic ones like space transportation systems. Cabal's depiction of competitive threats also feels very similar – a resurgent, militarily dominant, U.S. since 2004, committed to returning to the Moon prior to human Mars exploration; Russia as a (reinvigorated) major space actor, but now being superseded by China, alongside lunar and other programmes in India and possibly other powers. Lastly, the same fate is portended as earlier predicted if Europe fails to act – the loss of international prestige and sovereignty.

The report's title sums this all up. The price of ignoring such evils is (relative) *decline*. The prerequisite for conquering them is *daring*, by making space a political priority again and reopening the financial valves that fuel great power ambitions. Europe must thus take its place in the sun. Space is per se a means to avoid the shade.

Cabal's clarion call in all these respects echoes notes struck in the 1960 Price Report and the 1967 Causse Report, among other parliamentary and institutional reports of the 1960s.[257] To point out this affinity, possibly unintentional, is not to discount the message – such calls did contribute to a perspective that inculcated political will at the time. Figure 1 illustrates the kind of dynamic external factors have introduced.

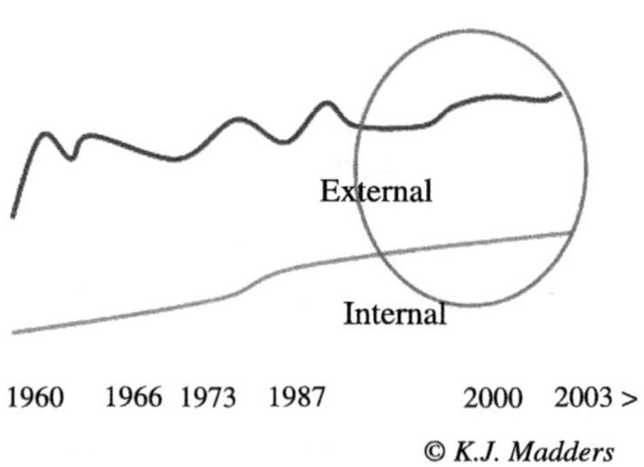


**Fig. 1:** *Long-term susceptibility of space investment to internal factors and external or competitive factors (e.g. ISS). An appeal to external factors does have a historical justification. But timing is everything.*

But of course, the context is not identical even if the technique may be.

On the one hand, Cabal recognizes that Europe is, whatever its shortfalls by comparison to the U.S., in particular, a mature space power. It has a baseline to be enhanced in some areas for the new challenges Cabal speaks of, but in others, as with Galileo[258] and the Global Monitoring for Environment and Security (GMES) programmes, the gaps are being closed. Space leadership – or at least substantial parity – is hence, in principle, not out of France and Europe's reach in significant areas like transportation, civil applications and space science. The report argues for investments to permit a substantial degree of leadership.

In pursuit of its goals, Cabal urges France and Europe to return to a policy of attaining major autonomy in critical areas such as space transportation augmented by cooperation without restriction in several areas of basic and applied science. This is a similar strategy to that which dominated the launcher, space science and applications programmes Europe undertook from the 1960s to the 1980s, including those under European Space Agency's (ESA) first Long-term Plan. But it contrasts with the cooperation-reliant approach ESA adopted in the early 1990s, as its autonomous space infrastructure ambitions unwound.

### 1.2.3. The vision – and about financing and propagating it

The type and scale of the programmes Cabal envisages surpass the ambitions fostered in the entire period of European space activities.

The report calls for Europe to join the – apparent – second moon race. The aim would be to return a European crew from the lunar surface by 2018 using a European transportation system. Cabal would also give full throttle to new military space programmes. These ambitions would be backed by a political commitment to make major new public financial resources available. In France, this would take the form of boosting the budget of the French space agency, the Centre National d'Etudes spatiales (CNES), by 8% annually and by creating funding reserves earmarked for French participation in new ESA projects. A space minister at cabinet level and a space council reporting to the President would underpin the new French political commitment and reverse the steady fading of space's profile within government over the past decade and a half.

The CNES budget increase would fuel stronger national programmes and foster change in ESA, as those programmes could act as progenitors for new ESA ones. The report does not go into the logic of the relationship between the CNES and ESA, but does condemn as "unacceptable" the fact that French participation in ESA is starting to grow at a faster rate than the CNES budget. The Cabal philosophy, rather, harks back to the manner in which progress in ESA was hastened by well funded initiatives led by CNES both in the early 1970s (Ariane, Meteosat) and the 1980s (Ariane 5, Hermes), as complemented by other national initiatives such as Columbus.

If the line of policy described were indeed to be pursued by France, it would mark a return to an exercise of French leadership within ESA not evident since the Hermes/Columbus crisis in the early 1990s, when Hermes and space infrastructure autonomy goals were abandoned and France retreated to ensuring Ariane 5s completion as its main commitment. The effects of such a change would be profound. It would extend French core commitment from its traditional domains, dominated by launchers, to exploration including human spaceflight with Moon and Mars missions in view. By CNES itself shouldering a significant part of the commitment, one may surmise that the new exploration programme might be less contingent upon European partners' ambitions (and hence, perhaps, be more achievable).

In short, Cabal urges France to apply the extra money and energy needed to kick-start the motor for European space activities, to regain that higher level of ambition to which France in particular, and Europe once aspired in space.

To accommodate the practical consequences of this ambition, various reinforcing measures are foreseen nationally, among them, a "space planning law" that would set a comprehensive framework for French space activities. At the same time, Cabal calls for action to "induce" Europe to adopt the report's vision, the aim being to secure its blessing from ministers within ESA and the European Union (EU) – at the level of the European Council – in 2008. Though fairly superficially, Cabal also recognizes the need to "popularise" the vision in order to attract public support. Part of this effort would include more attention to space education, an area Cabal in common with other commentators notes as requiring urgent action, even for the sector's current requirements.

The outcome hoped for in the successful pursuit of the report's vision would be to make Europe "the next world leader in the space domain", achieved by a judicious combination of strength in its own space exploration programme and by taking a lead in promoting cooperation, an area which the United States has indeed left open to others.

### 1.2.4. Joining the critics

Such a vision and approach obviously lifts the lid on the dismal lack of direction in European space policy in the last decade and a half.

Gone is the reproach to look at the other half of the glass; it is admitted that space policy in Europe has been emptied of major ambition. Nowhere is this more evident than in "the deafening silence of Europe on the subject of lunar exploration". Nor will worn-out excuses do any more: "France cannot justify its lack of ambition . . . by the shilly-shallying [*atermoiements*] of Europe". And nor should "the current flaccid consensus" (which one takes as referring to the basis for the 2007 European Space Policy) be allowed to parade as a real policy, since this expression of "minimum national or European space development provides no response at all" to the challenge of a world where every major power except Europe seems to be moving forward.

Speaking against a fundamentally utilitarian view which can be said to predominate currently, this is dismissed as an "error" of perspective, the correct way being to "go back to the aggressive policies which have led [Europe and France] to major achievements in the past". The consequence for France of doing otherwise would be to condemn it to "a descent into the cold darkness of anonymity, and a total disregard of [the space sector's] key importance for the future of the country". By extension, the same can be said for Europe and its security interests.

All this addresses the malaise Europe has been suffering in the area of space for many years now, which Cabal terms an "existential crisis".[259] Cabal does not go deeply into its causes, but the report clearly indicates that it is policy-makers who have not risen to their responsibilities. Many would agree.

## 1.3. Review of specific perspectives on European space policy in the report

This notwithstanding, Europe has collectively done a good deal of introspection on space policy over the past five or so years, since the quest for a "genuine" European space policy was started at the ESA Ministerial Council at Edinburgh in late 2001.[260] The report is obviously well briefed on ideas to have come out of this discursive process, and it cherry-picks the ones that suit it best. The following five main points selected from the report's discussion seem salient as guidance for elaborating future policy:

- The need for a simple and bold vision, comprehensible to governments, institutions and the public alike, anchored as a priority at the highest political level in France and in the EU, i.e. the European Council.
- The need for institutional reform both in the EU and ESA, but with ESA remaining the main instrument for the inception and execution of the major new programmes, in any case, the non-applications programmes.

- Identification of space exploration (Moon, Mars and beyond) as the political imperative around which a dynamic European space policy can be achieved, with benefits for strengthening European identity, stimulating interest in S&T including career choices, and with a view to long-term economic goals, but only once adequate techniques have been developed, possibly in some decades (as with mining). Among other things, Cabal in this context urges ratification by the European states of the 1979 Moon Agreement and its wider promotion.
- A combination based on an upgraded Ariane 5, the ATV[261] orbital transfer vehicle and the ARD[262] re-entry vehicle could provide a development concept that would be sufficiently attractive to negotiate access to lunar elements of the U.S.' Constellation programme. The precondition for any effective international cooperation, though, is a commitment to acquire substantial capabilities of one's own. If Europe were to such assets, it could place itself in a position to act as a "federator" of cooperation. For the time being, the International Space Station (ISS) should be exploited for its benefits in terms of in-space experience, multilateral cooperation, and microgravity environment access.
- Despite the emphasis on exploration and, elsewhere, on a toolbox of military space applications technologies, the report urges that existing applications should be promoted more intensively, such as for providing broadband over satellite to rural areas and in developing countries, while vigorous efforts should be undertaken to advance new applications with more comprehensive, cost-effective concepts (like the "system of systems").[263]

## 1.4. Assessment of the report's recommendations

Fifty recommendations are made, several of which have been mentioned above. A summary lists the main elements as a kind of recipe for a bold French and European policy, thereby separating the report's discussion from the actions it proposes.

Not a bad technique in itself, but the result here is so lopsided that the recipe is possibly unfair to the report itself.[264] Under the proposed new programmes, the list notably goes straight into launchers, followed by defence, with human spaceflight trailing after space services. Recommendation No. 50 is "... a first Moon landing by a European crew". A shopping list of launcher and military requirements making up a third of the recommendations is, moreover, unlikely to engender the wide support the report seeks.

Furthermore, some important recommendations seem inconsistent with others. This is the case with a call for qualified majority decision-making in ESA and the application of its industrial policy fair return rule only to groups of programmes. Superficially, these are appealing suggestions. But, if Qualified Majority Voting (QMV) and a "global" industrial policy were adopted, this would reduce ESA's distinctive intergovernmental, case-by-case cooperative character versus the EU, and the report is perfectly clear that this is just what it wants to preserve. Rather, when turning to EU-ESA relations, the report goes the other way in proposing that a 20-year plan on space services be "ratified by the European Union and placed under the aegis of ESA" – a proposal that seems, frankly, naïve, just like the proposal to commission the Galileo system in 2010.

While a proposal to make EUMETSAT GMES space segment operator is worthy of consideration, the recommendations generally overlook the outstanding performance of the private commercial space sector in Europe in satellite communication. Therefore, absent is any initiative aimed at better harnessing the market's potential to strengthen the sector. Instead, the secret of success – in contention with no less a power than the U.S. for space leadership – is revealed under the banner of "French genius at the service of Europe", primordially in terms of launchers and military missions.

This tantalizing sales pitch may perhaps work domestically, but not in Europe.

Better results would have been achieved if Revol and Cabal's own stricture were adhered to, that is, of advancing *only* "simple principles for a new space policy, the ambitions of which will be on a par with the position of Europe in the world". By contrast, their report fell into the familiar trap of going too far in *medias res*. What one is left with is a barely concealed national wish list.

The recommendations thus require further work to be fit for purpose, and should be regarded separately from the valuable discussion elsewhere in the report.

## 1.5. The report in the light of the 2007 European Space Policy (ESP)[265]

Where the ESP's content can be characterized as inward-looking, the Cabal perspective is global. And where the ESP is ascetic in its ambition, confining itself in its first issue to generalities, creation of a common EU-ESA-national framework, and a programmatic fact-finding exercise, Cabal provides a vision and broad indications of where new projects should be undertaken.

To this extent, the Cabal Report – if one extracts only the key messages – has the potential to complement the ESP well by offering substance that could be considered for inclusion in what is explicitly an iterative document (we shall return to this potential in the following section). Moreover, while the report

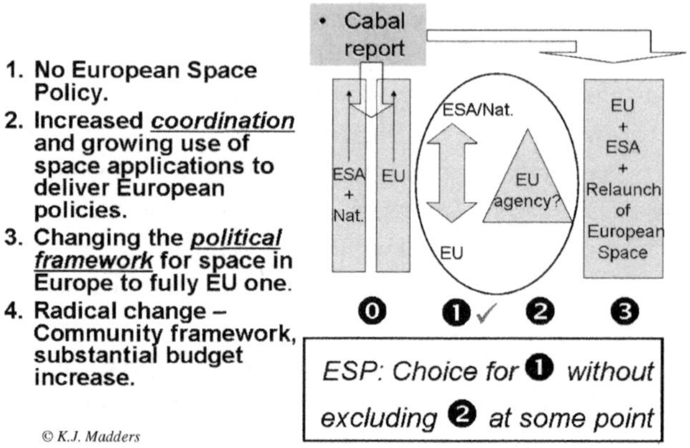


**Fig. 2:** *Relation to the 4 ESP scenarios. The Cabal Report is, on the one hand, more conservative and, on the other, more radical than the ESP.*

thunders that awaiting institutional reform before taking action on substance would be "suicidal", there is actually nothing in the ESP that restricts the kind of action the report advocates.

In other respects, whereas Cabal calls for a greater visibility of space within the EU institutions, the report remains highly conservative on institutional reform. The idea of ESA becoming an EC executive agency is specifically rejected, although the report did acknowledge the importance of the European Council for policy-making. Not least in these considerations is the modest contribution of the EU financially to total space expenditure today. The ESP, in fact, went further in terms of allowing for future institutional development (Cf. Figure 2), but maintained a gradual, consensual approach consistent with Cabal's concern not to cramp national initiative.

## 1.6. Evaluation and conclusions on the future utility of the report for the ESP

A report like this serves several purposes, as we have seen. Shorter-term and domestic ones must be distinguished from longer-term European ones. The key question is whether its content can contribute to a reviving the European space effort, and in particular, the ESP?

If the content is stripped to its essentials, the answer is a qualified yes. The report does mark a break with a past "flaccid" consensus and those essentials – contained in its vision and the proposed French action on this basis – are sufficient around which to build a more carefully considered initiative at the European level.

The qualification is threefold. First, in light of its political and financial magnitude, the French initiative on space that Cabal advocates would need to be taken up by the French President and championed bilaterally and in the European Council, ideally in 2008 during the French Presidency. Second, the planned injection of French financing would actually have to be forthcoming and ready for deployment to induce France's partners in ESA to engage in package-building around national programme objectives of their own, assuming these exist or can be promoted. Third, France will not be pushing at an open door. It will have to convince others of its analysis, particularly of the reality of the international competitive threat Cabal cites, and of their own concrete interests in joining in the spending of significant sums of public money in response. In other words France will have to move beyond Cabal's tissue of assertions to develop a coherent set of well-grounded and sufficiently persuasive arguments to accompany a Cabal-like vision.

Such persuasion has not proved impossible in similar rounds of European space policy formation over the decades. But it would be hard. France would need allies, and the report's brisk dismissal of the European Commission's role seems ill-considered here. Nor may antipathy towards including space in the EU sphere ultimately serve French interests. Preservation of Member States control in ESA and the benefit of secured industrial return are bought at the cost of complexity and extra volatility in decision-making in what is also the junior organization in terms of political weight. If the political will is there, EU structures are moreover more flexible than Cabal assumes and can be adapted to specific sectoral circumstances, as experience with other executive agencies has shown.[266] This part of Cabal's thinking should therefore be jettisoned; no strategy for a more ambitious European space policy should compromise its opportunities in the manner Cabal does.

A further defect that can and should be corrected in preparing a French (or other) initiative is the unmistakeable over-reliance Cabal places upon industry perspectives. The report could have been concocted by them. It is in this respect not enough anymore, as Cabal seems to believe, to consult a select group and then *inform* the wider public of what is for the best. Rather, it is mandatory to *involve* the public in the sense of the broadest sections of stakeholders and interested persons, notably present and future taxpayers. Indeed, while the world of the European space political/industrial elite may be cosy, it is precisely because that world is too small that it has often proved difficult to obtain broader and more heavyweight political support when it was needed. It also helps explain the lack of impact that past parliamentarians' output – including Senator Revol's – has had.

Lastly, while the report is to be commended for having vision, that vision is not sufficiently holistic for the multiple needs an ESP must serve. It fails, in particular, to incorporate a *process* that is self-reinforcing (Cf. Figure 3) for long-term success

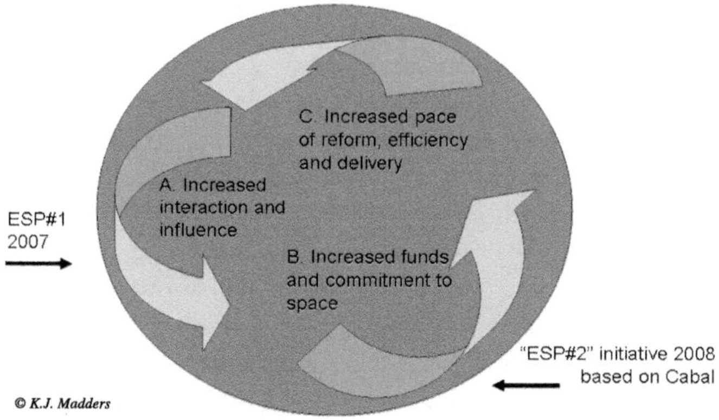


**Fig. 3:** *Creating a virtuous circle for the ESP.*

rather than subject to the vicissitudes Europe has so long endured as regards space issues. Again, a more elaborate initiative should correct this shortcoming.

In all these respects, just as in its too easy to resort to past rhetoric on threats (rather than giving positive emphasis to European values), the Cabal Report is old-fashioned, top-down and ultimately unsatisfying, despite its fundamental messages. The Cabal Report thus could serve as a basis for a French Presidency initiative, but it needs to be transformed and modernised, then debated and adapted to others' perspectives. And then adopted.

---

252 OPECST to the Assembly (Report No. 3676) and Senate (Report No. 223). "Rapport sur les grands domaines programmatiques de la politique spatiale du futur". Politique Spatiale: L'audace ou le déclin; Comment faire de l'Europe le leader mondial de l'espace 7–8 February 2007. http://www.assemblee-nationale.fr/12/rap-off/i3676.asp, http://www.assembleenationale.fr/12/rap-info/i3676-english.pdf. OPECST is drawn from both chambers in equal numbers reflecting political composition.

253 This report is available at: http://www.senat.fr/rap/r00-293/r00-2931.pdf.

254 The rapporteurs were conscious that the main messages of the report might be lost with a lengthy text. These are extracted in a brochure for wide distribution. See the English translation http://www.assemblee-nationale.fr/12/cr-oecst/05-06/synthese_polspatiale.pdf, http://www.assemblee-nationale.fr/12/cr-oecst/05-06/synthese_polspatialeeng.pdf.

255 Colloquium conclusion. Nov 2005. http://www.senat.fr/opecst/colloque_espace/colloque_espace.pdf.

256 Cabal himself uses that phrase in the concluding remarks just referred to.

257 Madders, Kevin. A New Force at a New Frontier. Parts I and II. Cambridge: Cambridge University Press, 1997.

258 The report was published before the 2007 Galileo crisis broke out.

259 Various indicators are referred to, the shrinking workforce being the most striking, down 16% from 2001 (and down considerably more if one were to compare it with the beginning of the 1990s).

260 The period witnessed, in particular, a European Commission Green Paper on European Space Policy and public consultations on it, a White Paper, and later, the establishment of a High Level Space Policy

Group of officials co-chaired by the Commission and ESA. Independent activities such as the European Space Policy Workshop series at the Catholic University of Leuven, alongside work by ESPI and others, complemented institutional ones. See http://www.eurospacepolicy.org for relevant documentation.

[261] Automated Transfer Vehicle. This ESA vehicle's maiden flight is due in early 2008.

[262] Atmospheric Re-entry Demonstrator. This ESA capsule was launched in October 1998. It reuses some elements from Hermes' development. As its name suggests, it is a technology demonstrator, but the ARD testbed could be used in developing new systems, notably in relation to a possible Crew Transport Vehicle and reusable launchers.

[263] Under this concept, distinct satellite projects serving different applications would be subsumed under an integrated approach from procurement to design through to deployment and operation.

[264] The recommendations are somewhat scattered across the report. In gathering them together, the summary failed to distinguish minor from major proposals.

[265] European Commission communication. "European Space Policy". COM (2007) 212. 26 April 2007: formulated with the ESA Director General and must be read in the light of Council Resolution 136. *OJ* 2007/C 136/01. 21 May 2007. That resolution "welcomed and supported" the policy but with qualifications. It was passed by the EU Council joined by the ESA Council to form the "Space Council" pursuant to the 2004 EC-ESA framework agreement.

[266] Hobe, Stephan, et al. Forschungsbericht ESA-EU: Rechtliche Rahmenbedingungen einer zufkünftigen kohärenten Struktur der europäischen Raumfahrt. Münster: LIT, 2006.

# 2. The new UK approach

Klaus Becher

How important is space to the United Kingdom? Are existing institutional arrangements for space in the UK Government adequate? Does the UK derive the right benefits from the European Agency (ESA) and the European Union (EU) activities? What roles must public space expenditure play? What is the proper balance between science and applications? How can the innovation and growth potential of the space sector be best realised? While it remains uncertain if a "new UK approach" to space policy exists or is about to materialise within the government, there has certainly been a remarkable, forward-looking debate on these questions in the UK in recent months. This ongoing process of re-evaluation of space policy in the UK should be of interest not only with a view to evolving UK policies, but also for the European space effort as a whole.

## 2.1. The UK experience

The UK can look back at a strong role in the 50 years of space history, which has always been achieved on a relatively limited budget. Britain's Ariel-1 satellite laid the foundation for international space collaboration in 1962. With Ariel-3 in 1967, the UK was the third country after the U.S. and U.S.S.R. to build and operate its own satellite. The launch of Prospero on Black Arrow in 1971 made Britain one of the very few states in the world with autonomous access to space able to launch their own satellite (from Australia). However, the launcher programme was abandoned at that same time, as a cost-cutting measure in a difficult economic phase based on the recognition that access to space was on its way to becoming a market commodity.

With the Skynet satellite series, the UK established itself as the third country with operational military communications satellites of its own already in 1974, more than a quarter century ahead of other Europeans. Subsequently, British military communications satellites were also provided to the North Atlantic Treaty Organization (NATO). The current Skynet-5 generation pioneers a successful private finance initiative (PFI) model for making secure military satellite communication services available to national and international defence customers through commercial enterprise.

**Fig. 4:** *Ariel-3 satellite (source BBC).*

Unlike in some of the continental European countries, hostility to space activities is not a factor in British political debates. The fact is, however, that the UK spends less than half per capita on space from public funds than other members of the European Space Agency (ESA). In 2006/2007, the UK's civil space budget was 218 million pounds (approximately 320 million euros), of which 72.5% have been channelled through ESA.[267] In addition, space-related spending by the Ministry of Defence, mainly for secure satellite communications, is a sizeable factor, representing a higher percentage of overall public space spending than in other European countries.

In 1987, Britain opted in ESA to abstain from manned space flight and involvement in the international space station. Given the later fate of Hermes and Columbus, this step can in retrospect only be seen as a fortunate decision, even if it may at the time not have been taken entirely for the right reasons. Staying away from some of ESA's most important optional programmes came at a price though. Not only did it limit industrial opportunities – it also moved the UK away from the core of ESA's decision-making in some respects. It allowed a mindset to grow in some quarters in ESA as well as in the UK that regards Britain as a marginal player in space – entirely inappropriate, given Britain's role as one of the main pillars of European strength in advanced space technology and the impressive vitality of public and private space activities in the UK.

Can Europe's space efforts in ESA and the EU take some pointers and rejuvenation from recent UK debates? Above all, this would imply a shift from an experimental science and technology development focus toward a mixed approach equally geared at satisfying operational requirements, for example, in the context of the strategic issue of permanent global climate change monitoring. Such a shift could also imply new funding approaches that reach beyond science and technology budgets, tap into the investment and operations funds of other user departments and get leverage from commercial markets and private investment.

## 2.2. The Case for Space

Today there is no functioning national supply chain in the UK space industry anymore. British small and medium sized enterprises cannot rely on business trickling down from a large national champion. They need to operate on a European-wide and global level to assure the desired flow of contracts. Likewise, British systems integrators are free to source beyond national borders, although regional development efforts to establish clusters of industrial and research excellence are designed to maintain a strong technology development network within the UK. This situation comes with all the costs and risks of globalised business, but opens up many opportunities to those who can generate sufficient agility.

The mindset of state subsidies that has been formative for the space industry, not only in Europe, is today giving way to an entrepreneurial, market-oriented approach that anticipates future demand and offers attractive products and services in a competitive environment. The UK sees its own future perspective in being the economic hub of globalisation, characterised by Gordon Brown as "a force for justice on a global scale" due to open markets, flexibility and free trade when duly combined with skills, education and infrastructure.[268] Against low-wage competition from new industrial powers such as China, wealth creation in the UK is expected to spring from globally operating knowledge and creative industries based on investment in science, innovation and technology, and from financial services. The goal is to turn advanced technology into profitable products and services quickly before it diffuses to others who can operate at lower cost.

Does this economic philosophy, combined with the increasing market strength of space services mean that Government has no role at all to play anymore in the space sector? Or does the increasing practical relevance of space technology and space-derived services rather demand increased government attention to space beyond the traditional science and technology development focus to make sure that best use is made in the public interest of the new operational opportunities space has generated across the full range of government responsibilities? For the 2007 Comprehensive Spending Review, a new round of the budget planning mechanism established by Labour in 1997, the Treasury identified certain major areas of change, such as climate change and intensified cross-border economic competition due to the rise of China and India, and suggested that spending should be reviewed and focused on addressing these "big challenges" over the next ten years.[269] In response, the point needed to be made that space-empowered applications and services provide essential contributions that underpin the UK's ability to cope with these challenges.

Outside the space community there is little awareness of the rapid advances of space technology and its fusion with other technologies as a key enabler of new public and commercial services that can empower better solutions. Too many casual observers are tempted to misjudge the space sector as one of the old metal-bending manufacturing industries whose time has gone by. Too little is known about the economic value created by the space sector and its role in driving and facilitating innovation in the geosciences and in low-energy, light-weight, high-precision, ubiquitous, always-on technology. Funding to foster innovation and growth flows to key technologies such as information technology (IT), sensors, new materials, robotics, or energy efficiency. A failure to realise that space is not just historically at the cradle of many of these cutting-edge technologies, but continues to drive and integrate them would put the space sector at risk. Government engagement in space could be crowded out by other priorities exactly at a time when space has more tangible benefits to offer than ever before.

**Fig. 5:** *Case4Space summary report.*

In order to create a wider, more solid understanding of the importance of space for the future prospects of the UK, a consortium of upstream and downstream UK space industry players launched the "Case4Space" campaign in 2006,[270]

It stressed the fact that space technology has moved from the sphere of early research, technology development and isolated specialist applications to widespread, routine operational use in communications, navigation and earth observation. Tools that depend on satellites are increasingly important for economic growth, quality of life and the successful delivery of policy objectives across a wide range of Government departments and public bodies. The most important and influential element of the campaign was a report by Oxford Economics on the economic value of the new space-enabled communications, navigation and earth observation markets using metrics and models very similar to the Treasury's own,[271] The report responded to earlier calls for an "evidence-based space policy" to justify public spending on space by putting hard numbers to the societal benefits of space policy, including the multiplier effect of targeted public spending on application-oriented advanced space technology development. Above all, the report substantiates the impressive catalytic and spill-over impacts of space technology that create capabilities and enhance productivity across the wider economy and even give rise to whole new sectors of economic activity that would otherwise not exist, such as satellite broadcasting and the Global Positioning System-based navigation industry.

## 2.3. UK Space Vision 2025

The Case for Space campaign suffered from its short-term focus on influencing imminent decisions on budget allocation in the Comprehensive Spending Review 2007 and Britain's Global Monitoring for Environment and Security (GMES) and Advanced Research in Telecommunications Systems (ARTES) subscriptions in ESA. In spring 2007, the trade association Ukspace[272] and EADS Astrium therefore initiated a series of consultations among key voices of the UK space policy landscape on a longer-term UK space vision. The results were presented to parliament, government and industry in June 2007 in the document "Vision 2025: A world of opportunities for UK space technology".

This non-official but representative vision places space in the big picture of a future world characterised by global interdependency, demographic pressures, climate change, scarce resources, relentless pace of innovation and risk of violent conflict. Space technology is seen as one of the major shaping forces of the time and a cutting-edge field where the fusion of information, communication and new, smart materials will drive innovation and trigger breakthroughs that feed into

**Fig. 6:** *Vision 2025: a world of opportunities for UK space technology.*

other fields of technology, science and economic development: "Space-age tools are essential for sustainability, good governance, security and defence. They are a world-wide engine of development and empowerment". Our quality of life and problem-solving capacities increasingly depend on tools with embedded space-based elements such as precise, ubiquitous timing and three dimensional positioning, resilient globally linked communications networks and a constant flow of world-wide current data measurements and imagery that provides a new evidence base and expands the limits of our understanding for addressing future challenges from climate change to counter-terrorism.

The vision paper stresses that space is the best teacher of green engineering and helps to support and catalyse low-carbon solutions in other sectors. New infrastructures enabled by satellites in conjunction with other technologies will render many of today's costly and energy-hungry terrestrial approaches obsolete. Satellites are indispensable for data collection and relay in the shared global effort to monitor global climate change.[273] In 2025, the vision foresees a real-time network of millions of sensors on the ground connected by satellites into a continuous Earth monitoring system, combined with continuous world-wide multi-instrument coverage from a low orbit for hyper-spectral change detection. Such a space-empowered network creates a new quality of disaster preparedness, scientific understanding, evidence-based multilateral policies and the enforcement of international agreements.

The world space market of 2025 is expected to be a worth a trillion pounds (approximately 1500 billion euros) annually, no longer dominated by governments but driven to a large extent by profit-oriented private investment even in areas such as space science and exploration. The vision suggests a new key role for the United Kingdom, building on its established strength in global capital markets and innovative financial engineering: "Britain holds the purse strings to space. London remains the place to go to finance your satellite business". The vision's target is to double the UK's world space market share by 2025, equal to a more than twenty-fold growth of annual UK space turnover to 150 billion pounds (approximately 270 billion euros). There are plenty of opportunities as space technology joins the ICT mainstream. The UK space vision group's message boils down to the slogan "Be in it to win it". Judged on the basis of first reactions by politicians, including within the Treasury, the message has essentially been understood.

## 2.4. "A Space Policy"

The House of Commons Science and Technology Committee held a major inquiry in the first half of 2007 leading to its report "2007: A Space Policy"[274] that provides an excellent overview of the issues. The committee's report endorses the established approach of being selective about what the UK does, concentrating on what it does well, such as environmental and planetary science, advanced communications satellites, small satellites and satellite finance. However, it recommends that the Government takes a more strategic approach to space, based on the realisation that UK involvement in the exploitation and exploration of space is crucial. The report recognises that a strong political lead is essential if the UK's research and industry are to play out their strengths in space, and therefore calls on

**Fig. 7:** *The inquiry initiated by the House of Commons Science and Technology Committee led to the report "2007: A Space Policy".*

Government to declare its ambitions. The report highlights Government's role in providing seed funding, risk-sharing and innovation-friendly regulation along with support for leveraging private funding for innovative products and services. It encourages the British National Space Centre (or BNSC) to take on the new role of explaining space technology to private investors and lending it the credibility it deserves.

The Committee criticises the absence of a long-term space vision and strategy, and refers to the "downward management of expectations" by BNSC headquarters as "extremely disappointing". It suggests that BNSC's next strategy document should include a long-term roadmap 2010–2050 that should provide "a flexible indication of where the space community is heading". BNSC's central staff should engage in horizon scanning and inform partners of emerging issues so that major new developments are identified early enough to benefit from them. The Committee weighed the arguments for and against replacing the BNSC partnership, whose current set-up is unsatisfactory in some respects, with a space agency but concludes that an agency would not improve matters at current funding levels. Instead, the report proposes that BNSC headquarters should have a small budget of its own to cover its own overheads as well as a modest national technology development programme.

The lack of funding arrangements within BNSC and Government as a whole for space-related programmes that involve more than one partner is a key weakness. The lead role and funding responsibility falls to one of the members of the partnership. This allocation is problematic where there is interest in end-use, but not in development. BNSC might provide a starting point for advanced budgetary solutions such as cross-departmental pooling of jointly administered funds or pooled purchasing that is better in tune with the specific nature of space activities and their operational importance for Government as a whole. The report quotes GMES and the Global Earth Observing System of System (GEOSS) as prime examples of cross-cutting relevance that would demand better interconnected thinking and organisation. The Committee urges that the lead on GMES should be shifted in the UK from the hesitant Department for Environment, Food and Rural Affairs (DEFRA) to BNSC headquarters. Meanwhile, however, the Natural Environment Research Council (NERC), a BNSC partner, established the new Centre for Earth Observation Instrumentation (CEOI) managed by a consortium of industrial and academic research partners to strengthen the UK's role in global climate research collaboration. Stressing the vital role of satellites for understanding global climate, the Committee expects that GMES should assure operational data continuity. In its current short-term project approach, however, GMES does not yet meet this requirement.

Britain is the second-largest contributor to ESA's Aurora space exploration programme, with a share of 17%. There is much enthusiasm in the UK for the manned aspects of international exploration of the Moon and Mars, but there appears to be no willingness to commit UK taxpayers' funds to participate in this journey for anything but unmanned missions. This does not preclude a British role in international programmes of human space exploration. The Science and Technology Committee recommends that the UK should fund the best science in space exploration, be it robotic or manned. There should also be no in-principle block on British funding for the future development of launchers. The message is to approach space without any ideological blinkers to seek out attractive opportunities and advantages where they exist.

## 2.5. Towards a new space strategy

In response to the Global Space Exploration Strategy, a BNSC working group reported in September 2007 on the opportunities and benefits of UK participation in space exploration,[275] The report advocates a full and active UK role to answer "scientific questions of great interest that can only be answered through the continued exploration of space", with some of them requiring human presence.

This exploration report is expected to have an impact on BNSC's new strategy that is to replace the current civil space policy documents "UK Space Strategy 2003–2006 and beyond" (November 2003) and "Earth Observation Programme Board Strategy 2003–2006" (January 2004), complemented by the individual space policy objectives of the members of the partnership. Publication of the new strategy for 2007–2010 is now expected at the end of 2007. A public consultation had been launched in January 2007.

With the split-up of the Department for Trade and Industry, BNSC's host department, into two new departments, BNSC's central staff, as an element of the Office of Science and Innovation (OSI), has moved to the new Department for Innovation, Universities and Skills (DIUS), not to the new Department for Business, Enterprise and Regulatory Reform (BERR) as the desired market-orientation of space might have suggested.

## 2.6. The security and defence dimension

In military space applications other than communication, the UK enjoys privileged access to US-operated technical means under its close post-1945 alliance and cooperation arrangements with the United States. The UK continues to adapt and update its established policy of combining autonomy in some of the most critical defence technologies with trusted access to US capabilities and defence markets. Generally speaking, it tries to pursue the most cost-effective strategy that avoids relying entirely on others.

In this context, the Ministry of Defence (MoD) is re-evaluating its approach to space in the light of the Defence Industrial Strategy (December 2005) and the Defence Technology Strategy for the Demands of the 21st Century (April 2006), the first open publication of the MoD's research and development priorities.[276] Specifically, the MoD has begun to work with UK industry and academics to provide national niche ISTAR capabilities (Intelligence, Surveillance, Target Acquisition, Reconnaissance) through small satellites, following the Topsat optical demonstrator satellite launched in 2005. This development is regarded as a necessary element of network-enabled capabilities (NEC) and flexible situational awareness, particularly also for alliance and coalition operations.

In spite of the strict separation between civil and defence space in the UK, there must be an effective effort to make technology development funding in military space useful for the whole space sector and vice versa. This has always been the main rationale of the MoD's presence in the BNSC partnership. The technology directorate at BNSC headquarters is traditionally held by a civil servant from the MoD with awareness of both worlds. The 2007 Science and Technology Committee report found that there is an increasing overlap between the uses of space for civil and military purposes today. The Committee indicated that it may

**Fig. 8:** *Topsat satellite (source Qinetiq).*

return to the issues of military use and dual-use systems soon and suggested that BNSC includes dual-use technologies in its new strategy. Dual-use could prove to be an important catalyst for change in UK's space policy. On the one hand, the government has identified dual-use space technologies as a cost-effective way to deliver better security. On the other hand, it is not clear how it can build and operate dual-use systems that cut across departments. Open issues of dual use also keep affecting the UK's positions on the European level, including Galileo, ESA's "peaceful purposes" mandate and EU involvement in civil security and counter-terrorism, but not defence.

## 2.7. Key messages

For French observers in particular, it is apparently hard to grasp that their habitual key arguments for substantial government spending on space usually fall on deaf ears in the UK, or are plainly counterproductive. Essentially this phenomenon reflects diverging notions of the role of government and the state. The difference in style is obvious, but it must not be overrated. There are plenty of civil servants with a statist, central-planning mind in the UK, and there are not just a few growth-oriented, market-minded innovators in France.

Still, perceptions have a life of their own. The fact in the UK today is that advocates of European cooperation and pooling of efforts in space must first of all confront the concern that a penny thrown at Europe is a penny lost. This is an unfortunate attitude not least, because the UK conducts such an exceptionally high proportion of its own civil space activities through ESA. The UK is industrially more integrated with Europe than other big space players. It has achieved a good return on investment from ESA, both over time and in most recent years – although the latter point is somewhat obscured by ESA's changed reporting format.

From a UK perspective, it is generally assumed that the European integration process is not a desirable thing in itself as "l'art pour l'art". This simply means that European programmes must always be judged on their merits from case to case with a good dose of scepticism and an open mind. One must also acknowledge in this context that the bureaucratic machinery of European civil space cooperation, influenced by multiple national science and industry lobbies with little transparency, parliamentary accountability and political leadership, does not always come across as the perfect vehicle for deriving the best economic and political value. In sum, it is not very easy to present European space programmes in a positive light to a UK audience, including key decision-makers.

This does not mean, however, that it is any easier to mobilise their support for national space programmes.

International space policy, as institutionalised as it may be on the working level, remains an element of foreign policy. The weightiest single element of global science policy by budget volume, it can even be regarded as a key factor in today's international system, setting the political and legal stage for crucial new areas of world politics such as climate change cooperation. Accordingly, in its proper pursuit, the diplomatic art of pressing the right buttons should not be neglected even if this requires the non-trivial effort of observing and respecting multiple national discourses in a multinational context.

In the UK, the buttons for others to press if they wish to strengthen the UK's space involvement today include climate change, African development, market-spawning new technologies, quality of life and – if pursued in an appropriate, capable framework and without attempting to undermine NATO – security and defence.

In dealing with their British partner, other Europeans also ought to be mindful of the specific nature and limited mandate of the BNSC, designed in 1985 as a partnership between departments and agencies. It is not an executive agency. It is responsible for international and cross-government representation in civil space and leads efforts to define an overall civil space strategy. It has no political weight, makes no decisions and cannot take initiatives of its own. Decisions on UK space policy are made, or not made, in cabinet departments and ultimately, for practical purposes, at the Treasury.

For bringing the UK's space policy more vigorously behind joint European efforts in ESA and the European Union (EU), both organisations and their other members would be well-advised to present their work in a more results-oriented, accountable, value-for-money manner to impress and attract the UK. There is currently hope that the planned new ESA centre in the UK, expected to cover climate change, exploration and space applications, could potentially make a real difference. While it is unlikely at this time that the UK Government is going to release additional funds for space as such in the foreseeable future, it is quite possible that UK spending on other priority areas of public policy such as climate change or security and resilience can help to drive Europe's future space agenda.

In addition to such important operational uses of space by governments, the emerging new UK approach to space stresses the global market strength of advanced space technology as a core component of the information and communication technology (ICT) revolution. This suggests a likely gradual shift in the role of governments, as well as ESA and the EU, away from operating large technology development programmes in the space sector to becoming intelligent lead customers in the marketplace, as has been the case in other parts of the IT sector since the 1980s.

In telecommunications and meteorology, ESA has a strong record of generating operational and commercial services already. The task is now to make the institutional framework of European space cooperation agile and dependable enough for bringing the effective, globally networked use of space technology to reality in order to confront the big political, social, economic and environmental challenges which governments will be expected to cope with in the next decades. From a UK perspective, the operational provision of climate change information is likely to be the test case, and time is indeed very precious.

[267] United Kingdom. British National Space Centre (BNSC). UK Space Activities 2007. London, July 2007: 67.

[268] Mayer, Catherine. "Gordon Brown: The Interview". TIME Magazine (10 May 2007).

[269] The UK. HM Treasury. Long-term opportunities and challenges for the UK: analysis for the 2007 Comprehensive Spending Review. London. Nov. 2006.

[270] UKspace. Case4Space. Summary Report. Oct. 2006.

[271] Oxford Economic Forecasting. The Case for Space. The Impact of Space Derived Services and Data. Oxford: Oxford Economics Ltd. Nov. 2006.

[272] UKspace is the new short name of the United Kingdom Industrial Space Committee (UKISC), jointly sponsored by the aerospace industry trade association SBAC and the technology industry trade association Intellect. It represents over 75 percent of the UK space industry by both turnover and people employed, with member companies engaged in all aspects of space activity.

[273] For the high priority assigned by the UK to climate change, see the Treasury-sponsored report: Nicholas Stern. Report on the Economics of Climate Change: The Stern Review. Cambridge 2007.

[274] The UK House of Commons Science and Technology Committee. A Space Policy – 7th report of session 2006–07. London, 17 July 2007.

[275] The UK Space Exploration Working Group. Report of the Space Exploration Working Group. London, 13 Sep. 2007.

[276] The UK House of Commons Science and Technology Committee. Memorandum 76: Submission from the Ministry of Defence – A Space Policy. London 2007: 291–296.

# 3. What's the problem with Europe's flagships Galileo and GMES?

Serge Plattard

The general impression today is that the two European space flagships initiatives, Galileo and the Global Monitoring for Environment and Security (GMES) programmes, are progressing quite slowly, if not even stalled for the first one, since they were started nearly ten years ago; they are still far from entering an operational mode for delivering the projected services. Indeed, the idea of having a European positioning, navigation and timing system based on a constellation of satellites comparable to GPS (Global Positioning System) was floated in the nineties and pushed by the transport commissioner Neil Kinnock (1995–1999) in 1998 mainly for security and sovereignty needs in Europe. The GMES concept was aired through the Baveno declaration in May 1998, calling for Europe to have its own system for monitoring environment and security worldwide on a 24/7/365 basis. Both initiatives would give Europe the appropriate strategic tools to participate fully in the information revolution that would ensure Europe's grip on geopolitical issues; economic competitiveness and sustainable development.

Yet, what looked so promising and structuring for the European Union (EU), in line with the Lisbon goals, has encountered a series of difficulties of political, organisational and industrial nature that could endanger the leverage that Europe stands to gain from these endeavours.

## 3.1. The continuing difficulties of Galileo

Early May 2007, the German Transport Minister Wolfgang Tiefensee indicated that "Galileo is going through a deep and serious crisis", referring to the current situation as "a dead-end street" and acknowledging the "need to find an alternative solution",[277] echoing a letter of the Commission's Vice-President Jacques Barrot sent to the German Presidency of the EU on 14 March 2007 listing a series of identified problems and fixing a 10 May deadline for incorporating the Galileo Operating company (the future concessionaire) and appointing a Chief Executive Officer (CEO) for that company, and a target date for the signature of the heads of terms by 15 September 2007.[278] After nine years from launching Galileo and five years of public funding, how could such a dramatic situation occur?

From the very beginning, the Galileo endeavour has been challenging and complicated, since it is the first large-scale infrastructure project of operational nature launched by the EU, not based on a multilateral organisation and aimed at delivering a variety of services to European citizens. It had to start from scratch, overcoming technical challenges, organising a framework in which the Commission, the European Space Agency (ESA), National Space Agencies and industry would be involved, and preparing the most efficient mode of distribution of the relevant services. It had also to face the existing GPS, which has been providing free-of-charge signals for navigation since 1993, and an accuracy better than 10 metres since the discontinuation of the selective availability by the U.S. Department of Defense in May 2000. One should bear in mind that the development phase followed by the Full Operational Capacity (FOC) should be such that by the end of the current

## THE GALILEO SYSTEM

The fully deployed Galileo system consists of 30 satellites distributed in three circular Medium Earth Orbit (MEO) planes at 23 222 km altitude above the Earth, the orbital planes having a 56° inclination with respect to the equatorial plane. Each plane contains nine active satellites plus one spare. Each satellite houses four atomic clocks providing extremely accurate timing. Each point on the Earth surface receives signals from at least four satellites at all times. It is planned for Galileo to offer five basic services:

- Free, open-service providing location and timing to all users, comparable to the free-of-charge GPS signal;
- Commercial service that will offer unprecedented accuracy, about two meters, for paying subscribers;
- Public regulated service (PRS) serving the needs of police, safety/security, defence and intelligence agencies and using encrypted signals;
- Safety-of-life services for functions such as air-traffic control;
- Search and rescue (SAR) assistance.

The total cost for bringing the system to full operational capability (FOC) is estimated at 3.4 billion euros for the 2007–2013 period including the exploitation of the European Geostationary Navigation Overlay Service (EGNOS) (2008–2013).

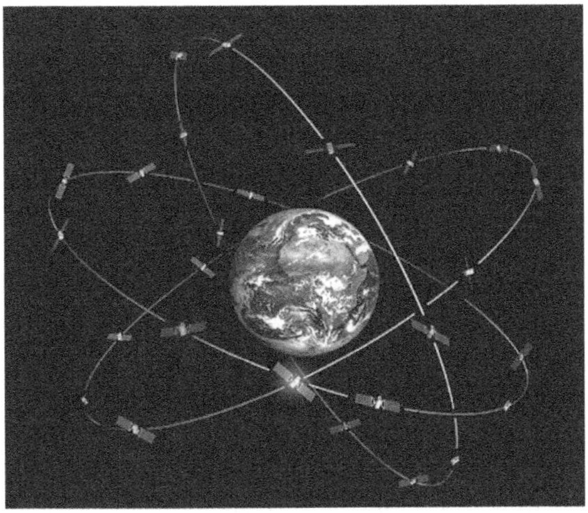

**Fig. 9:** *Galileo Constellation (source Telespazio).*

decade, the five proposed services would be operational providing sufficient lead time to capture a large chunk of the market and retain a significant competitive edge before an upgraded version of GPS (GPS-III) will pose new challenges that is scheduled to be introduced in the 2014–2018 period. When the European Council of 5 April 2001 approved in a resolution to launch the Galileo programme, the start of the operational phase was scheduled to be in early 2008.

### 3.1.1. The troubled Galileo Joint Undertaking (GJU)

A Galileo Joint Undertaking was set up in 2002 with the objective of managing the development phase of the programme and finalizing the procedure for selecting the future operating company of the 30 satellites of the Galileo constellation. This entity, the only one suited in the eyes of the Commission able to channel public funds coming from the EU and ESA was supposed to select the partners that eventually would form a concessionaire for running the system over a period of 20 years. At the same time, ESA had to deal with a consortium called Galileo Industries to be the maître d'ouvrage for the first four satellites of the constellation and be responsible for the in-orbit validation (IOV) of the first elements of the system before transferring the responsibility of having the constellation completed under the Global Navigation Satellite System (GNSS) Supervisory Authority (GSA). It was anticipated that Galileo would be operational by 2009.

The first sign of slowdown was noticed in the excessive time it took for the GJU to be set up and, in particular, to have its director selected and approved, a person knowledgeable with industrial management but with no space experience. The GJU launched a tender for the deployment/exploitation phase on 15 October 2003 which received four initial bids in December 2003. In September 2004, the two remaining bidders submitted detailed offers, and a month later the GJU requested additional information. The GJU concluded that neither of the two consortia had made a sufficiently attractive proposal satisfying the criteria needed for the development and the exploitation of the constellation. Due to strong political pressure underlining the importance of nationalistic positions within the industry, even in an EU led project, the GJU was unable to come to a rational conclusion, which would have been to select one of the consortia as the winner. Instead, contrary to the current rules of the Commission when selecting bids, the two consortia were asked in 2005 to make a common offer merging the two bids that was accepted by the GJU under three conditions: i) the merged bid should lead to an improvement of the offer; ii) the merged consortium should create a single entity as the sole interlocutor of the GJU; and iii) the merged offer should not lead to additional delays, in particular, the signature of the concession contract by the end of 2005. Putting aside the competitive nature of the offer and introducing a disguised sort of "juste retour" proved to be dramatic for the project: inevitably, the worm was introduced in the fruit. Indeed, although the above conditions were accepted by the companies forming the merged consortium, known as MC (AENA, Alcatel Alenia Space, EADS, Finmeccanica, Hispasat, Inmarsat, TeleOp, and Thales), only an incomplete and partial version of the heads of terms, i.e., the core elements of the concession contract, were signed on 23 November 2006, therefore leaving some major issues open, noticeably the design risk and the market risk. Internal industrial disagreements requiring mediation with regard to the division of roles and responsibilities as well as the location of major ground installations introduced an additional one-year delay. For the first five months of 2007, the negotiation came to a stop.[279]

### 3.1.2. A five-year delay

Reasons for such a delay in the GNSS programmes Galileo and EGNOS (European Geostationary Navigation Overlay Service) with regard to the initial schedule need to be looked at in more details.

- First of all, the political momentum underpinning the launch of the project has considerably diminished because the negotiations have lasted for years, not giving to the politicians concrete results for them to dwell upon in order to get additional funds. For some of them, Galileo could turn into a lame duck for which fighting for has become secondary when compared to other costly projects, i.a., renewable sources of energy or environmental conservation. This insufficient political support is also connected with the difficulties Europe is experiencing in building its own identity: a clear vision for the next 20 years spelling out the role Europe should play in organizing and participating into the world governance would definitely underpin the strategic nature of Galileo, and therefore, the importance for Europe in obtaining early results.

- Second, there has been a constant discomfort about introducing a public regulated service (PRS) for governmental/public authorities' use and hence the possibility for military personnel to benefit from such a service. Although the Galileo operating company of the system is civil and Galileo is a civilian programme, there should be no incompatibility, for instance, for France, to use such a signal for non-civilian applications. Quite contrary to the UK government, which maintains that a civilian programme cannot, under any circumstances, be derived into military applications; a rather hypocrite position since no one complains about Arianespace launching dedicated military satellites, the Helios and the Syracuse series, for instance, not to mention the Skynet and Sicral satellites, while the Ariane programme was developed by ESA using European civil funds. In addition, there is still a considerable degree of uncertainty as to the extent the public authorities will use the PRS. Some estimates predict over 20% of the available signals, but it is hard to tell for sure.[280]
- Then, there is the central question of risk-sharing under a Public-Private-Partnership (PPP) scheme where the "in-orbit-validation" infrastructure including the first four operational satellites and the associated ground segment are completed by an industrial consortium, ESNI for the matter (European Satellite Navigation Industries) and subcontractors under the IOV contract from ESA that would be followed by the deployment phase concerning the remaining 26 satellites to be procured with two-thirds of private money and one-third coming from public funding from the Commission and ESA. Nine blocks of risks have been identified[281] for which solutions have been found for seven of them, but differences remain concerning the risk-sharing associated with the conception of the system, a major concern, and those related to the commercial revenues and the size of future markets. As of mid-2007, no satisfactory solution has surfaced to bridge these divergences. The transfer of the design risk and the completion risk, cost overruns risk and performance risk to the MC has not proven possible at reasonable conditions. The EU from the outset, and more specifically the GJU, has underestimated the technological complexity of EGNOS and Galileo, assuming that the system could be deployed in a much shorter time and with less public funding that it took the U. S. for GPS (a high-ranking Pentagon official

explained in May 2000 to the Centre National d'Etudes spatiales (CNES) that with the planned performances for Galileo, there was no way it could be completed with the anticipated funding, and a rough guess would be to at least double the amount, i.e. 10 billion U.S. dollars).

- Last is the concern about political and industrial governance. The current industrial organisation lacks leadership. For instance, for years there has not been any recognized interlocutor from the MC having the full power to negotiate the concession contract with the GJU, and later with the GSA. Furthermore, the difficulty in reaching decisions has been due largely to disputes regarding roles, responsibilities and programme work shares. The intrusion of politicians from a country whose industry is substantially involved in the project have complicated the decision-making process, in particular, when negotiations were finally settled and it still tried to get an "over return" for its national industry in terms of workload regarding a ground infrastructure for instance. Germany and Italy, which from the outset had separately wanted to be the sole leaders of Galileo have certainly not contributed in smoothing the progress of this complex project, to say the least.

### 3.1.3. Public funding: better later than never, or how to waste time and taxpayers money...

After the EU Council of Ministers of Transport met in March 2007, it was clearly recognised that the current scheme had induced too many delays to continue on the same track with reasonable chances of success by 2011. Another approach had to be found favouring a scenario based on purely public funding for both the IOV and deployment phases, and de facto abandoning the PPP funding scheme. It also kept the possibility of the private sector to manage the constellation once operational, and gave it the exclusivity of delivering services for a 20-year period.

Three weeks before the Council of the Ministers of Transport was held on 7–8 June 2007, the Commission issued a communication[279–281] where six possible scenarios were presented with their strengths and drawbacks, ranging from pursuing the current scheme, to a complete stop of Galileo leaving only EGNOS to be implemented, hence relying more or less entirely on GPS. Of course, none of the extremes were politically sellable. The salvaging scenario, no. 4, as described above, seemed to be the only one that was acceptable and presumably feasible within a 2013 horizon.[281]

At the aforementioned Council, held in Luxemburg, a resolution requested to stop the current funding scheme and rely only on public funds for the IOV phase and the deployment of the constellation amounting to 3.4 billion euros, meaning an additional 2.4 billion euros to the 1 billion already provided for the project.[282] The total costs when taking into account a 20-year exploitation by a private consortium is currently estimated at 10 billion euros. No decision has been made concerning the funding modalities. Nonetheless, it seems that a dividing line has started to complicate matters once again: on the one hand, the Commission together with the European Parliament and most of the Member States would favour the supplementary funds coming from the European budget; on the other

## GALILEO KEY DATES

**1998** Sir Neil Kinnock expresses the strong need for Europe to have its own GNSS (Global Navigation Satellite System) for sovereignty and security purposes

**2001** April, EU Council approves the launch of Galileo. A 2008 operational phase is foreseen

**2003** 1st September, formal establishment of the Galileo Joint Undertaking (GJU)

**2004** September, the two remaining short listed bidders submit detailed offers

**2005** 4 July, the GJU agrees to the creation of a merged consortium, MC, comprising AENA, Alcatel Alenia Space, Finnmeccanica, Hispasat, Inmarsat, Thales, TeleOp

28 November, launch from Baikonour of Giove-A (Galileo In-Orbit Validation Element)

**2006** 20 November, first version of the heads of terms in initialled

**2007** Early in the year the negotiations are stalled

14 March, Vice-President Barrot, in a letter to the EU Transport Ministers, lists the problems and fixes

10 May Deadline for incorporating the Galileo operating company and appointing a CEO

16 May, a Communication of the Commission proposes alternative scenarios and recognises the difficulties of the PPP scheme

7–8 June, the Transports Council endorses the full-public funding scheme for construction and deployment of the infrastructure. No decision on funding modalities

1–2 October, Ministers at the Council of Transports could not agree on a public financing scheme. Germany opposes drawing funds from the EC budget, whereas a large majority of Member States is in favour

hand, Germany, the Netherlands and the UK would prefer a system in which each country would provide a direct contribution to ESA. The latter would induce greater advantages for their industry since the "geo return" is directly in line with the level of input funds. This would be a way for Germany to enjoy a dominant status that it has always tried to gain since the beginning. This delicate question, which should have been answered at the Transport Council of 1–2 October 2007 is still pending.[283] At any rate, the negotiations for the privately-owned concession would then only need to be completed by 2009.

The June Transport Council also requested that ESA be retained as the procurement agent and designing authority on behalf of the EU, which will imply that ESA will have to exercise its technical authority under EU rules under the overall EU management of the programme. It also recognised that EGNOS will achieve operational capability in early 2008 requiring immediate action to implement its services as a pre-cursor to Galileo.

As of mid-2007, it is amazing to read how the Commission is advocating full public funding of the whole infrastructure – as it has been and is still the case for GPS – as the solution to unstuck Galileo, whereas six years earlier the only way to go ahead was through a PPP scheme. This was probably to make sure that some lukewarm governments would support the Galileo programme from the outset.

Most likely, as far as one can tell, in particular, after the late July 2007 agreement concerning the workload sharing, management, and responsibilities between EADS-Astrium, Finmeccanica and Thales Alenia Space, Galileo looks in better shape for deployment of operational services by 2013 or early 2014. In our view, this will require: i) reaffirmation of a strong political commitment at the highest level of the Member States; ii) definition of a strong industrial leadership able to interact smoothly with the designing authority and procurement agent (ESA); and iii) the obligation to guarantee operational services available on time to avoid the risk of having a significant slice of the market captured by GPS-III, Compass/Beidou (the Chinese satellite navigation system under development) or Glonass (the Russian system which will regain full operational capacity in the early next decade).

## 3.2. The slow, but steady growth of implementing GMES services

GMES represents a concerted effort to bring data and information providers together with end users so they can interact better with each other, and thus make environmental and security-related information available to those who need it through enhanced or new services.[284] The nature of the second European space

**Fig. 10:** *GMES Sentinel-1 artist's impression (source ESA).*

## TOWARDS A EUROPEAN STRATEGIC CAPACITY FOR EARTH MONITORING

**1998** The 19 May Baveno Manifesto calling for a true European strategy in the field of global environment monitoring

**2001** A 100 million euros budget line for research and development (R&D) at the EU level is created together with a 100 million euros ESA Programme (GMES Service Element)

**2004** The Commission proposes an Action Plan (2004–2008) establishing a GMES capacity by 2008

**2005–2006** The fast-track services are selected: Emergency response, Land management and marine services

**2005** ESA creates a programme called *Space Component of Galileo*, the 1st phase of which is subscribed up to 258 million euros and approved by the December Council held at Ministerial level

**2006** The EU 7th Framework for R&D (2007–2013) displays a 1200 million euros budget line for GMES: 650 million euros as the EU contribution to the ESA programme and 550 million euros for services development

**2007** Launching the 2nd phase of the ESA programme on a 430 million euros basis

**2008** Implementation of the first operational services and of the governance structure. The ESA Council at Ministerial level planned for November should approve the schedule and the level of funding for the continuation of the programme

flagship, GMES, is quite different from Galileo in many respects. First of all, GMES is not a space programme *per se*, rather a system of land, sea and space-based probes for gathering environmental data to be assimilated to deliver added-value information, which eventually will become services for end users. The space component relies on both: existing Earth observation satellites, optical as well as radar, operated by ESA, national agencies or companies and satellites to be designed by ESA (the Sentinels 1 to 5) or by European space agencies that will become operational in the next decade. Hence, with the current components already functioning, the availability of some of the services exists. Like for Galileo, the ultimate goal is to guarantee the continuity of services over decades and find an operator capable of handling the enormous quantity of data to be assimilated and delivered.

A poor judgement would be to consider that the progress of GMES is too slow because of insufficient political support and/or improper management. Indeed the complexity of the programme is related to the integration of data from space-based and *in situ* Earth observation capabilities into user driven operational application services. The programme managers realised early on that the capacity could only be built up progressively based on clearly identified priorities and using existing elements whenever possible. From this perspective, three services have been selected for Fast-Track treatment (called also GMES Service Elements in the language of ESA): Emergency response, land monitoring and marine services. The objective being to develop and validate pilot operational services based on selected R&D by 2008.

### 3.2.1. A decisive year 2006 for GMES

2006 has been an important year for GMES in many ways. The decision of the December 2005 ESA Council at Ministerial level in Berlin cleared funds for the first phase of the programme paving the way for putting on firm grounds the development phase of the Sentinel family of satellites which should become operational in 2010 and beyond; gathering Synthetic Aperture Radar (SAR), optical/hyperspectral data, focusing also on operational oceanography and atmosphere chemistry. In March 2006, a co-decision of Vice-President Verheugen, Commissioners Potočnik and Dimas created the GMES Bureau to be effective on 1 June.[285,286] Although the first objective of this Bureau, embedded within the structure of the Directorate-General for Enterprise, is to deliver the Fast Track Earth observation services in 2008, its mission is to become the management structure of the GMES programme contributing to the long-term sustainability of GMES and enabling it to move gradually from R&D towards a more user-driven and user sustained approach. The Bureau will be the focal point for the coordination of GMES-related services of the Commission and shall identify priorities for the future GMES services, with air quality, and security being certainly two likely candidates. The GMES Bureau will eventually manage GMES services in addition to the Commission, including EU institutions and bodies, Member States, intergovernmental organisations such as ESA, EUMETSAT or the EUSC (European Union Satellite Centre). After one year of existence, with no control over funds, it seems that the GMES Bureau has difficulties in exercising full authority on moving the programme ahead. Its position vis-à-vis the space applications unit of the Directorate for Industry, Aerospace, Security, Equipment and Defence is not as easy as had been anticipated. This governance issue is definitely a hindrance for smoothing the management that could be easily solved by giving to the Bureau full responsibility for conducting the programme.

2006 saw also an interesting development in the downstream aspects of GMES, in particular, at the Graz conference organised under the Austrian Presidency of the EU on 19–20 April, and that focused on a market approach for GMES in Europe and its regions. A roadmap for GMES service development was proposed, insisting on the importance of having a relevant governance scheme, the necessity for tailored services, data access policies and standards definition, strong involvement of the users and a clear definition of their needs, active participation of the Member States to enhance the awareness on GMES advantages.[287] The regional dimension was considered as essential in the definition and use of GMES services. Eventually, in 2007 it initiated the organisation, under the purview of the Committee of the Regions, of a standing conference of the space applications user regions in which GMES services are, of course, included.

### 3.2.2. The governance question

Despite these initiatives showing progress in the development of GMES, the governance issue was still pending due to the variety of actors involved from one end to the other of the programme. A working group under the authority of the GAC (GMES Advisory Council) was set up and delivered a report whose conclusions has not won the consensus of the Member States[288], partly because there is no GMES recognised authority yet. Two levels of governance are foreseen: i) a general governance scheme aiming at establishing a sustainable institutional and financial framework for GMES; and ii) governance principles for the three GMES components (observation infrastructure-space based and in situ based, core services and downstream services, information management and dissemination[289,290]).

In the first half of 2007, from a space programmatic view point, the situation looked brighter since the second phase of the ESA programme was launched.

On April 17, the EU German Presidency organised a symposium in Munich on "The way to the European Earth observation system GMES – The Munich roadmap". This roadmap, resulting from a consensus among the GMES stakeholders and the GAC, summarizes the agreed-on architecture and proposes principles for the operational implementation of the European Earth observation services including milestones for the way forward. More specifically, concerning the architecture of the aforementioned services, it distinguishes the **Core Services**, which provide standardized multi-purpose information common to a broad range of EU policy-relevant applications areas leading to sizable economies of scale, and the **Downstream Services**, which are more targeted to (trans-) national, regional, or local information needs. The GMES observation infrastructure is complemented by systems whose data would become accessible via international mechanisms such as GEOSS (Global Earth Observation System of Systems). More innovative on this roadmap are the governance and funding principles that are outlined, reflecting real progress compared to the year before, although the part concerning the Core Services is somehow still fuzzy:

- Dedicated GMES satellite missions are developed by ESA and operated by the relevant European-level operating institutions, such as EUMETSAT or ESA, depending on the type of mission;
- Because of ESA responsibility in implementing the overall GMES space component, it also coordinates the contribution of space data and other elements coming from the Member States, Eumetsat and other GMES partners;
- The development, operation and governance of the in-situ infrastructure may be contributed by the operating bodies at European, national and regional levels;
- The Core Services are supervised and generally managed by dedicated governance schemes on the basis of the specific characteristics building on existing coordination mechanisms;
- The operation and management of downstream service is entirely driven by their specific users.

The Roadmap also reaffirms the capability of delivering operational and autonomous Fast Track services by the end of 2008. The Council is invited to decide on the programmatic, management and financial schemes for GMES to be implemented if possible by 2012. ESA and the Member States are also invited to ensure the implementation of the first generation of the dedicated GMES space component missions, if possible by 2012.

The Munich Roadmap clearly highlights the importance and the difficulty of the governance issue which has to be addressed fully and in depth today to guarantee an efficient and timely development of GMES. Otherwise, the programme might turn out to be a technical success, in particular, through its space component, used mostly for R&D purposes but not a response to the critical needs for sustained information services expected by the European users irrespective of who they are: decision-makers at the various levels, private industry or services, and citizens.

## 3.3. Lessons learned

Galileo and GMES, two programmes of a very different blend, but both aiming to provide added-value services for end users ranging from ordinary citizens to operational entities, are experiencing painful, if not rough developments essentially because of:

- Insufficient political drive from the top, disrupting the schedule for critical decisions;
- A tendency to "re-nationalise" European programmes to maximise the national return and stature instead of strengthening European institutions and the economy on a broader scale;
- Enormous difficulties to obtain a consensus for a workable, efficient and lean governance scheme. Indeed, it is striking to note that generally speaking, when a programme is based on existing elements and compelled to incorporate new ones by involving a large variety of actors, public and private, including end users and suppliers, there is a tremendous difficulty to move ahead in a coherent way, unless the agreed-on rules of governance are recognised that clearly define the roles and responsibilities of the actors involved. This drawback lies genuinely with European programmes and becomes a major impediment with the larger ones, is and therefore, a persistent weakness for Europe.

There is no silver bullet for managing such complicated projects involving so many actors. But one thing is sure: unless a recognized leader able to commit a significant amount of time and money comes forth, the momentum will be lost and Europe will lose its soul. Today, and maybe even tomorrow, this weakness will be the Achilles' heel of Europe. Indeed times have changed, but it is hard not to feel nostalgic when remembering the drive of the European space projects 20 or 30 years ago.

[277] "Galileo in "dead-end-street" after partners pull out". EurActiv. 8 May 2007. http://euractiv.com/en/science/galileo-dead-street-partners-pull/article-163633.
[278] Barrot, Jacques. "Letter to the German Presidency of the EU". EC Press release, 14 Mar. 2007.

[279] EU Commission. Staff working document. Galileo at a crossroad: the implementation of the European GNSS programmes. COM (2007) 261 final. Brussels, 16 May 2007.
[280] Nardon, Laurence. "Strengthening Galileo's Business Case". Space News (7 May 2007): 15.
[281] European Commission. Staff Working Document. Accompanying Document of Reference cclxxix SEC(2007)624. Brussels, 2007.
[282] EU Council. "Transport, Telecommunication and Energy". Press Release 10456/07 (Presse 133). June 2007: 6–8 http://www.consilium.europa.eu/ueDocs/cms_Data/docs/pressData/en/trans/94576.pdf.
[283] Mason, Jeff and Wissenbach, Ilona. "Germany rejects EU satellite funding proposals". Reuters (20 Oct. 2007).
[284] Brachet, Gérard. "From Initial Ideas to a European Plan: GMES as an Exemplar of European Space Strategy". Space Policy 20 (2004): 7–15.
[285] European Commission. decision creating a Bureau of Global Monitoring for Environment and Security (GMES). C673. Brussels, 11 Mar. 2006.
[286] "Mission of the GMES Bureau". http://www.gmes.info/165.0.html.
[287] "A Market for GMES in Europe and its Regions. The Graz Dialogue". 19–20 Apr. 2006. EU Council 9182/06-Rech 118, Compet 113. Graz, Austria. 15 May 2006.
[288] Working Group of GMES Advisory Council (GAC). "Future GMES structure of governance". 2006.
[289] Data policy and management aspects are addressed in the frame of INSPIRE and the European Spatial Data Infrastructure.
[290] European Commisssion. Directive establishing an Infrastructure for Spatial Information in the European Community (INSPIRE). EU Official Journal L 108. Brussels, 25 Mar. 2007.

# 4. The new European Space Policy as seen from across the Atlantic

John M. Logsdon

The new European Space Policy is the latest step in a long-running process through which advocates have sought to link the potential contributions of the space sector to the broader political, social, economic, and security objectives of the "European project." As such, it is at a minimum an important symbolic act. In giving their approval to the policy, the members of the European Space Council asserted in their resolution that "the space sector is a strategic asset contributing to the independence, security and prosperity of Europe and its role in the world".[291] Such a statement lays the foundation for an enlarged and more assertive European presence in space.

Whether such rhetoric can be turned into reality is yet to be seen, but the new European Space Policy is a potentially important step in that direction. The very fact that the policy received the unanimous approval of twenty-seven European governments is in itself politically significant. While previous European Commission (EC) space documents such as the 2003 "Green Paper" and "White Paper" reflected primarily the ambitions of the EC in space, the new policy can legitimately be seen as a statement endorsed by the governments of European countries.

**Fig. 11:** *The Fourth Space Council held in Brussels on 22 May 2007 (source ESA).*

## WHAT IS THE EUROPEAN SPACE POLICY?

What is called "the European Space Policy" in this essay is actually a composite of several documents. The proposed policy was set forth in the "Communication from the European Commission (EC) to the Council and the European Parliament – European Space Policy ." [COM(2007)212, 26 April 2007] A similar communication was sent to the Ministerial Council of the European Space Agency (ESA) by the ESA Director General. Those communications were accompanied by an "Impact Assessment" prepared by the staff of the European Commission [SEC(2007) 505, 26 April 2007] and by a proposed "European Space Programme" [SEC(2007)504, 26 April 2007] that resulted from stakeholder consultations among the various public and private space entities and individuals, working together under the guidance of the High-level Space Policy Group established on the basis of the 2004 ESA-EC Framework Agreement. Finally, The Competitiveness Council of the European Union (EU) meeting together with Ministerial Council of the European Space Agency as the Fourth European Space Council on 22 May unanimously approved a resolution adopting the 26 April proposal and adding some new interpretations to the policy's meaning.

This action gave a formal government stamp of approval to the new policy.

The following discussion draws primarily on the policy statement as set forth in the 26 April Commission communication. It will be made clear in the text when one of the other documents listed above is referenced.

Beyond its symbolic character and its statement of European aspirations in space, it seems to this long-time observer of European space activities that the new policy is largely an affirmation of the current space *status quo* in Europe. It proposes no new programmes, does not call for increased budgets, and accepts that the best that Europe can do organizationally at the current time is to manage its space activities through a complex process of coordination among the EU, and particularly the European Commission (EC), ESA, and national space agencies. No dramatic institutional innovations related to space governance in Europe are proposed, although in the run up to the policy they were apparently considered and rejected as politically infeasible.[292] In its essence, the European policy is an advocacy document making the case for why space capabilities should be recognized as vital to Europe's future and suggesting some steps that should be taken in the short- to mid-term to take make sure space contributions to that future are maximized. It will be of great interest to observe the policy's implementation; there are plans already underway for significant space initiatives to be proposed during the French presidency of the EU in the second half of 2008.

This is not to say that the symbolic aspects of the new policy are unimportant. Indeed, the policy sends important messages both to European leaders and to other countries active in space with respect to European aspirations as a global actor, with space capabilities as an important element in pursuing those aspirations. In this way, the European policy mirrors (probably not consciously) the preamble of the most recent statement of U.S. National Space Policy, released in October 2006, which declared that "In this new century, those who effectively utilize space will enjoy added prosperity and security and will hold a substantial advantage over those who do not".[293] Through its new policy, Europe clearly hopes to be one of those "who effectively utilize space".

This analysis thus suggests that the primary audience for the policy is the political leaders in Europe who will consider in coming years the appropriate priorities and funding levels for space activities at the EU, ESA, and national levels. While the policy states in general terms European aspirations to become even more a leading space power, the immediate competitive challenge to the United States is rather muted. This policy is not a statement that Europe intends to mount an across-the-board challenge to U.S. space leadership, but rather that Europe aspired to be seen as one of the leaders in global space activities to more significant degree than heretofore has been the case. It remains to be seen whether the implementation of the policy will lead to any meaningful changes in the transatlantic space relationship, with Europe, based on its increasing space capabilities, taking on a more assertive role.

## 4.1. Space and Europe as a global actor

Indeed, one of the more interesting aspects of the new European Space Policy is the explicit link made between space capabilities and the European Union's ability to exert influence on a regional and global scale. The policy suggests that that Europe cannot achieve its global ambitions without increased reliance on space assets.

It is beyond the scope of this paper to discuss in depth the EU's plans for increasing its global involvement and influence in coming years; those aspirations include becoming the world's leading knowledge-based economy; a global leader in dealing with issues of sustainable development, climate change, and environmental management; and an actor able to project military power in areas beyond its borders. The new policy argues that having world-class space capabilities based on a solid foundation of a well-trained workforce, a healthy industrial base, and innovative science and technology is a key to European "hard power" in the economic and security spheres, and that having Europe centrally involved in the inspiring adventure of space exploration will contribute to Europe's "soft power", i.e., the ability to convince others to follow European leadership. It declares that having a coherent policy is "a strategic choice for Europe, if it does not want to become irrelevant", that an effective European space policy is essential if Europe hopes "to exert global leadership", and that "space systems are strategic assets demonstrating independence and the readiness to assume global responsibilities".[294]

These claims are in many ways similar to those made to justify large space budgets in the United States; the difference is that the United States has for many years made such investments, while in Europe there remains a significant gap between such stated aspirations and governments allocating the resources needed to achieve them. Compared to the United States, Russia, and perhaps China, European leaders have not yet made a major commitment to the full development of comprehensive space capabilities in both the civil and the security sectors. The new policy argues that it is Europe's interest to make such a commitment.

The words "autonomous" and "independent" appear several times in the policy statement. The contention is that Europe must have its own independent means to access space and to carry out vital activities there. Some of those activities would gather and communicate the data needed for Europe to make its own decisions on key global issues such as the character and sources of global climate change. The implication here is clearly that Europe does not want to depend on any other space faring country, and most particularly the United States, for those space capabilities or the information derived from them needed to exercise power in international political, economic, and security relations, and to otherwise pursue European interests in the world. As the United States concerns itself with the challenge to its currently leading position in space from China and possibly a resurgent Russia, it would do well to also pay attention to whether Europe will turn the ambitions sketched in its new space policy into the ability to compete with the United States for important commercial, political, and security payoffs from space, and whether Europe will make its capabilities one of the foundations of its enhanced power in global relations.

## 4.2. Space and the European Union

One question that might be asked of the new policy is "what is the added value of increased EU involvement in Europe's space efforts?" This question is germane because to this observer, the need for such added involvement is the primary institutional innovation suggested in the policy statement. (Although the topic is beyond the scope of this essay, it is worth noting that giving the European Union a formal "competence" in space to complement the roles of ESA and individual European countries is an element of the European Reform Treaty to be voted on in coming months. This would legitimize EU involvement in space matters; in formal terms the EU currently does not have a recognized role in space policy or programmes.)

The new policy assigns to ESA its current responsibilities, including developing space technologies and systems, exploring space, pursuing a robust space science effort, and guaranteeing access to space. It also notes that ESA will provide the R&D capabilities needed to implement space programmes sponsored by the EU. It seems as if there is little new for ESA in the policy. So for ESA, the policy is largely a reaffirmation of the value of what it is already doing.

By contrast, the policy notes that "the EU will use its full potential to lead in identifying and bringing together user needs and to aggregate the political will in support of wider policy objectives".[295] For most civilian space efforts that might be undertaken under EU leadership, carrying out these roles would be the responsibility of the European Commission. One way to understand the new statement, and indeed the lengthy process that preceded its approval, is to see the policy as proposing that the lead role in European space policy should shift from Paris to Brussels. For the past three decades, European countries have used ESA as their primary means for pursuing Europe's collective ambitions in space. Now the priority setting and policy development role for at least space applications, the policy suggests, should be assigned to the EU, with the EC in the lead role, and with ESA acting as the implementer of EU decisions.

There are at least two reasons for considering such a shift. Over the past few years, space leaders in the largest ESA Member States became somewhat disenchanted with the returns on their investments in the agency. One concern was that the combination of the voting system within the ESA Council and the procurement rule of "*juste retour*" resulted in the smaller Member States of ESA having disproportionate influence over the agency's efforts. One result of this attitude was a tendency towards "renationalisation" of some space efforts; France, Germany, and Italy decided to carry out through their national space programmes some efforts that previously might have been implemented through ESA.

Another, more fundamental, reason for proposing the move of space policy authority from ESA to the EU is the belief among European space leaders in government and industry that the more explicitly political context of EU decision making might result in convincing their national leaders to give additional support to space programmes, and particularly to user-oriented space initiatives such as Galileo and the Global Monitoring for Environment and Security (GMES) programmes. Rather than treating space purely as an R&D area, such a move, it has been argued, might put space on the agenda of European leaders in areas such as transport, environment, communications, and industry, and eventually foreign policy and military security. The result, European space advocates hope, would be larger space budgets, more effectively used in pursuit of core European interests. Whether such a shift can in fact lead to sounder decision-making for space appears problematic, if the complex decision process through which ESA moves forward is replaced by the even more complex way the EU reaches its decisions and carries them out.

It is fair to ask whether the EU, and particularly the EC, is up to such a central role in space. Jurisdiction over space policy within the Commission was shifted in 2005 from the Research Directorate to the Directorate General for Enterprise and Industry, with a small unit in the Industry Directorate having day-to-day responsibility for space matters. This shift recognized that the primary reason for EU involvement in space was linked to using space for broader purposes, not sponsoring space research. But within the General Directorate for Enterprise and Industry, space issues must compete for policy and political attention with other areas of Directorate activity, and it is not yet clear how they will fare in that competition.

When the European Commission published its White Paper on space in 2003, there was hope that the EC would obtain significant new resources to support new space initiatives. At the time, EC spending on space from 2003–2006 was estimated to be 755 million euros, an average of 189 million euros a year.[296] For the 2007–2013 period, the new policy indicates that EC space spending will be 2.8 billion euros, an average of 467 million euros per year.[297] While this represents a more than 200 per cent increase in EC space spending, it compares unfavorably with an ESA budget for 2006 of 2.9 billion euros;[298] clearly the EC will remain a relatively minor funder of European space programmes compared to ESA and the larger national space agencies. If the EC is to lead space in Europe, it will have to be through other means than its budgetary power.

If the European Commission is to become a major presence in space policy, it will have to gain that position through its actions in bringing together user needs and aggregating political will in support of enhanced European space efforts. One recent indication that the Commission can play a significant role in this dimension is the July 2007 decision to use EC funds (rather than funds provided through ESA) to pay the costs of the Galileo programme for at least the next year. This decision was made necessary by the collapse earlier this year of the prior plan to fund Galileo through a public-private partnership. In the aftermath of this development, the EC was able to convince key Member States to commit to funding Galileo development totally with public funds.[299] Whether this success can set a precedent for EU funding of GMES and subsequent space initiatives will be interesting to observe.

## 4.3. The European emphasis on applications

One thing that strikes a U.S. observer of the new European Space Policy is its emphasis on space applications – the use of space capabilities to provide direct services to people on Earth. The policy notes that "the key to securing the maximum political, economic and social return from investment in space technologies lies in the development and exploitation of space applications".[300] This emphasis stands in rather stark contrast, at least in the civil sector, with the new focus in the United States on space exploration. Current U.S. policy makes advancing "U.S. scientific, security, and economic interests through a robust space exploration programme" the central goal for the National Aeronautics and Space Administration (NASA), which controls by far the largest share of U.S. civilian space funding.[301] In the United States, the use of space systems for communicating various types of information is the responsibility of the private sector and the Department of Defense. Observing the Earth from space is an activity lead by the National Oceanographic and Atmospheric Administration, part of the Department of Commerce, the Department of Defense and the National Reconnaissance Office, and various private sector satellite operators; NASA's role is limited to science-driven Earth observations. Providing basic positioning, navigation, and timing services is the responsibility of the Department of Defense, overseen by an interagency management board. This complex structure tends to downgrade the priority of most civilian space applications as a government responsibility, at least in budgetary terms.

The preliminary "European Space Programme" submitted along with the space policy lays out in some detail current and planned European space activities; it does not propose significant new initiatives. Its goals are to enable "major space actors" to move towards "increased transparency" thereby "reducing unnecessary duplication and enhancing complementarity." The goal is to move towards a "coordinated joint European space effort".[302] What is new about this effort is a listing of all European space activities at the EU, ESA, and national levels in one document, one that can be adapted as new space efforts emerge and can serve as a

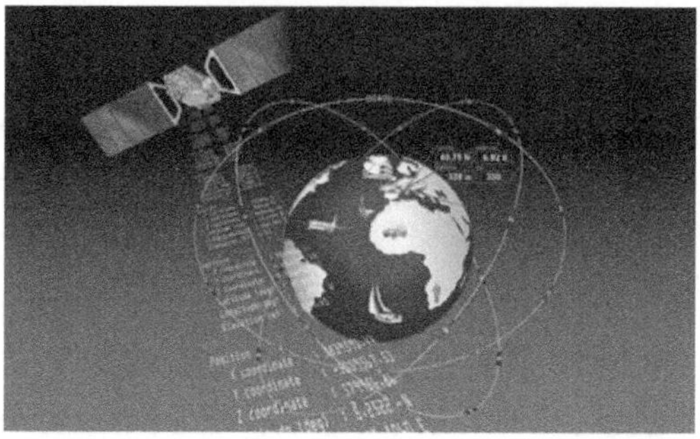

**Fig. 12:** *The Galileo System (source ESA).*

common and comprehensive frame of reference for discussions ofl European space activity.

If the new European policy is implemented, management of space applications efforts would be centralised under the EU. In principle, it would be user demand that would set the requirements for various space applications, with ESA in charge of managing the development of the appropriate systems to meet those requirements. Primary funding for space applications programmes would come from national governments to the EC, which would then transfer the bulk of that funding to industry through ESA. The controversy over how best to fund the Galileo system foreshadows how difficult it may be to make such a coordinated arrangement work in practice.

More to the point, there seems to be a mismatch in space priorities between the United States and Europe. The new European Space Policy gives very terse

**Fig. 13:** *Artist's impression of a lunar outpost (source NASA).*

treatment to the European role in space exploration, noting that exploration "has a significant political appeal in a vision of European identity". But it also seems to accept the notion that Europe will continue to be a junior partner in future space exploration activities, noting that "Europe needs to remain an indispensable international partner providing first-class contributions to global initiatives". Europe may seek leadership in "selected domains in accordance with European interests and values", but it does not appear that exploration beyond Earth orbit is one of those domains.[303] A relevant question is whether the policy indicates a potential split in the transatlantic space partnership that has been one of the hallmarks of the past half century in space, with the United States focusing on exploring "the Moon, Mars, and beyond" while Europe concentrates of using space to deliver concrete benefits to its citizens.

This potential divergence seems to have been recognized on both sides of the Atlantic Ocean. While NASA and ESA have a long tradition of consultation and cooperation, only in the past two years have the EC and the U.S. Department of State initiated discussions about U.S.-EU government-to-government space cooperation. This is a positive development. As the EU pursues a different path in space than NASA, the leading U.S. civil space agency, the creation of a new mechanism for U.S.-European discussions on topics other than space science and exploration is a positive development.

## 4.4. Space and security

As a product of the EC and ESA, both entities focused on non-military activities, it is not surprising that the new policy gives only limited attention to security applications of space capabilities. That is not to say that such applications are totally ignored. First of all, while the United States may have the resources to develop separate systems with largely duplicative capabilities, the policy recognizes that this is not an option for Europe. The policy contains only a brief discussion of the use of European space capabilities for security and defense, recognizing both that such capabilities will primarily "continue within the remit of Member States".[304] The Space Council resolution approving the space policy makes explicit, but cautiously so, that military users will be able to take advantage of the capabilities of the Galileo and GMES systems, at least as such uses are consistent with the policy that "GALILEO and GMES are civilian systems under civilian control".[305] Even such a careful statement is an achievement, given the sensitivity of space security issues in some European countries.

Europe is at a very early stage in its recognition that dependence on space systems for key economic, societal, or security purposes comes accompanied by vulnerability, should an adversary want to interfere with the use of those systems. Thus the policy makes only a cursory mention of the reality that Europe's space systems "must be protected against disruption".[306] This single mention stands in vivid contrast to the 2006 U.S. National Space Policy, which has an overriding focus on U.S. freedom of action in space and the likely U.S. reaction if that freedom is threatened with disruption. If the ambitions of the new European policy are achieved, Europe will also have to pay increased attention to how best to protect its space systems from threats of disruption or even destruction.

As Europe moves forward with the development of space systems with dual-use capabilities such as Galileo and GMES, there is a need for continuing and enhanced transatlantic space discussions. While the interaction between the U.S. space research and development agency NASA and ESA, and between individual European space agencies and NASA is well established, policy dialogue between the U.S. government, represented by its Department of State, and the European Union, represented by the EC, is a recent development. The EC-Department of State interaction could become an important forum for dealing with sensitive policy issues such as export control and U.S.-EU cooperation in space situational awareness.

## 4.5. Developing the space policy

There is a striking difference between the way the United States develops its space policies and the process pursued in the creation of the new European Space Policy. Successive statement of U.S. space policy, and there have been many since 1960, are products of discussions that take place internal to the U.S. government, with no formal channels for stakeholder input. These policies are developed through a process of negotiation and compromise among various executive agencies (NASA, Department of Defense, State Department, Transportation, Commerce, etc.) This process is managed by some White House organization. The Office of Management and Budget is always deeply involved, and thus an approved policy statement implies a budgetary commitment (not always honored) to its implementation. Only after agreement is reached on all or most issues is the policy is sent to the President for his approval. When approved by the President, a U.S. space policy becomes an authoritative statement that can be used by the Chief Executive and his staff to enforce compliant behavior by the various agencies of the U.S. government.

While the formal process of developing space policy is limited to government officials, the United States has an open political system, and those officials consider a variety of stakeholder interests. There is no practice of studying and analyzing alternative space policies themselves prior to government decision. Instead, the people involved in the policy development process bring their own knowledge, backgrounds, contacts, and experience to the process.

The process by which the European Space Policy was developed appears remarkably different, and may indicate that the resulting policy statement is of a different character than that which emerges from the U.S. policy development process. The new policy is a step, albeit a particularly significant one, in a long process that has included: various EC communications; publication of an EC Green Paper that was then discussed in a variety of public settings; issuance of a subsequent EC White Paper; studies by non-governmental groups on various aspects of the policy; creation of an EC-ESA Framework Agreement that provided the terms of reference for the two entities to work together in crafting the policy and the creation of a High-level Consultation Group that was the venue for (sometimes tense, it is reported) discussions between officials from the EC, ESA, other multinational organizations such as EUMETSAT and the new European Defense Agency, and Member States; invention of a "Space Council" to accommodate the differences between the EU and ESA memberships and to provide government guidance to those drafting the policy ; consultations between ESA and the EC and private sector stakeholders; an impact assessment of the policy prepared by the EC; and finally agreement on the content of the policy. This elaborate process reflects the complexity of Europe today. One must also wonder whether the process can produce an output that is more than the lowest common denominator among all who have had inputs along the way.

There are some interesting similarities in the language of the 2006 U.S. National Space Policy and the new European Space Policy. Table 1 summarizes some of those similarities.

Oversight of policy and programme implementation is crucial to policy success. In the U.S. context, the White House and associated Executive Office of the President agencies such as the Office of Management and Budget, the Office of Science and Technology Policy, and the National Security Council provide continuing oversight of how the various agencies of government are implementing presidential policy decisions. They have the budgetary and other levers of executive power to enforce policy compliance. There appear to be in Europe no similar mechanisms for making sure that the guidance in the new policy shapes future decisions and actions. Perhaps

**Tab. 1:** *Comparing Space Policies.*

| | U.S. National Space Policy (NSP) | European Space Policy |
|---|---|---|
| Leadership | A fundamental goal is to "strengthen the nation's space leadership and ensure that space capabilities are available in time to further U.S. national security, homeland security, and foreign policy objectives". | "Europe needs an effective space policy to enable it to exert global leadership in selected policy areas in accordance with European interests and values. To fulfill such roles the EU increasingly relies on autonomous decision-making, based on space based information and communication systems. . . . The EU, ESA, and their Member States have continued to invest strongly to maintain leadership in space-based science". |
| Freedom of Action in Space | "Freedom of action in space is as important to the United States as air power and sea power. . . . The United States considers space capabilities . . . vital to its national interests. Consistent with this policy, the United States will: preserve its rights, capabilities, and freedom of action in space. . . . The United States will oppose the development of new legal regimes or other restrictions that seek to prohibit or limit U.S. use of space".<br>The NSP recognizes that with reliance on space comes vulnerability, and makes very clear that the U.S. will defend its ability to benefit from the investments it has made in space. | "Independent access to space capabilities is therefore a strategic asset for Europe." One goal is "to secure unrestricted access to new and critical technologies, systems and capabilities in order to ensure independent European space applications." |
| Protection of Space Assets | The United States will "dissuade or deter others from either impeding those rights or developing capabilities intended to do so; take those actions necessary to protect its space | "Space-based capabilities . . . must be protected against disruption." |

*(continued)*

**Tab. 1:** (*continued*)

| | U.S. National Space Policy (NSP) | European Space Policy |
|---|---|---|
| | capabilities; respond to interference; and deny, if necessary, adversaries the use of space capabilities hostile to U. S. national interests." | |
| International Cooperation | "The United States will seek to cooperate with other nations in the peaceful use of outer space to extend the benefits of space, enhance space exploration, and to protect and promote freedom around the world."<br>A fundamental policy goal is to encourage international cooperation with foreign nations and/or consortia on space activities that are of mutual benefit and that further the peaceful exploration and use of space, as well as to advance national security, homeland security, and foreign policy objectives." | "Europe needs to remain an indispensable partner providing first-class contributions to global initiatives and exerting leadership in selected domains in accordance with European interests and values. With an open attitude towards cooperation, Europe must take judgments on when to rely on partners and where to retain independence." |
| Competitive Space Industry | "It is in the interest of the United States to foster the use of U.S. commercial capabilities around the globe and to enable a dynamic, domestic commercial space sector." | "A competitive European space industry is of strategic importance.... To achieve this goal it is essential that European public policy actors define clear policy objectives in space activities and invest public funds to achieve them. This public investment could help create a critical mass stimulating further public and private investment. A focused industry policy for space will also stimulate companies competing throughout the full value chain..." |

the fact that the policy does represent a wide consensus among those concerned with the European space sector will suffice to support its implementation. That, of course, is a very optimistic outlook.

## 4.6. Governance issues

The complex process through which the new policy was developed reflects the fragmented character of space activities in Europe, with individual countries, the European Space Agency, EUMETSAT, and now the European Commission and perhaps other elements of the European Union all involved in space policy and programmes. In addition, the space industry and operators of space systems are active participants European space policy debates. It appears as if the new European Space Policy does little to rationalize this fragmented system except to note that more coherence would be desirable. The very interesting "Impact Assessment" which accompanied the proposed policy makes it clear that from a rational perspective, a more sweeping change in the political framework for space in Europe, moving most or all space activities under European Union management, would be a desirable step, but that "current political considerations constrain the options which could be pursued in the immediate future." The best that could be recommended at this point, a reality reflected in the new policy, is "increased coordination and growing use of space applications to deliver other European policies".[307]

The policy recognizes that "the different approaches, separate legal processes and divergent membership of the EU and ESA can lead to cumbersome decision processes." That much is obvious, but this observation is followed immediately with the weak statement that "The [EC-ESA] Framework Agreement has provided significant advances in the working between the EC and ESA, and with the Member States, in policy development. The Agreement will be assessed and improved if required".[308] One is tempted to substitute "if possible" for "if required"; the Impact Assessment makes it clear that most European countries are not yet ready to transfer significant space programme responsibilities from ESA to the EU.

## 4.7. A step along the way

As the first of its kind, the European Space Policy has both strengths and limitations. It most certainly is not the final word on Europe's future in space.[309] It is, however, an achievement to be welcomed by those in Europe who believe in the importance of space capabilities to the European future. What will be interesting to observe is what happens next. The new European Space Policy lays out a positive path for the space sector over the next few years, but it needs to be implemented to produce the hoped-for results. Will those governments that gave unanimous approval to the policy be willing to abide by its priorities, provide the resources needed for carrying them out, and respect the lead role of the EC in subsequent policy-making for space? Only if this happens to a significant degree will the policy be evaluated as a success.

---

291 EU Council. Resolution on the European Space Policy. 22 May 2007. http://www.consilium.europa.eu/ueDocs/cms_Data/docs/pressData/en/intm/94166.pdf.
292 European Commission. Commission Staff Working Document: Impact Assessment of the European Space Policy. SEC (2007) 505. Brussels, 26 Apr. 2007. An interesting discussion of the pros and cons of significant changes in the European governance framework for space.
293 United States. The White House. US National Space Policy. 26 Aug. 2007. http://www.ostp.gov/html/US%20National%20Space%20Policy.pdf.
294 European Commission. European Space Policy. Brussels, 26 April 2007: 4.
295 European Commission. European Space Policy. Brussels, 26 April 2007: 11.
296 European Commission. White Paper: Space: a new European frontier for an Expanding Union. COM (2003) 673. Brussels, 11 Nov. 2003: 46.
297 European Commission. European Space Policy. Brussels, 26 Apr. 2007: 10.
298 European Space Agency. Funding. Paris, 19 Oct. 2007. http://www.esa.int/SPECIALS/About_ESA/SEMNQ4FVL2F_0.html.
299 Baker, Simon. "Governments Agree to Fund Galileo from Existing Budget". European Voice. (10 July 2007). http://www.europeanvoice.com/archive/article.asp?id=28544.
300 European Commission. European Space Policy. Brussels, 26 Apr 2007: 5.
301 United States. The White House. A Renewed Spirit of Discovery: The President's Vision for U.S. Space Exploration. Jan. 2004.
302 European Commission. Commission Staff Working Document. European Space Programme – Preliminary Elements. SEC (2007) 504. Brussels, 26 Apr. 2007: 3.
303 European Commission. European Space Policy. Brussels, 26 Apr. 2007: 12.
304 European Commission. European Space Policy. Brussels, 26 Apr. 2007: 7.
305 EU Council. Space Council Resolution. Brussels, 22 May 2007: 6. http://www.consilium.europa.eu/ueDocs/cms_Data/docs/pressData/en/intm/94166.pdf.
306 European Commission. European Space Policy . Brussels, 26 Apr. 2007: 7.
307 European Commission. Commission Staff Working Document: Impact Assessment of the European Space Policy. SEC (2007) 505. Brussels, 26 Apr. 2007: 22–26.
308 European Commission. "European Space Policy". Brussels, 26 Apr. 2007: 11.
309 Peter, Nicolas and Plattard, Serge. "The European Space Policy: Europe's New Compass". ESPI Flash Report #1 (May 2007). http://www.espi.or.at/images/stories/dokumente/flash_reports/flash-report1-espi-europe _in_space.pdf.

# 5. The U.S. missile defence programme

Tomas Valasek

## 5.1. Introduction

Later this year, the Czech Republic and Poland will decide whether to host components of the United States missile defence system on their territory. Talks with Washington have been underway since January 2007. The United States would like to station radar on a former military base in the Czech Republic, while Poland would host ten interceptor missiles. The purpose would be to defend the United States mainland and Europe against missiles launched from Iran and elsewhere in the Middle East. The Czech and Polish sites would enlarge the architecture of the Ground-Based Midcourse Defense System (GMD) protecting the United States. This system currently consists of a network of sensors and missiles stretching from Japan to Alaska.

After months of progress, negotiations on the Czech and Polish sites have lately run into difficulties. In the Czech Republic, public opposition to the bases runs strong. The Polish government collapsed in the summer of 2007 and elections were scheduled for the fall. In the meantime, the United States. Congress cut funding to the Polish base. Washington as well as the Polish and Czech governments have come under Russian pressure. Moscow is running a vehement sticks-and-carrots campaign to stop the plans for new bases in Central Europe.

None of this may stop the construction of the Czech and Polish sites. The Czech opposition is unlikely to garner enough votes for a referendum. A Polish government of any stripe is likely to support the project and the United States Congress may restore its funding cuts at later date. However, the sudden visibility given to the planned Czech and Polish sites has rekindled a debate in Europe on the costs and benefits of ballistic missile defences.

Lurking in the background is the question of the use of space in missile defences. Some proponents of missile defences argue that only space-based missile defences can reliably protect the United States homeland. If space-based interceptors were to become reality, they would constitute the first use of space for offensive military purposes. Already, current plans call for use of space-based sensors – satellites that will both detect the launch of enemy missiles and guide interceptors to their targets. The use of space for GMD guidance may raise the risk of anti-satellite attacks in the future.

## 5.2. Missile shield development

The United States government has pursued missile defences since the 1950s. The immediate predecessor to GMD is the National Missile Defense (NMD) programme created by the US Department of Defense (DoD) during the Clinton Administration. The original NMD design consisted of a network of missiles as well as land- and space-based sensors. It was intended to intercept warheads aimed at civilian areas in the so-called midcourse stage of the ballistic curve; that is, at or near their highest altitude, after they had separated from the booster rockets. NMD co-existed with a number of separate missile defence programmes such as THAAD (Terminal High-Altitude Area Defense) and shorter range theatre systems intended to protect troops in the field. However, early in his presidency, George W. Bush reorganised these efforts, eliminating the barriers between the various programmes and creating one unified Pentagon agency, the Missile Defense Agency (MDA). Today's missile defences are thus a system of systems, incorporating components of the original NMD, but blending them with other plans aimed at intercepting missiles in other stages of flight (either the initial 'boost' phase or the terminal stage).

In practical terms, the backbone of today's GMD relies heavily on former NMD components. The United States has over twenty midcourse interceptors deployed, based at Fort Greely, Alaska and Vandenberg base in California. Several sea-based radars of various ranges and two types of land-based radars are either already connected to the grid, or will be soon. Work on other parts of the programme such as THAAD and the Patriot missile continues: some are far more technologically mature than the midcourse interceptors. However, not all are yet integrated into the seamless system.

The proposed sites in Poland and the Czech Republic have their roots in the NMD system. The X-band radar considered for the Brdy military base in the Czech Republic is designed to track and illuminate incoming missiles. It also has to assess whether they have been successfully destroyed. The radar (which has already been built and is currently stationed in the South Pacific) uses a long and narrow beam to stay with its distant, small and fast-moving targets.

The interceptor to be used in Poland would be a brand-new, as yet undeveloped two-stage missile similar to those in Alaska. Its job would be to defend against long and intermediate range ballistic missile attacks. Mid-course interceptors of the type slated for Poland are designed to destroy incoming missiles in the middle portion of their ballistic curve, at high altitudes outside the Earth's atmosphere. The interceptor missile's booster launches a small 'kill' vehicle into space, which smashes into the incoming warhead and destroys it through kinetic energy. It is not clear how different the flight characteristics of the new missile would be from the

**Fig. 14:** *Interceptor lowered into silo at Ft. Greely base (source: US Missile Defense Agency).*

existing interceptor. It is to have two stages instead of three on the Fort Greely missiles, which may suggest a shorter range.

## 5.3. Technical shortcomings

Silo-based GMD is technically not yet operational. Unusually for a weapons system, it was deployed before the full testing and evaluation cycle had been completed. GMD has never undergone exercise involving all its components in a realistic scenario.[310] While the MDA carried out numerous tests on the different components of the system those brought mixed results. GMD missiles have made only six intercepts out of twelve attempts. Many of the successful intercepts were notable for their simplified scenarios which, critics say, fail to simulate a real attack.

The Center for Defense Information (CDI), a private military research group, has been particularly critical of the 'scripted' nature of tests conducted to date. In their comprehensive overview of past GMD tests[311] CDI recounts cases when the interceptor was given prior information on the location of the mock 'enemy' missile.

(This made it easier to score an intercept.) The tests also did not appear to simulate the full range of deception measures likely to be used by an enemy missile. For example, decoys used in testing were designed to seem three to six times larger than the oncoming warhead. This vastly simplified the task of distinguishing between the real warhead and decoys. A real enemy missile is unlikely to extend similar courtesy.

Simplified scenarios are not unusual for early stages of testing. Nor, given the complexity of the task of shooting down a missile with a missile, should the high rate of failure surprise. But it is rare in peacetime that a weapons system is deployed on the basis of incomplete test data; even more so if half the attempted 'kills' of simulated enemy missiles failed. President William J. Clinton declined to authorise deployment of national missile defences in September 2000, arguing that he did not "have enough confidence in the technology, and the operational effectiveness of the entire NMD system, to move forward to deployment".[312]

Nevertheless, President George W. Bush 2002 instructed the DoD to field the available missile defence assets. He argued, in essence, that it is better to have unproven defence against enemy missiles than none. That may be true, with two very important caveats. It assumes that the costs incurred by the deployment are smaller than the benefits of possessing untested missile defence capability. And given Moscow's threats (see "Russia's Concerns", below) it is not yet evident that the threat to the United States has not actually increased. If Russia, for example, withdraws co-operation on ending Iran's nuclear programme, or if it deploys new missiles in response to GMD, the US may find itself in more, rather than less, danger.

There is also a risk in making optimistic assumptions about the missile defence abilities, and in basing national security decisions on them. No one knows for certain whether GMD really works. It has never been used in anger and never tested 'end-to-end', with all its components involved. The Congressional Research Service (CRS) concluded as recently as July 2007 that "ongoing technology problems have not yet demonstrated the operational capability of [the] deployed system".[313] Yet, the head of MDA, Lt Gen Trey Obering argued in a July 2007 article that the United States has "operational capability to defend against a missile launched at the United States".[314] The worrying scenario for Americans is that the government will pursue policies based on the assumption of little to no vulnerability to a missile strike – and that that assumption turns out to be false.

## 5.4. The links between missile defence and space

Current missile defence architecture calls for satellites to detect the enemy launch and to cue the interceptor missile to the target. This is clearly not the first use of

**Fig. 15:** *'Kill' Vehicle (source: U.S. Missile Defense Agency).*

space for military purposes – space is the key arena for U.S. military communication, surveillance, observation and positioning. However, use of space for missile defence creates new incentives for an anti-satellite strike. Furthermore, the dual role planned for U.S. satellites – they serve as tools of launch detection as well as for guiding the interceptor to the target – creates its own challenges. If satellites are struck to deny the U.S. the ability to use missile defences, Washington would also lose the capacity for detecting enemy strikes. This raises the likelihood of a nuclear launch by the United States. Washington would have lost both the ability to defend against an enemy launch as well as the ability to detect such launch, presenting it with a choice of launching its own arsenal or risk losing it.

It is unclear what space debris would be created through an interception. Midcourse interceptors destroy enemy missiles near the highest point of their trajectory, well outside the atmosphere. No publicly available studies confirm or reject the possibility of debris remaining in the orbit after the intercept. However, the Pentagon is said to have deliberately designed previous intercept tests so that interceptors are destroyed below the highest point on their trajectory. One possible explanation is that the Pentagon sought to avoid leaving space debris after a successful intercept.

Lastly, the Pentagon is exploring the possibility of basing interceptors in space. The idea is attractive – space-based interceptors would focus on intercepting ballistic missiles in their 'boost' phase; shortly after launch, while they are still moving relatively slowly and while their brightly-burning boost engines leave large infrared radar signature. This makes detection and interception easier. But the risks of space-based interceptors (SBIs) are as large as attractions. Critics of missile defences warn that SBIs can be destroyed with relatively cheap and rudimentary technology.[315] The attack on space assets, depending on the technology used could easily incapacitate other satellites as well. The US military, by far the most dependent of all powers on use of space for military purposes, has the most to lose from a conflict in space.

## 5.5. Politics of missile defence

All discussions of weapons systems are political. The United States Congress frequently overrules the Pentagon and the White House on funding requests for new weapons. The legislators may want to steer a greater share of the work to their electoral districts; they may view particular weapons as unnecessary or even provocative. But even by U.S. standards the missile defence system has attracted unusually close attention from the nation's political classes.

President Ronald Reagan famously made homeland missile defence the centrepiece of his national security strategy. The Republican Party has maintained staunch commitment to it ever since. Missile defences were a top priority in the 1994 Republican election manifesto "Contract with America", which marked the beginning of nearly decade-long Republican control of the United States Congress. Under their leadership the Congress passed several acts designed to force the reluctant Clinton Administration into deploying missile defences as soon as possible. The Republicans also worked to increase the perceived need for missile defences, by highlighting the ballistic missile threat to the United States. In the late 1990s, Donald Rumsfeld (who later became the Secretary of Defense) chaired a Congressionally-mandated commission to assess missile threats to the Unites States. The 'Rumsfeld Report' accused the U.S. intelligence agencies of deliberately underestimating the danger which ballistic missiles pose to United States homeland.

The Democratic Party has had a more ambivalent relationship to missile defences. A number of arms control 'doves' in the party have opposed it, but some party conservatives joined the Republicans in voting for missile defence acts. President William J. Clinton walked the middle road. For most of his term he continued to fund missile defence research, while resisting Republican calls for its early deployment. He even vetoed the 1996 Defence Authorization bill because it contained an amendment mandating the deployment of a missile defence system. By 1999, however, President Clinton relented. He signed the National Missile Defense Act of 1999, partly because of pressure from the Republicans and partly because the technology advanced while he was in the office. The legislation mandated the deployment of missile defences "as soon as is technologically possible".

The Republican takeover of both the White House and Congress in 2000 temporarily muted the political discussion of missile defences. Both the executive and the legislative branches of the United States government were solidly in favour. President Bush's national security team came to the office with a programme that clearly stated that the United States would deploy missile defences as soon as possible. Donald Rumsfeld moved up from a private missile defence advocate to the post of secretary of defence. Before joining the Pentagon, Rumsfeld's deputy, Paul Wolfowitz co-authored a book on U.S. foreign policy that made missile defences a cornerstone of US security policy.[316] The new president quickly obliged: he ordered the existing capability to be deployed in spring 2002.

But the controversy was to flare up again. After the Democrats regained control of Congress in 2006, missile defences came back to the centre of U.S. political discourse. The Democrats first tested the president's resolve with a number of bills critical of missile defences. Then, in the summer of 2007, both houses of the United States Congress cut funding for the Polish interceptors (albeit by different amounts; the bills have yet to be reconciled). Even if the funding is restored at a future date; the cuts represent the strongest signal yet that U.S. commitment to European sites in particular, and to missile defences in general, may be more wobbly than generally assumed.

U.S. presidential elections will change the outlook for the European sites yet again. Hillary Clinton has voted against missile defence funding four different times. She may feel compelled to take a more supportive stance if elected; presidents do change their mind once in the office. But her congressional record to date suggests serious misgivings. Republican candidates are in favour; several made supportive statements on the campaign trail and those who sit in the Senate (John McCain, Fred Thompson) voted repeatedly for missile defences. The simple rule of thumb is that a Republican president is nearly certain to keep expanding the missile defence system. A Democratic commander-in-chief is somewhat more likely to slow down the expansion, but not likely to kill it altogether.

## 5.6. Europe and NATO on missile defences

Several European governments expressed worries about U.S. missile defence plans for Europe. The former French President Jacques Chirac said that Europe must be careful to prevent new dividing lines on the continent.[317] Foreign Minister Frank-Walter Steinmeier, a German Social Democrat, warned that U.S. facilities must not start a new arms race in Europe.[318] But overall, there are few signs of a U.S.-European divide on the subject. The debate is intra-European and often intra-national rather than trans-Atlantic. Moreover, Europe's North Atlantic Treaty Organization (NATO) allies are gradually moving toward linking up with U.S. missile defence plans.

Germany's position is divided along party lines. The left and the right have traditionally disagreed on Russia and the United States, and missile defence discussions mirror this division. While the foreign ministry criticised U.S. plans, the German defence minister Franz Josef Jung, a Christian Democrat, spoke out in favour of building a NATO-wide missile defence system. Chancellor Angela Merkel mostly stayed off the topic of missile defences other than arguing, in early April, that the issue needed to be dealt with through NATO. In France, missile defences apparently ceased to be an issue when the new government of President Nicolas Sarkozy came to power.

Most European governments seem more concerned about connecting to, not opposing, US missile defences. Their plans focus on NATO. The alliance has long explored the possibility of fielding missile defences of its own. A study launched in 2002 concluded that it was technically feasible to construct a missile shield covering all allied territory against a limited strike from a rogue nation. And in 2006, NATO heads of states tasked the Secretary General Jaap de Hoop Scheffer to produce a study assessing ballistic missile threats to NATO (work on study is still in progress).

While these are NATO studies and considerations, technically separate from U.S. missile defence plans, a close link to the American system has always been assumed. Finally, in the summer of 2007, the allied defence ministers explicitly linked the two. They decided at a 14 June 2007 meeting to study the implications of U.S. missile plans for Europe to NATO's own plans for protecting allied territories. In the words of John Colston, NATO's Assistant Secretary-General: "this is [about] recognising that the United States' system would be likely to provide a very substantial degree of coverage of the European continent and therefore it does make sense for us to examine the United States system alongside possible potential future NATO elements". The U.S. plans are gradually blending into NATO plans. In the future, we may see a dual-control system emerge: NATO would make limited use of the United States system, but Washington would retain the ability to operate it independently.

## 5.7. Russia's concerns

Even though the Polish and Czech sites would represent a relatively modest expansion of GMD, and even though they are not the first missile defence sites in Europe,[319] the U.S. request caused a stir in Russia. General Yuri Baluyevsky, chief of the general staff, called U.S. plans "inexplicable" and hinted that Moscow might withdraw from the Intermediate Range Nuclear Forces (INF) treaty in response. Russia's Commander of Strategic Missile Forces, Colonel General Nikolay Solovtsov, went a step further, stating that his forces were capable of targeting the Czech and Polish facilities. In July 2007, President Vladimir Putin withdrew Russia from the Conventional Forces in Europe treaty citing, among other things, his displeasure at U.S. missile defence plans in Europe.

The statements by Putin and his generals reflect a trend in Russia toward a more confrontational foreign policy. In the first seven months of 2007 Russia also skirmished with its neighbour Georgia, fought a diplomatic conflict with Estonia, engaged in a war of words and tit-for-tat expulsion of diplomats with Great Britain and unilaterally claimed ownership of North Pole's mineral resources. Russia is clearly trying to raise its foreign policy profile, and the row with U.S. over missile defences is a part of this broader trend.

But Moscow also has very specific problems with the Czech and Polish bases. It objects to their location and their intended target. Moscow views the planned Polish and Czech sites (together with proposed new U.S. military bases in Romania and Bulgaria) as violation of earlier promises not to enlarge NATO military infrastructure to new Member States. Those bases may be puny compared to U.S. presence in Europe during the Cold War, but fears of encirclement resonate very strongly in Russia's domestic politics.

Russia also worries because it assumes that the European missile defence sites are meant to contain Russia, and not Iran as the United States claims. Moscow has repeatedly questioned the White House's assessment of the missile capabilities of Iran. It stated that Iran simply does not pose a missile threat. And it also accused the US of upsetting the nuclear balance.

Washington says that not only is GMD not aimed against Russia, it is also not capable of limiting Russia's ability to launch a mass nuclear strike. The ten interceptors to be deployed in Poland, MDA says, are no match for the several thousand missiles on the Russian side. That is true, but not without qualification. At the moment, Russia's nuclear arsenal would overwhelm any missile defence system the United States possesses. Russia's missiles are too sophisticated and too numerous for the GMD. But that could, under some circumstance, change. The life expectancy of its arsenal is limited. The country is retiring old weapons faster than it produces new ones. Recently, the Putin government put more money into missile production and tested several new missiles. But that may not be enough to arrest the decline in Russian missile holdings. Some estimates put the level to which Russian nuclear arsenal could drop to as little as 150 missiles.[320]

Even so, that still appears to be far too many for the proposed interceptor missiles in Poland to effectively shoot down. Moscow points out that technology evolves, and if the United States keeps adding missiles to its defensive system in Poland, it may at some point build an effective wall against Russian missiles.[321] If so, that point is very far off in the future. After a decade of development, U.S. interceptors are barely capable of distinguishing between a single warhead and a simple decoy. Under any scenario Russia will posses as a minimum dozens of far more capable, multiple-warhead missiles.

Russia seems reconciled that the United States will deploy some form of missile defences. Its efforts are switching to rerouting U.S. plans away from the sites in the Czech Republic and Poland, which Moscow views as too close to the likely trajectory of Russian ballistic missiles. At a July 2007 meeting with his U.S. counterpart, president Putin suggested upgrading and using the Russian early-warning radar in Gabala, Azerbaijan instead of the proposed Czech site.

The proposal met with a mixed reaction – there is little agreement among experts on the technical capabilities of the radar in question. Broadly speaking, early warning radars such as Gabala use wide beams to scan large areas for signs of missile launches. The X-band radar sought by the United States in Europe uses a narrow beam to track a missile in flight. Whether Gabala could do the same is unclear. Some Russian and U.S. experts have lauded the radar's abilities. "At that location, the proposed missile defences can 'defend' all of Europe, including south-eastern Europe, [whereas] the Poland/Czech Republic arrangement cannot cover all of Europe ",[322] wrote Phil Coyle, former chief of testing and evaluation at the U.S. Department of Defense and now one of the leading critics of GMD. The US Missile Defense Agency disagrees. MDA chief Lieutenant General Trey Obering said that Gabala is too close to Iran to serve as mid-course radar, and that it would be only useful if working as a complement to the Czech radar.[323] And Pavel Felgenhauer, Russian military expert at the Novaya Gazeta newspaper, said that Gabala was not designed to guide interceptor missiles to their targets. He added that it does not cover all of Iran.[324]

This and other topics will be the subject of a U.S.-Russian dialogue on missile defences, which the two sides launched in the summer 2007. At the time of this writing (October 2007), the two sides were far apart but possibly closing the gap. Russia insists that the Gabala site be in lieu of the Czech and Polish sites, and that the US stop its work on space-based missile defence. Washington has not accepted either of the conditions. However, it made a potentially important proposition in October 2007 – the United States offered to make the deployment of missile defences conditional on both the US and Russia agreeing that Iran represents a missile threat. Presumably, this would allow the United States to build the Czech and Polish sites, but postpone their activation until such day when the United States and Russia agree that Iran possesses a missile capable of threatening the United States and Europe. Russia has not officially responded to the details of the proposals. However, in October 2007, President Putin, during a visit to Tehran, said that there was no evidence that Iran poses a missile threat. This suggests that U.S. and Russian views on Iranian capabilities differ substantially, which will no doubt complicate the task of crafting a joint threat assessment.

## 5.8. Conclusion: Missed opportunities to reduce nuclear tensions

As one observer pointed out, while Russia (and parts of Europe) think of missile defences as a Cold War hangover, the opposite is truth. Missile defences, in their current reincarnation, are a part of a broader attempt beginning in 1990s to reduce the nuclear rivalry between the United States and Russia.[325] Two subsequent U.S. nuclear reviews (1994 and 2002) gradually de-emphasised the Russian nuclear threat. By 2002, the plans congealed into a two-pronged strategy. First, the United States and Russia would radically cut their nuclear weapon holdings and intensify joint effort to stop nuclear proliferation. Second, the United States would also build missile defences as a means to make nuclear weapons less attractive to countries such as Iran. However, the strategy never fully came to fruition. The United States gradually lost Russian trust and co-operation on its new nuclear strategy. Attempts to reduce nuclear arsenals stalled. And that, rather than missile defences, should be the real cause of concern.

President George W. Bush came to power promising to end Cold-War nuclear targeting. The United States and Russia did indeed de-target the strategic missiles (by removing components carrying the co-ordinates), and agreed in 2002 to dramatically slash their nuclear arsenals. They kept some nuclear weapons, in part as deterrence against aspiring or new nuclear powers such as Iran and North Korea.

And because these powers were viewed as unstable and unpredictable – which means they may not be discouraged from using nuclear weapons in the event of a threat of counter-attack – another layer of defence would be added: missile defence. The two-layer approach, the United States believed, would dramatically decrease the utility – and hence desirability – of nuclear weapons. All in all, the greatest threat to mankind – that of nuclear confrontation between Russia and the United States – would be completely eliminated. The smaller, but still catastrophic danger of nuclear exchange with a smaller 'rogue' state would be lessened as well.

Alas, the plan is now essentially on life-support. The United States and Russia put only parts of the vision in place. Both committed to slash nuclear holdings (but by far less than the logic of the plan required) in 2002. The Anti-Ballistic Missile (ABM) Treaty, was abolished (as it should have; it is the legal codification of deterrence and needed to go to make room for new positions). But the United States and Russia fell far short of implementing the entire package of reforms. They failed to keep a lid on proliferation on nuclear technologies. Despite much rhetorical commitment to fighting the spread of WMD, North Korea has since 'gone nuclear' and Iran seems to be on the verge of doing so. US efforts to find new uses for the nuclear arsenal (such as the bunker-buster bombs) weakened its own credibility in fighting proliferation. U.S. interventions in the past decade also seem to have genuinely frightened Russia – so much so that Russia began to think of the US as a threat again. Trust between the United States and Russia has evaporated, and without trust it is impossible to move away from deterrence. And last but not least, Russia has changed, too. President Putin has in recent years stepped up confrontational anti-American 'trong-arm' rhetoric as a means of solidifying power.

The end result is that U.S.-Russian co-operation on nuclear arms reductions is all but frozen. Russia now actually threatens to build new categories of nuclear weapons if the United States deploys missile defences. Iran, concluding that without nuclear weapons it will surely be attacked, ratcheted up its weapons programs. Despite good news from North Korea (which in September 2007 vowed to dismantle its nuclear programme), the non-proliferation regime looks weak.

That is the context for today's discussion about new bases in Poland and Czech Republic. The case for them – assuming the United States can make the underlying technology work – remains strong. That is why European NATO allies are more interested in adding to U.S. missile defence plans rather than defeating them. Missile defences are not the problem: the real concern lies in the failure of the other components of the security architecture sketched out six years ago. Without the whole underlying structure – stronger non-proliferation regime, constructive and co-operative relationship with Russia, continued nuclear arms cuts in both Russia and the United States – missile defences seem almost irrelevant. They will not stop a large-scale exchange of nuclear weapons between the United States and Russia. The risk of such exchange remains unnecessarily high. More countries are developing nuclear weapons now than six years ago, and U.S.-Russian tensions (partly because of missile defences) mean that neither side is doing all that can be done to stop proliferation. And that, rather than missile defences, is the real worry.

## CHRONOLOGY OF KEY MISSILE DEFENCE DEVELOPMENTS

**2007.** The United States launches formal negotiations with Poland and the Czech Republic on missile defence cooperation, including the construction of a missile defence facilities.
**22 July 2004.** The first ground-based missile interceptor is placed in an underground silo at the missile defense complex at Fort Greely, Alaska, USA.
**14 May 2003.** Denmark and Greenland sign an agreement in principle to expand Greenland's Thule Base as a link to U.S. missile defence.
**5 February 2003.** The United Kingdom agrees to allow the United States to upgrade the ballistic missile early warning radar at RAF Fylingdale Moor.
**13 December 2001.** President George W. Bush serves notice to Russia that the United States was withdrawing from the Anti-Ballistic Missile (ABM) Treaty.
**January 2001.** President Bush affirms his plan to deploy a robust National Missile Defense (NMD) system.
**1 September 2000.** President Clinton decides not to proceed with deployment of the NMD system, citing the status of technology and concerns among the U.S. allies and opposition from Russia and China.
**23 July 1999.** President Clinton signs the National Missile Defense Act of 1999, but lists four criteria he will use to make an ultimate deployment decision: threat, cost, technological status of NMD, and adherence to a renegotiated ABM Treaty.
**31 August 1998.** North Korea launches a Taepo Dong 1 missile over Japan, but the third stage fails to put its payload in orbit.
**July 1998.** A commission chaired by Donald Rumsfeld finds that the threat of a ballistic missile attack could emerge sooner than predicted in the 1995 intelligence estimate.
**24 June 1997.** First fly-by test of the 'kill vehicle' for the NMD system.
**November 1995.** A report from the U.S. intelligence community declares that no country could threaten the United States with a ballistic missile attack in the next 15 years.

**29 January 1991.** President Bush announces the Global Protection Against Limited Strikes (GPALS) system to counter unauthorized, accidental or limited attacks.
**14 June 1989.** President Bush decides to continue the missile defence programme even as the U.S.S.R. collapses.
**October 1986.** President Reagan and Soviet President Gorbachev discuss eliminating nuclear weapons, but talks collapse when President Reagan refuses to agree to limitations on missile defences.
**March 23rd, 1983.** President Reagan announces that the United States will start an expanded research and development program of missile defence system.
**January 1976.** The full Congress approves shutting down Safeguard, and Secretary of Defense Rumsfeld announces the system's termination.
**1 October 1975.** Safeguard system begins operating in Grand Forks, North Dakota, USA.
**May 1972.** U.S. and Soviet Union sign the ABM Treaty, banning nation-wide missile defences. It is amended in 1973 to allow only one limited missile defence site to each side.
**August 1969.** United States Senate votes to deploy Safeguard missile defence.
**1966.** United States Secretary of Defense McNamara announces that the Soviet Union has deployed its Galosh missile defence system.
**1957.** U.S. begins work on its first major missile defence effort, the Nike-Zeus system.

*Sources: The United States Missile Defence Agency, Union of Concerned Scientists, U.S. Embassy to the Czech Republic, the Nuclear Threat Initiative*

[310] Coyle, Phil. What about the most recent missile defense test? Nieman Watchdog. 11 Sept. 2006. http://www.niemanwatchdog.org/index.cfm?fuseaction=background.view124.
[311] Black, Sam and Samson, Victoria. "Missile Defense Flight Tests". Center for Defense Information. (18 June 2007). http://www.cdi.org/program/document.cfm?DocumentID=1984&StartRow=11 &ListRows=10&appendURL= &Orderby=D.DateLastUpdated6index.cfm.
[312] Hildreth, Steven A. "Ballistic Missile Defense: Historical Overview". Congressional Research Service. Updated (July 9 2007): 5.
[313] Ibid.
[314] Obering, Trey. "Missile Defense Hits Mark". Defense News (23 July 2007): 21.
[315] Hitchens, Theresa and Victoria Samson. "Space-Based Interceptors: Still Not a Good Idea". Georgetown Journal of International Affairs Summer/Fall (2004).
[316] Kagan, Robert and Bill, Kristol, eds. Present Dangers: Crisis and Opportunity in American Foreign and Defense Policy. San Francisco: Encounter Books, 2000.

[317] Dombey, Daniel, Bertrand Benoit and Guy Dinmore. "Chirac Hits at US missile plans". Financial Times (10 Mar. 2007).
[318] Schulte, Sebastian. "U.S. Missile Defense Plan Splits German Leaders". Defense News (2 Apr. 2007).
[319] The GMD system also uses radar facilities in Britain and in Greenland (Denmark-administered territory).
[320] Kimball, Daryl G. "Missile Defense Collision Course. Arms Control Today". (Aug./July 2007). http://www.armscontrol.org/act/2007_07-08/focus.asp.
[321] Kozin, Vladimir. "NATO and Russia square off over US BMD". Jane's International Defence Review. (July 2007): 28–29.
[322] E-mail correspondence with author. 17 July 2007.
[323] Mannion, Jim. "Russian radar site doesn't fit US missile shield needs: general". Agence France Press, 16 Aug. 2007.
[324] Galpin, Richard. "Inside Russia's missile defence base". BBC News (2 July 2007). http://news.bbc.co.uk/1/hi/world/europe /6262220.stm.
[325] Becher, Klaus. "Prepared remarks for the IISS/CEPS European Security Forum". Brussels (2 Apr. 2001). http://www.eusec.org/becher.htm.

# 6. Controlling the freedom of using space: the White House Space Policy dilemma

Xavier Pasco

Despite the relatively short history of the use of space – the $50^{th}$ anniversary was celebrated in 2007 – this first decade of the twenty-first century may mark a new departure in the occupation of space by mankind, or more precisely by the military. For some years now, projects for developing space weapons to control space have nourished international controversies. A Chinese anti-satellite test performed in January 2007, has only served to reinforce the general feeling that there is "weaponisation" of space. This feeling has quietly become widespread as a common consideration in the space community, as for the first time since the beginnings of space programmes, such perspectives might benefit from a more favourable political context. As an example, the new U.S. White House Space Policy unveiled in October 2006, while not breaking with earlier security-oriented policies, openly legitimates radical options should these become necessary for the security of the United States and its space assets. It is the open character and the general tone of the declaration, rather than its content that does not differ much from earlier documents. Up to now, such issues had been kept low profile on the political agenda and had never been promoted to the point of feeding possible domestic and international political arguments. This article intends to propose, as seen from the European perspective, an analysis of the reasons behind this development, to better judge the realities behind the political stance, and to provide some perspectives of possible consequences.

## 6.1. The evolution of political and military interests in the use of space

First, a brief review is necessary to understand why the weaponisation of space has never been translated into real programmes. Certainly, the political bodies in the United States have been no less interested in the use of space than the currently. Historical data provide ample evidence of the permanence of public investments with a constant and active interest in the military use of space. In fact, if one considers the continuity of the funding, the strongest relation between the space sector and governmental activity has clearly been in favour of the military sector over the civilian sector. This relation started very early on in a situation where the two superpowers were confronted with the advent of a totally new era as regards strategy due to ballistic and nuclear weapons. Even though a number of space activities in the United States have not depended solely on military activities, in comparison they have remained very fragmented and subject to frequent changes. In addition, this link between military space and governmental action has been characterized by a profound continuity that has been structured in successive layers based on complementary military user communities. This structure, which will be detailed below has allowed military space to be regularly renewed, thus avoiding the "identity crisis" that has regularly plagued the civilian space programme and NASA, the agency in charge of it.

### 6.1.1. Strategic space 1958–2007: the historical link between nuclear technology and space

In the United States, just like in the U.S.S.R., now Russia, space activity basically started parallel to the ballistic and nuclear arsenals. It is because of the competition between the U.S. and the U.S.S.R. to equip themselves with those armaments between 1945 and 1953 that the governing bodies in the two countries developed an interest in the use of space.

This relation between the worlds of space and nuclear technology has not been based only on the similarity of the technologies useful for developing missiles and rockets. It has also been derived from the necessity (perceived very early on in the United States and made official in 1995 in reports published as early as 1946) to possess monitoring and possibly targeting capabilities that are both permanent and invulnerable. While aerial capabilities would rapidly prove to have limitations in this field,[326] detaining satellites with early-warning surveillance, reconnaissance and targeting capabilities was to become a priority given the rapid evolution of the offensive weapons. The so-called MAD doctrine (Mutual Assured Destruction) was to give such assets almost the rank of a national life insurance in the U.S., but also in the U.S.S.R. later on, making space seem like a mutually recognized sanctuary at the time.

Even though this relationship between space and nuclear technology has supposedly made satellites a way to improve the efficiency of terrestrial and naval nuclear forces, satellites have never been considered a possible long term substitute. As demonstrated by a number of U.S. historical documents, space programmes aiming to weaponise space have regularly proven unable to attract sufficient attention from military and political authorities. From a political point of view, the global costs (including the political costs) incurred by such programmes have always by far exceeded the benefits. In the context of the MAD, the governments were better off accepting the mutual use of spy satellites allowing a precise count of offensive arsenals rather than run the risk of a renewed confrontation, which might have ended with reduced observation capabilities and a generally decreased mutual deterrence capacity. Moreover, those countries – among which was the United States – were highly confident of the efficiency of their nuclear delivery means as early as in 1960. This confidence contributed to discarding any complicated programme, aiming, for example, to put missiles into orbit.

Today, this historical link has endured and remains the basis for the military space effort in the U.S. The continuing development in the United States of efficient space technologies for gathering information on missile arsenals and other vectors throughout the world shows this very well. Moreover, there are dynamic research and development (R&D) programmes related to innovative sensing techniques (such as infrared, hyperspectral, etc.) supported by the current project anti-ballistic "missile defence". In this regard, it seems legitimate to consider space R&D financed by the Anti-ballistic Missile (ABM) effort as a continuation of this "space-nuclear" historical link in new domains.

### 6.1.2. "Theatre level" or "force multiplier" space (1991–2007): adapting the space systems to the new strategic environment (first try)

The end of the Cold War was a first far-reaching change in the relationship between space and military activity. In the aftermath of a series of conflicts that started in 1991 with the Gulf War, and further illustrated by the Kosovo conflict, space capabilities suddenly appeared as a "must-have" if one wanted to win a conventional (i.e. non-nuclear) war. In particular, space capabilities were soon associated with the use of innovative air strategies with increasing global efficiency, and thus almost giving birth to a new paradigm of the "art of war" according to promoters. This revived importance given to space in the very conduct of military operations constitutes one aspect of a general adaptation of the military tool to the new strategic conditions that took place in the U.S. at the time. Space has drawn some benefits from the reshaping of the U.S. military apparatus underway at the time, and has been assimilated as a genuine "Revolution in the Military Affairs" (RMA).

From this general point of view, military authorities have perceived a dual logical approach to space applications: on the one hand, space was soon viewed as the cornerstone of future defence architectures around which forces and doctrines would have organise. Becoming part and parcel of the "Battlefield Awareness" concept,[327] information from space should be used at the lowest level on the battlefield, down to the soldier who will have at his/her disposal the most advanced communication means. On the other hand, as seen from the organisational perspective, space could also appear as one of the most flexible sectors of activity given its recent history, an a priori less entrenched area than other military domains supposedly more heavily structured. Thus, military space activity could envisage the use of very diverse resources stemming from civilian, commercial or dual-use programmes (with the exception of hardcore intelligence or typical military programmes) or from extended international cooperation. This evolution implies a redefinition of the requirements strictly connected with the conduct of military operations, a distinct form with more general purpose needs that can be satisfied by commercial operators with satellites already on-orbit or to be developed in the near future. This increasingly important domain is the subject of numerous international discussions, notably trans-Atlantic discussions on standards relating to procedure or to interoperable techniques.

### 6.1.3. Space as a "security enabler" (1994–2007): adapting the space systems to the new strategic environment (second try)

The global political answer in the United States and in Europe to the accelerated change of the strategic landscape (with the concept of "enlarged security"), as well as a better understanding of what space technology could add to this security during the nineties has brought about further changes in the relations between space and defence. The increased use of space observation, telecommunication and navigation/localisation/timing techniques during the military operations has gradually resulted in introducing more and more space technologies at the very heart of the military and security systems. Of course, the precise guiding of missiles and ammunitions by the Global Positioning System (GPS) system is what first comes to mind.

Generally, a priority military objective seems to have been the development of innovative intelligence systems as soon as the end of the first war in Iraq, as well as during the Kosovo conflict. This choice reflects the difficulty of dealing with new military (or security) "targets" that are more elusive, mobile, and to put it briefly, less characteristic than the Soviet missile silos. Based on the experience gained during the nineties, the idea progressively imposed itself to adapt existing space assets, which had been developed for decades, to monitor the Soviet Union. Space systems must be omnipresent, capitalising on technical progresses made in the field of sensors with projects of more comprehensive space systems destined to improve both early-warning and intelligence capabilities, along with observation satellites that are both capable of increased resolution (geometric and spectral) and of enlarged field of views. Such systems can also benefit from more efficient "small" satellites that, despite reduced performance compared to traditional military systems, can usefully complete national means. Last but not least, all these capabilities can be networked to build a genuine space architecture that can be interconnected with other information tools, air-, sea- or ground-based. In brief, in the eyes of numerous military strategists in the United States, mastering the information technologies is becoming a pre-requisite if one wants to win wars. Very progressively then, space techniques have quickly become more than a simple "force multiplier" to represent a real strategic or security enabler, but more and more a central component of current and future defence and security systems.

This position, basically shared by most of the military establishment in the United States, has found a more political translation that has helped sustain this effort until now. Indeed, these security space applications have somewhat come to reflect the new political analyses of the new threats facing the United States before and after 2001, and stand to benefit from the ubiquity and the new performance levels provided by these new collective capabilities. Again, the permanence and the performance offered by enlarged networks of collaborative sensors, potentially both military and civilian, could theoretically allow one to deal with a large spectrum of missions, ranging from police and law enforcement type missions to the direct use in theatre-level weapons systems and operations in high intensity conflicts. This ability to basically improve the knowledge of the general "security" environment may be legitimately presented as providing the United States with strategic and political gains in a world embroiled with more and more elusive threats.

This general context is contributing to the reinforcement of the trend towards more and more sophisticated networks of space sensors and means, creating dynamism for space not to be found anywhere else in the world. It is precisely this latest evolution of the military uses of space in the United States seems to be directly linked to its strong political meaning and impact. Indeed, space can also be considered a powerful tool to reinforce the presence of the United States on the international scene from the military point of view as well as from the high technology standpoint. Any space architecture has a natural federative impact, as it can be considered "systems of systems" that merge military and civilian satellites for intelligence or for communication purposes, for example. From an extreme perspective, it could be maintained that the stress is not so much on the individual missions devoted to such or such system, but rather on their ability to integrate a larger, cooperative and more flexible architecture able to answer military needs when they arise. The rapid technological evolution would then make such "smart" connectivity possible. Even the civilian industry would provide new space assets usable by the military communities given the broad progress made in the "high-tech" information world.

This evolution has translated into several concrete steps with the implicit goal of highlighting the value of the political and industrial investments made during the Cold War. This is illustrated by the well-known efforts made by the most recent administrations to promote Earth observation or navigation/localisation techniques that were previously considered as sensitive capabilities and kept at the classified or "military use only" level. At the same time, U.S. administrations have actively invested in the international space launch domain, also with the global objective to better organize a crucial capability for supporting these developments in a highly competitive environment.

As "security enablers", space choices made in the United States have reached the point where they can be considered as participating by essence, consciously or unconsciously, in a wider political picture that will gradually transform military space orientation into a dominant strategic driver for the aerospace industry in the decades to come, with both domestic and international dimensions. This background must be well understood, as it has paved the way for new political considerations, formulated during the Clinton Administration and endorsed by the Bush Presidency, aiming at reintroducing perspectives to achieve a better control of space.

### 6.1.4. "Controlled" space? (1995–2007): making space a dominating factor

This new situation has indeed spawned new interpretations of the importance of controlling space. Up-to-now, projects aiming to put defensive or armed systems in space had never rallied real political support, even if proposed several times in the United States since the sixties and despite sometimes strong lobbying effort. The highest political authorities seldom endorsed making space a possible new battlefield as suggested by well-known U.S. Air force Generals in the fifties.

Today, reconsiderations about the military role of space have started, bringing about international discussions fora in charge of strategic disarmament, notably the Conference on Disarmament (CD) in Geneva, Switzerland. As seen by most of the U.S. strategists, the increasingly diverse and numerous functions assumed by exiting or coming space systems make those "a vital national interest": space is becoming "a center of gravity, both economically and militarily".[328] A new mission for space techniques can then be crafted around the general notion of "space control", usually compared to the strategy that led the major commercial powers to seek dominance of the seas in the past centuries. A very important directive signed in 1999 by William Cohen, then Secretary of Defence in the Clinton administration announced that space must be considered as

> "a medium-like land, sea and air within which military activities will be conducted to achieve U.S. national security objectives. The ability to access and utilize space is a vital national interest because many of the activities conducted in the medium are critical to U.S. national security and economic well-being."

As a consequence,

> "purposeful interference with U.S. space systems will be viewed as an infringement on our sovereign rights. The U.S. may take all appropriate self-defence measures, including, if directed by the National Command Authorities, the use of force, to respond to such an infringement on our rights."

In particular, it is considered vital that

> "an adversary can not obtain an asymmetric advantage by countering our space capabilities or using space systems or services for hostile purposes".[329]

These new strategic objectives have been translated into the definition of radically new military requirements in space. This new expression of needs is currently shaping the budgeting of nascent R&D programmes, as well as consolidating older defensive concepts such as, for example, powerful jamming systems directed at space, operational since September 2004 in the United States. Indeed, the "space control" doctrine, as defined in the middle of the 1990's can be either defensive or more offensive. As early as in 1995, the space control concept was to include technical and non-technical measures.

Non-technical measures relate to traditional diplomatic and political decisions aiming to protect the development of strategic space techniques from undesired interferences produced in any international forum or by any human activity. This general political position cannot be seen as directly related only to the space control concept, but rather can also be interpreted as a traditional general political goal of any nation in any national interest domain. Concerning space security issues, such a general position has been expressed quite openly since the beginning of the decade in forums like the CD, for example. It is also the case when negotiating frequency reservations in the International Telecommunication Union (ITU), not only in the case of the United States, but for most of the national players. It may also be claimed that general or specific aids or subsidies offered to the national space industry with the goal of sustaining and even enhance competitiveness are part of this defensive space control exerted by most space-faring countries. More specifically, in the United States it also translates into the setting up of forces specialized in "space combat", with the goal to better thwart any hostile foreign use of space assets.

Technical measures can be passive and add to this defensive mode of space control. For example, they include the hardening of satellites, making them more robust against malfunctions, irrespective of whether these are a result of intentional or non-intentional causes. Electronic components more resistant to possible electro-magnetic interferences (as those generated by sun bursts, for example) or platforms capable of sustaining the shock generated by tiny space debris are fully part of research programmes destined to feed this strictly defensive posture.

Beyond such research, some level of military planning exists in the United States with the associated R&D orientation that aims to develop future programmes among which some are more offensive in essence. Obviously, they are also more controversial. There are three principal domains:

- Space surveillance, or space situational awareness;
- On-orbit satellite protection;
- Possible active space weapon systems.

Space surveillance, today more frequently referred to as space situational awareness (SSA), envisions the use of terrestrial as well as space-based sensors to better identify orbiting space systems. Efforts would be focused on observation domains perceived as insufficiently studied. This activity implies the reinforcement or the development of both optical and radar ground-based assets simultaneously with existing observation satellites or future dedicated platforms devoted to the inspection of low-Earth orbit or geostationary satellites.

On-orbit satellite passive protection focuses especially on the hardening of the electronic component of the satellites, both civilian and military. It must be noted that networking the space systems is also expected to bring some kind of passive protection due to a natural redundancy resulting in a better resiliency of the space architecture.

Last, but not least the control of possible "hostile uses of space" is underlined by multiple experimental activities in the field of anti-satellite techniques. These techniques comprise both ground-based projects including powerful lasers able to disrupt, may be even destroy, optical satellites. Such programmes, along with other technologies usable either from Earth or directly from space have the clear mission to "deny the use of space by any hostile force, if necessary". Such assets are many, ranging from anti-satellite platforms to attacks conducted on vital ground control stations by specially trained units, including also satellite jamming or spoofing.

Such efforts are nascent in the United States, even if real large-scale threats on space assets remain largely virtual. Expressed in 2004 in a first military doctrinal document (see below), these views were also reaffirmed as a politically plausible perspective in the last presidential space policy made public in October 2006. Still, the recent Chinese laser "painting" of a U.S. observation satellite and the voluntary destruction by Chinese authorities of a national meteorological satellite by the use of a ground-based missile equipped with a kill vehicle has largely reinforced these perspectives.

## 6.2. From space control to space weaponisation? A tougher rhetoric introduced with the Bush Administration

Over the last few years, activities related to space control have steadily increased and have become the most influential perspective permeating the latest presidential directive. A first publication from the U.S. Air Force may be considered as having marked the first decisive step in this direction. The document published in August 2004 defines possible future "Counterspace Operations".[330] For the first time, this doctrinal document translates the general orientation already adopted by the Clinton Administration in the July 1999 defence directive (already quoted) signed by William Cohen, then Secretary of Defence into technical and procedural perspectives. Until then, such an orientation, explicitly referred to by the 2004 "Counterspace Operations" doctrine, were the only official references in broad terms to the notion of space control. The new military text is significant as five years later it demonstrates the importance attached to this general framework by the Pentagon. In particular, it is the first time that such a notion as space weaponisation is associated with the Department of Defence Joint doctrinal *corpus*, using notions such as "transformation" or "effect-based Approach" which imply a wide-open inter-service and administration cooperation (beyond the Pentagon itself) to attain political objectives and implying some high-level interoperability.

The 2004 document envisions four different types of activities that almost mirror the structure of "space control" evoked in 1995, adding to the three categories of activities already described a fourth one called "planning and execution". This supplement clearly aims to reinforce the operational side of mainly experimental space control efforts. Very significantly, the three categories of efforts envisioned by the space control doctrine have been renamed to appear as the following list:

- Space Situational Awareness;
- Defensive Counterspace;
- Offensive Counterspace;

As expressed in the 2004 documents, SSA covers diverse applications related to general environment monitoring, reconnaissance and intelligence, with a direct interaction with command and control functions. Operationally, the SSA is conceived as supporting different stages of counterspace operations: It is used to "find, fix and track" any space event, fulfilling a detection-and-alerting mission. It must also help "target and engage" the threat and assess the effect produced on this threat. The so-called "space events" include "orbital manoeuvres, anticipated and non-anticipated launches, atmospheric re-entries, laser emissions, sun bursts, and adverse electromagnetic pulses."

The Defensive Counterspace operations (DCS) are aiming to protect, maintain and repair U.S. and allied space and ground-based equipment. DCS operations envision two types of distinct operations, some addressing the detection of a possible attack on those systems as well as the monitoring of hostile operation before and during the crisis. It also addresses the active preservation of the space systems integrity during the attack (as demonstrated during the recent Iraqi war, at the time when GPS systems were jammed). It is clearly stated that DCS operations could be implemented either to deter any attack (that would be discouraged by protective systems) or to defend the space segment using both passive and active measures. These last measures include the detection of an attack as well as a positive attribution of its origin, either intentional of not. It also implies the use of manoeuvrable platforms as well as systems that would be able to afford defence against anti-satellite attacks, jamming or interference devices. Last, but not least, DCS operations also include some "recovery" functions, allowing the replacement of any crucial element of the space chain, leading incidentally to the development of highly reactive launching systems.

The Offensive Counterspace operations (OCS) aim directly to deny any adversary the use of its own space systems for military purposes. OCS operations rely on the so-called "5 Ds doctrine" which defines the following graduation as a possible implementation plan for such operations: Deception, Disruption, Denial, Degradation and Destruction. OCS operations imply a gradual overdependence on space systems by any possible adversary. In particular, the increased dependence of Command and Control (C2) systems regarding satellite networks is highlighted in the document as making space systems a necessary military target. OCS operations, unlike DCS, fulfil combat functions, going beyond just the interdiction of an adversary to stop it from using its own space system for war purposes. It implies temporary or definitive annihilation to make "space superiority" a prevailing factor from the start.

A first overview of the current R&D programmes associated with such functions can be traced from unclassified literature. Some OCS-related programmes have already been translated into ground-to-space systems, among which, at least one, dubbed the Counter Satellite Communication System (CSCS) and intended to jam hostile telecommunication satellites, would have been deployed since September 2004 by the 76th Space Squadron. A second programme, called Counter Surveillance Reconnaissance System (CSRS) with the objective to disturb the function of remote-sensing satellites, has been slowed down following budget reductions.

Other programmes such as RAIDRS (for Rapid Attack Identification Detection and Reporting System) made for on-board threat analysis and reporting, or for

**Tab. 2:** *FY2008 budget of main U.S. space control associated projects (in millions U.S. dollars) As known in summer 2007 (source: Center for Defense Information)*

| Programme | Administration Request | House Authorization | Senate Authorization | Conference Committee |
|---|---|---|---|---|
| Space Test Bed | 10 | 0 | 0 | TBD |
| NFIRE | 36 | 36 | 36 | TBD |
| XSS | 28.9 | 28.9 | 28.9 | TBD |
| ANGELS (Nano) | 2.5 | 2.5 | 2.5 | TBD |
| Starfire Optical Range | 44 | 44 | 44 | TBD |
| Airborne Laser | 548.8 | 298 | 348.8 | TBD |
| Kinetic Interceptor | 227.5 | 177.5 | 227.5 | TBD |
| Multiple Kill Vehicle | 271.2 | 229.2 | 271.2 | TBD |
| TICS (Tiny Independent Coordinating Spacecraft) | 6 | 6 | 6 | TBD |

experimental inspection and protection satellites (as demonstrated by the XSS satellite series launched from 2003 to 2005, and continued through the nano-satellite ANGELS programme) comprise most of the current federal efforts in DCS-related projects.

At this stage, SSA programmes are among the best-supported space control-related activities with a particular focus on the modernization of existing space surveillance ground components (the Spacetrack programme) as well as on the development of space-based surveillance systems with possibly four to eight dedicated satellites on orbit on the horizon of 2012–2013.

## 6.3. Managing dilemmas: What is collective security from the U.S. perspective?

The August 2006 presidential directive provides a political framework for the continuous development of these different projects.[331] While the nationalistic tone of the directive has been widely held as a premiere for such a document, attentive reading shows how much it confirms a space strategy that had already been announced in the past, albeit with more diplomatic words.

If the Clinton Administration had to pay attention to the preservation of U.S. space power, as clearly stated in the 1996 space policy directive, it never transformed the notion of space control into a political top priority and made sure that any possibility to involve some level of international coordination would be explored. The overtly benign stance towards space control, which became official in the 2006 text, symbolizes the obvious increase in political support for U.S. space dominance that must be undisputed. Not much different as regards the principles, the two texts differ mainly in their form.

Since August 2006, space has become a medium comparable to others where the "national vital interests" of the U.S. must be defended by all means, including possibly dedicated space-based programmes that had remained confined at the R&D level up to now, considering their political sensitivity. The idea of developing space weapons is gradually becoming more acceptable, by referring, for example, to possible "protective zones" surrounding satellites, with the perspective of considering any suspicious movement in such zones as hostile and thus granting the right to use these new "defensive/offensive" systems. While different from the "softer" space control doctrine as advocated by the Clinton Administration, such a demonstrative policy is still being hailed as promoting the defence of space for the common good. As put recently[332] by Robert Joseph, Undersecretary for Arms Control and International Security,

> "(...) not all countries can be relied upon to pursue exclusively peaceful goals in space. A relatively small number of countries are exploring and acquiring capabilities to counter, attack, and defeat U.S. space systems. These capabilities include jamming satellite links or blinding satellite sensors, which can be disruptive or can temporarily deny access to space-derived products. Anti-satellite weapons – whether kinetic or conventional – or Electro-Magnetic Pulse weapons – can permanently and irreversibly destroy a satellite. (. . .) In view of these growing threats, the United States must close the gap between its stated space policy and its deployed capabilities. Our space policy instructs us to increase our ability to protect our critical space assets and to continue to protect our interests from being harmed through the hostile use of space. This requires us to remain at the forefront in space, both technologically and operationally, just as we have in the air, on land, and at sea. (. . .) let me say that what is clearly in our interest is to continue to be the leader in expanding the use of space for peaceful purposes. Our advances in space in the fields of communication, medicine, and transportation, as well as many others, have come to benefit not just Americans, but all of mankind, including citizens of countries that have not yet ventured into space. For the United States, that means continuing our tradition of pursuing diplomatic efforts to gain the broadest possible appreciation for the benefits that all nations receive from the peaceful uses of outer space".[333]

According to this logic, this notion of a "common good" is the clear legitimisation of the fact that the United States has positioned itself to "develop capabilities, plans and options to ensure freedom in space and, if directed, deny such freedom of action to adversaries".[334]

However, it must be noted that this predominantly unilateralist approach does not represent the exclusive path for space security followed by the United States military and political authorities. Indeed, maximizing tension in space cannot appear as the preferred solution for a country that relies so much on its space assets. More and more frequently, discussions of international "rules of the road" or "codes of conduct" are surfacing in the internal U.S. debate. The necessity is evoked of exploring more diplomatic means that may help regulate and decrease a full range of threats, from space debris to space traffic management issues, with all issues possibly leading to security breaches that cannot be dealt with by space weapons, either offensive or defensive. In this respect, it is highly remarkable to note the recent speech made during a congressional hearing by General Kevin Chilton, Commander of the United States Strategic Command, referring to the possible usefulness of a code of conduct:

> "I think as a government, we should examine the potential utility of a code of conduct or "rules of the road" for the space domain, thus providing a common understanding of acceptable or unacceptable behaviour within a medium shared by all nations".[335]

A clear sign of the persistence of a lively and open debate in the United States, this declaration also gives credit to critics arguing about the limited efficiency of a strategy based only on what appears as unilateral weaponisation considering the existence of many other pending security issues that need to be addressed by the international community. However, any renewed interest for a collective approach to space security will come about only in win-win solutions, especially if they demonstrate that such international "rules of the road" provide solutions in the best U.S. national interest.

In any case, long-term security trends in space will necessarily involve new actors, whether private operators or an increasing number of emerging countries. It also seems to be in the interest of the United States and of the international space-faring community to help manage these trends and keep a number of transformations under relative collective control. This means that the United States will have to more actively enter collective security-oriented processes, also meaning that positive moves will have to be made on either side, namely, in the United States and in China, and to a lesser extent in Russia to allow scope for some possible agreement. To which extent will the domestic U.S. debate allow such moves? This may very well be the next strategic dilemma the United States will have to face, balancing between its military space dominance and vulnerability in space.

---

[326] The U-2 spy plane piloted by Gary Powers would be shut down in 1960 by the USSR anti-aircraft defense. In August 1960, the first images transmitted by satellites were on the US presidential desk.
[327] Perry, William. "Battlefield Awareness (...) is the edge which gives our forces unfair competitive advantage in any combat they're involved in".
[328] Cook, Donald. The Military Utility of Space. Rusi Conference, London: Sept. 1999.
[329] Cohen, William. Memorandum for Secretaries of the Military Departments, accompanying the Defense Space Policy Directive #3100.10 (9 July 1999). www.dtic.mil/whs/directives/corres/html/310010.htm.
[330] Air Force Doctrine Document. "Counterspace Operations". Aug. 2004: 2-2.1, 2.
[331] United States. US National Space Policy. 31 Aug. 2006 http://www.ostp.gov/html/US%20National%20Space%20Policy.pdf.
[332] In fact, on 11 January 2007, the very day when the Chinese ASAT test was performed.
[333] Joseph, Robert G. "Remarks to Center for Space and Defense Forum". Colorado: Colorado Springs, 11 Jan. 2007. www.state.gov/t/us/rm/78679.htm.
[334] United States. US National Space Policy. 31 Aug. 2006. http://www.ostp.gov/html/US%20National%20Space%20Policy.pdf.
[335] Chilton, Kevin. Speech of the General, Commander of the United States Strategic Command. Committee on Armed Services, Advanced questions (27 Sept. 2007). www.senate.gov/~armed_services/statemnt/2007/September/Chilton%2009-27-07.pdf.

# 7. China's ASAT test – A warning shot or the beginning of an arms race in space?

Götz Neuneck

## 7.1. Introduction

On 11 January 2007, the People's Republic of China (PRC) conducted its first successful anti-satellite (ASAT) military test, becoming the third country proving that it can destroy satellites in low-earth orbit (LEO). This event raised international concerns about the worsening of the space environment and the potential of inspiring other countries to work on future hostile space activities. The artificial collision is considered to be "one of the worst manmade debris-creating events in history"[336] threatening civilian and military satellites in the future and, possibly, accelerating the weaponisation of space. Over the long run, this could lead to a costly and dangerous arms race in outer space which might draw in other space-faring nations. China has long been an advocate of arms control in space and has criticised the United States, above all, for working on "weaponising space". Within

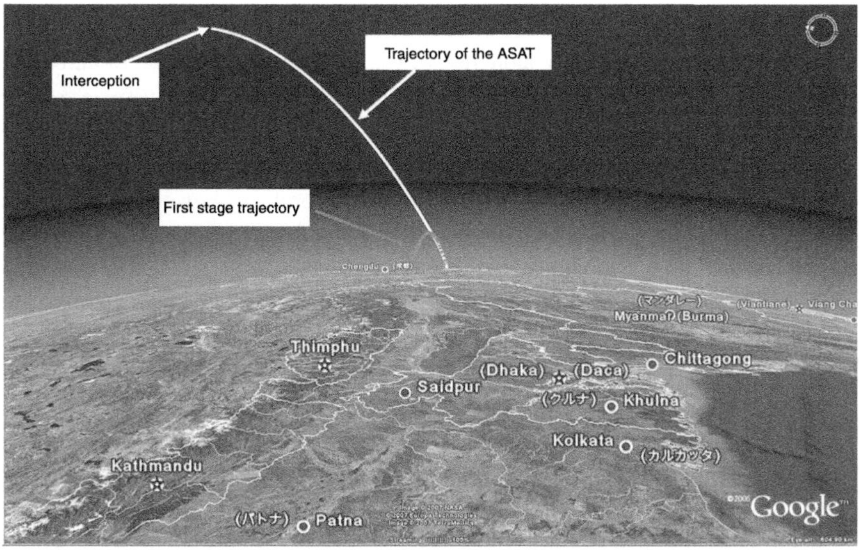


**Fig. 16:** *Analysis of the ASAT trajectory (source G. Forden).*

the Prevention of an Arms Race in Space (PAROS) talks at the Conference on Disarmament (CD) together with Russia, in 2001/2002 it proposed a draft treaty for banning weapons in space, though the January 2007 test event came as a surprise for many. Several questions have emerged from this situation: How developed is China's ASAT capability? What does this event say about China's technical and political intentions and motivations? How will other states and the international community react in the current situation? The paper examines the ASAT-tests, the Chinese explanations, the international reactions, and the influence on the debate on space security.

## 7.2. China's FY-1C destruction – technical analysis and the space debris issue

The Chinese Military launched a direct-ascent anti-Satellite device in the early sunlight of 11 January 2007, which destroyed the aging 800 kilograms. Chinese weather satellite Feng Yun 1C (FY1C) at an altitude of 865 kilometres at 6:26 local time thus creating thousands of pieces space debris.[337] The People's Liberation Army (PLA) used a medium-range ballistic missile, probably a two-stage, solid-fuel ballistic missile "Kaituozhe-1" KT-1,[338] a version of the DF-21/CSS-5, to transport a kill-vehicle to directly hit the south-flying satellite, which had been orbiting the Earth since 1999 on a polar orbit. The ASAT payload was launched northward from near China's Xichang Space Centre in the Sichuan province from a mobile transporter-erector launcher (TEL) and hit the satellite directly at a distance of 1.150 km from Xichang. It is very likely that the launch vehicle and the collision was observed by a ground-based radar. Details of the ASAT collision payload are not known. Some argue that it is not too complicated to hit a satellite which is flying on a predictable orbit and probably emitting a well-known signal. Technical analysis, however, shows that China's ASAT could be a sophisticated kinetic interceptor, which might also be equipped with an optical tracking device.[339] It is reported that the test in January was not the first attempt, but this was the third test, the previous three tests might have been fly-by tests or perhaps also failures.[340] Due to the high collision speed greater than 8 km/s, the satellite has broken up into thousands of fragments reaching up to 3.500 kilometres and perigees reaching below 200 kilometres. This ASAT cloud pattern with orbital periods from 95 to 132 minutes and inclinations between 96.8 and 102.6 degrees is typical for a high-energy collision. Because the atmospheric density is low at that altitude, a large fraction of the fragments will remain in orbit for centuries. The United States Space Surveillance (SSN) in Colorado Springs,

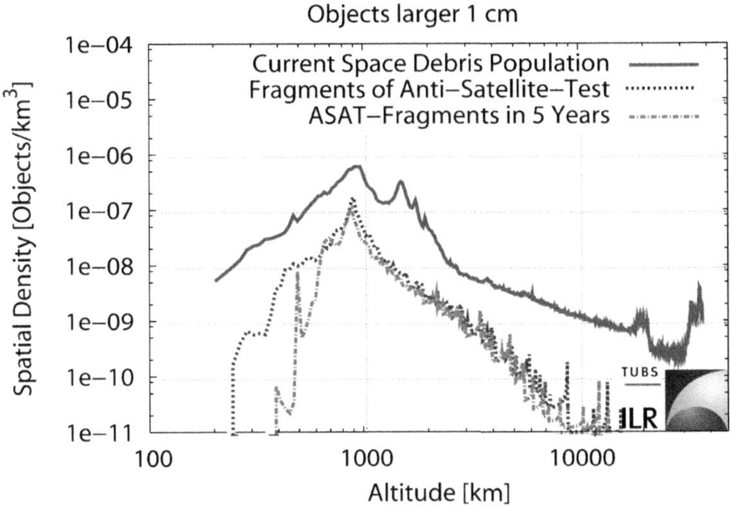


**Fig. 17:** *Comparison of the current space debris population with the debris cloud created by the Chinese ASAT Test from January 2007 for objects larger than 1 cm (ILR/TU Braunschweig see Footnote 343).*

which can track objects in LEO with sizes larger than 5 to 10 centimetres with its radar and optical sensors, has tracked about 2 000 pieces. The tracking of all fragments by ground-based radars is ongoing. Models show that the number of debris greater than 1 millimetre ranges from 1 to 2 million.[341] One should have in mind that due to the high velocity of objects in space, even small orbiting fragments can destroy satellites. An effective shield against pieces larger than 1 cm is nearly impossible. The test increased the number of cm-sized space debris by 8% (see Figure 17).[342] Together with another fragmentation event, the break-up of a Russian upper-stage on 19 February 2007, these two events are responsible for a drastic increase of almost 30% as compared with the space debris, which was tracked at the beginning of 2007, thus creating a long-lasting hazard for satellites in LEO. An analysis by ESA scientists reveals that these two events increased the collision risk of the International Space Station (ISS) with objects larger than 1 centimetre by 80%.[343] The European Space Agency (ESA) has around 60 active and non-active satellite payloads below 900 kilometres. A future Chinese ASAT capability might also endanger new surveillance satellites of local competitors such as Taiwan, Japan, South Korea and India.[344] The United States maintains a lot of civilian and military satellites in LEO, which might be threatened.[345] If detected early enough, the ISS and some other satellites can manoeuvre to bypass a catastrophic collision.

Simulations also show that more than 50% of the debris with a size greater than 1 centimetre will remain in orbit for more than 20 years, and perhaps considerably

**Tab. 3:** *Estimated amount of space debris in different altitudes by size based on the ESA MASTER 2005 Debris Environment Model and estimates (source: Footnote 341).*

| | 0.1–1 cm | 1–10 cm | >10 cm |
|---|---|---|---|
| Total debris at all altitudes | 150 000 000 | 650 000 | 22 000 |
| Debris in LEO | 16 000 000 | 270 000 | 14 000 |
| Debris from the Chinese ASAT-Test | 2 000 000 | 40 000 | 2000 |
| Debris from a potential break-up of a 5–10 tons satellite | 8–14 000 000 | 150 000–250 000 | 3000–5000 |

longer, because the altitude of the break-up is so high. National Aeronautics and Space Administration's (NASA) chief scientists for orbital debris Nicholas Johnson said: "This is by far the worst satellite fragmentation in the history of the space age, in the last 50 years".[346] Table 3 shows the estimated amount of space debris by size for all altitudes and for LEO, including an estimate for a break-up of a heavy satellite in LEO. It can be shown that the catastrophic disintegration of a 5 to 10 tonnes satellite can roughly double the amount of space debris larger than millimetre currently located in LEO.[347]

The international space community reacted to the growing problem of space debris by developing space debris mitigation guidelines. Introduced by the United States, which elaborated and released its own national guidelines in 1997, such voluntary measures will help to design missile stages or satellites to reduce the release of artificial fragments when placing objects in outer space.[348] The International Inter-Agency Space Debris Coordination Committee (IADC) adopted debris guideline in 2002, which formed the basis for voluntary guidelines of the United Nations Committee on the Peaceful Uses of Outer Space (UNCOPUOS).[349] It should be the goal that all space-faring nations legally adopt the UNCOPOUS guidelines. After the Chinese ASAT test it would be an important step forward if China were to fully abide by these guidelines. A discussion should be started regarding the establishment of a consultation mechanism or ad-hoc meetings to discuss large "space debris-producing events" within the scope of an IADC Subcommittee in the future.

The direct international reaction of governments, media and non-governmental organisations (NGOs) was highly critical of the Chinese space weapon test. The United States Government, but also other countries, such as Australia, Japan, South Korea, Taiwan and the European Union, responded with critical comments to China.[350] Australia's Foreign Affairs Minister, A. Downer, complained about the potential economic costs and the possible damage to other satellites due to the generation of space debris. The Japanese Government criticized that China should have given advanced notice and Prime Minister Abe said that the test would not be "in compliance with basic international rules, such as the Outer Space Treaty." Many Taiwanese believe that Beijing's test is directed primarily against Taiwan, but Taiwan's Deputy Defence Minister, Ko Cheng-en, speculated that the ASAT test is part of a warning to both, Japan and the United States, to avoid interfering in Taiwan-related security issues.[351] The response of another regional competitor, India, was relatively mild, but Indian analysts pointing out that the test opens a new vulnerability to India's space activities. In Russia, government officials, media and analysts blamed mainly the United States Government for having failed to negotiate an international agreement to ban space weapons.

Given the fact that the United States government knew of the ASAT testing series in advance, the White House only responded by issuing a formal protest. National Security Council spokesman Gordon Johndroe said: "The U.S. believes China's development and testing of such weapons is inconsistent with the spirit of cooperation that both countries aspire to in the civil space area."[352] Neither the White House put pressure on China to cancel one of the previous tests nor was China démarched for conducting the test. Vice President Cheney said that the "anti-satellite test, and China's continued fast-paced military build-up ... are not consistent with China's stated goal of a "peaceful rise". The cooperation between NASA and the Chinese space agencies, which started during the Chinese-U.S. summit in April 2006, was frozen. The United States Air Force (USAF) chief of staff, Gen. T. Michael Moseley said: "It makes space astronomically more dangerous it was before".[353] The Chairman of the Joint Chief of Staffs, General Peter Pace, didn't learn much more about China's intentions during his visit in China in March 2007. NGO representatives see China's move as a "signal of failed U.S. policy" (Center for Defense Information Director Theresa Hitchens) or a "big mistake", which could lead to a space arms race. (Michael O'Hanlon, the Brookings Institution).[354]

## 7.3. China's reluctant explanations

Despite the emerging international critique, the Chinese Government did not release an official statement until 23 January, twelve days after the event. Foreign Ministry spokesman Liu Jianchao said that "this test was not directed at any country and does not constitute a threat to any country".[355] On 1 February 2007, another Foreign Ministry spokesman insisted that China opposes the weaponisation of space and reaffirmed China's proposals to avert an arms race in outer space. Beijing's reluctance to clarify its motives and its delay in explaining the event raised concerns that the Chinese leadership might not have approved the test or might have underestimated the international response. There are also rumours saying that the test was not coordinated between the Ministry of Foreign Affairs (MFA) and the PLA. The uncoordinated statements showed a lack of a clear communication strategy and a careful preparation, but given the fact that China's President is the head of the political entities as well as Military Commander in Chief, it is very implausible that the political leadership did not know about the test. The Director of the World Security Institute's China Programme concluded that this indicates "that the military's action to develop space weapons during China's diplomatic offensive were a separate and perhaps independent hedging track rather than a deliberate design to develop space weapons".[356] Another analyst for Chinese space activities follows that: "It is not unlikely that compartmentalisation within Chinese institutions and bureaucracies, still rampant, played a part in the debacle as well, leaving the Foreign Ministry to twist in the wind when international protests began pouring in".[357]

It is also reported that on 12 February 2007, the Chinese Defence Minister Cao Gangchuan said in a meeting with a former Japanese defence chief that China has no plans to conduct another ASAT test.[358] On the other hand, articles in the Chinese media are quoting Chinese analysts who believe that the use of space for military activities is inevitable.[359] According to China's 2006 white paper on space activities the country is participating actively in the IADC by initiating a Space Debris Action Plan, increasing international exchange on space debris research. On the other hand, IADC member China has not yet formally adopted debris mitigation guidelines and it is believed that no national policies have been put into place as yet. Nonetheless, in April 2007, the Chinese delegation cancelled the 25th IADC meeting shortly before it was scheduled to begin. Obviously, there were concerns that the meeting would turn into a forum to criticize China for its ASAT test and the debris created. Since then, no further official explanations were released. It is obvious that an international as well as national discussion on the event is being kept silent by the government.

## 7.4. The international discussion and China's future space capability

The Chinese test triggered a discussion in the United States and worldwide on how to handle the emerging dangers of new weapons in space. Concerning

China's deeper motivations, there are two schools of thought. Some think that China wants to prompt the United States to engage in arms control negotiations for space, other argue that China is developing a wide range of counter space-capabilities such as kinetic kill vehicles or anti-jamming techniques to raise the costs for an U.S. intervention on behalf of Taiwan. This would be consistent with the Chinese doctrine of "asymmetric capabilities" to exploit U.S. vulnerabilities.[360] Such capabilities are an inexpensive response to the growing domination of space by the United States. Another motive might be the concern over the development of several ballistic missile defense systems by the United States, partially together with Japan. If Ballistic Missile Defence (BMD) ever works effectively, such a system could negate the Chinese limited deterrence arsenal by intercepting all of the some 20 ballistic missiles DF-5A with intercontinental range. The strategic BMD systems are also very dependent on the space-based sensor platforms such as Defense Support Program (DSP) or Space-Based Infrared System (SBIRS)-Low, which might be threatened soon. With the ASAT test, China demonstrated that it has a limited capability to destroy LEO satellites under restrained testing conditions. Some believe that China has been developing its ASAT capabilities for decades.[361] Further indications are alleged attempts by China to blind or "paint" U.S. satellites. In late 2006, National Reconnaissance Office (NRO) Director Donald M. Kerr said that China fired with High-Power Lasers at U.S. imaging satellites, thus exposing that China is developing a comprehensive ASAT capability by using kinetic and jamming technologies. The United States satellites were not physically damaged or disrupted. Details of these incidents are not known, but it is unclear if this incident was simply orbital satellite tracking or an attempt to test its dazzling capabilities.[362] Phil Meek, director of space law for the USAF General Counsel's office is quoted as having said that at least ten cases of foreign, mostly Electronic Warfare-type attacks against space assets occurred in the past.[363] Three annual Pentagon Reports on the PLA (2004, 2005, and 2006) concluded that the PLA could only destroy or disable satellites by a ballistic missile "armed with a nuclear weapon" Only the 2003 report said China was developing a direct-ascent ASAT system, which could be deployed between 2005 and 2010 [364].

The Chinese space policy goals are very ambitious and expansive.[365] The PRC has a civilian and military space programme, and is investing in manned space flight including missions to the Moon and Mars.[366] Its "Long March" launcher families (CZ) are derived from its military ballistic missile programme. China operates three launch sites and is building a brand new space port on the Hainan Island. The PRC is developing a wide range of satellite capabilities and is cooperating with other countries in areas such as navigation (Galileo/ESA),

**Tab. 4:** *Important achievements of the Chinese space programme.*

| Date | Activity |
|---|---|
| 1956 | The PRC launches it's space programme |
| 10 January 1970 | First Launch CZ-1 satellite launcher |
| 24 April 1970 | First successful launch of a Satellite (DFH-1) |
| 3 March 1971 | First research satellite in orbit |
| 26 November 1975 | First remote photo sensing satellite in orbit (FSW-1) |
| 8 April 1984 | First Geostationary satellite (DFH-21) in Orbit |
| 7 September 1988 | First experimental meteorology satellite (FY1A) in a polar orbit |
| 7 April 1990 | First commercial communication satellite Asiatsat-1) |
| April 1993 | Foundation of "Chinese National Space Administration (CNSA) |
| 10 June 1997 | First geostationary meteorology satellite (FY-2) |
| 20 November 1999 | First unmanned test of the spacecraft Shenzou 1 |
| 31 October 2000 | First navigation satellite (BNTS-1) |
| 15 October 2003 | First manned spaceflight of Yang Liwei with Shenzou 5 |
| 11 January 2007 | ASAT-Test |

communication (DFH/Germany), Remote Sensing (ZY/Brazil, RADARSAT / Canada). Beijing also stated that it plans to launch 100 civilian and military satellites in the next five to eight years. The country is also interested in the research and development of small satellites. Despite the fact that China's space activities are expanding rapidly, its space technology capabilities are far behind those of the United States or Russia. Militarily, China is working on "space enhancement capabilities", but does not have the knowledge, the resources or the political orientation to strive for the range of space activities the US already has.

The question is first, whether China is willing to deploy its ASAT System in the future. One precondition is the reliability of the kill vehicle and the missile. Neither the DF-21 or other missiles are very reliable systems. One may assume that it would require a larger stockpile of missiles to guarantee success in a confrontation.

Certainly, some intelligence analysts concluded that China might have an "operational ASAT capability" that could threaten U.S. LEO imaging satellites in the future. Pentagon officials are concluding that the Chinese military shows the ability to blind U.S. imaging satellites and hamper U.S. military operations if there would be a conflict over Taiwan. A report of a meeting at the "National Defense University" (NDU) concluded that deployed Chinese ASATs "could threaten a range of U.S. military capabilities that rely on space assets and might have significant consequences for a Taiwan contingency".[367] Some other experts argue that "the weapons system was used against a satellite that was much harder to hit than more strategically important satellites such as communications and early-warning satellites in geostationary orbits".[368] China has already showed the capability to launch satellites into geostationary orbit with powerful satellites, but such a geostationary orbits ASAT-technology has to be developed and tested too. The test also showed that the United States and Russia no longer have a monopoly over ASAT capabilities. Other nations such as India might follow or consider investing more in comparable missile defence programmes. A key problem here is that a ban of ground-based ASAT capabilities is harder to verify than weapons deployed in space.

After the ASAT test in January 2007 and the harsh public, but relatively mild reactions by governments, possible immediate and long-term technical as well as political consequences were not only discussed in the United States and in Europe, but concrete actions have been proposed. Conservative politicians such as Senator Kyl (R-Arizona) who called the test a "wake-up call" are demanding for an increased defence bill for "space control" and the implementation of a "space-based test bed".[369] Democrats seem to be less worried and are calling for a ban of future tests.[370] Some U.S. voices argue that "the test has also cast doubts on China's reliability as a global partner".[371] Given the political facts in Washington, there is the danger of an increased space-control spending in the United States, but also in other countries. There is a wide spectrum of unilateral, military and political answers. The NDU round table of space policy experts discussed costly and risky aggressive means such as a direct attack against Chinese ASAT facilities on the ground before their use or the introduction of space-based weapons to defend the United States space-based infrastructure. More moderate reactions are the substitution of damaged satellites, the use of foreign satellites or the introduction of "stealthy satellites" or a clustered constellation of satellites. Another way to prevent China and others from investing or deploying space weapons or capabilities against satellites is to pursue diplomatic solutions such as arms control talks or the development of an international code of conduct. An important assumption here is that a ban on kinetic ASAT weapons is feasible and in the interests of all countries.

## 7.5. The necessity for new approaches to ban ASAT-technologies

In 2007, the international community celebrates the $50^{th}$ birthday of Sputnik and the $40^{th}$ anniversary of the Outer Space Treaty (OST) in a climate where governments and the public are realising that the likelihood of the "weaponisation of space" is increasing drastically. The reason is the changing global world order, new published national Space Policies (esp. by the United States), the 11 January 2007, Chinese ASAT test and the increasing space competition. These developments are underlining the necessity to introduce new regulations to prevent the introduction of weapons into space.[372] The benefits for the increasing use of space applications such as communication, navigation, earth observation are obvious. Space-based services are crucial for the functioning of modern societies and the breakdown of critical infrastructures would have numerous societal consequences. An increasing number of countries are developing and operating satellites with military applications. Early warning, verification or monitoring from space platforms are more or less stabilising technologies. Today, existing satellites have "passive functions", i.e. they are not capable of directly eliminating adversarial satellites. To this day, however, no state has deployed active reliable "space weapons" permanently. The introduction of space weapons would not only mean the violation of an international taboo, but it could trigger a whole chain of new developments: new threat scenarios, costs for additional means of protection and the danger of an arms competition between space-faring nations in space.

The Chinese ASAT-test could serve as a wake-up call for the space-faring nations and all other members of the Outer Space Treaty and related regimes to start serious negotiations to prevent more countries from investing in destructive ASAT-technologies. The lessons from ASAT-test are obvious and are again highlighted by the Chinese ASAT-event: Testing is necessary for developing a reliable space weapon, but ASATs will create persistent space debris, which will significantly affect space traffic, especially in LEO. If weapons are deployed in space, huge numbers of such devices are necessary. They have to be operated, maintained and repaired. Accidents, failures and the potential use will threaten the space environment significantly and permanently. ASATs have little if any military value. They are complicated to operate, costly and ineffective to defend other satellites.

International law and multilateral institutions can play an essential role in prohibiting or restricting activities in space to ensure that outer space is used only for "peaceful purposes" and for the "benefits of all countries". If applied, political and legally binding regulations can foster predictability, transparency and trust in the space environment among space-faring states. The number of space-faring nations is still limited and therefore easier to handle than in the future where more and more nations might develop access to space and manoeuvre in space. The 1967 OST and related conventions, the UN Charter, the (not legally binding) UN space principles and other arms control treaties represent a good platform for improving space security in the 21$^{st}$ century. Given the current developments, this network of regulations needs constant review, improvement and expansion. For example, Article 4 of the OST prohibits the place nuclear weapons and other weapons of mass destruction in space. Unfortunately, the OST lacks some clarity in defining exactly where outer space begins or what "peaceful purposes" are. Some argue that all national military activities are permitted in space unless they are banned by other treaties. This "grey zone" of clarity should be filled by exact definitions, additional interpretations of current space law provisions and the introduction of verification procedures. The Chinese ASAT test, which was not "illegal" according to international space law in a strict sense, clearly challenges the spirit of cooperation in space matters for peaceful purposes between space-faring nations. Unfortunately, important international fora which could be used to review, expand or reform the existing regulations are being blocked by some nations.

Space security issues have been often debated within the UN General Assembly (UNGA) First Committee (Disarmament and International Security). For more than 20 years, the overwhelming majority of nations within the UNGA are holding the view that an arms race in space must be prevented. In 2005, at the 60$^{th}$ session of the UNGA, the United States voted for the first time against the "Prevention of an Arms Race in Outer Space" (PAROS) Resolution and repeated this stance in 2006.[373] The UNCOPUOS which was created in 1958 and which works by consensus is very active in studying legal, technical and commercial problems in the space environment, but was unable to reach the agreement to include military issues. An UNCOPUOS subcommittee, the IADC is very concerned about the increasing space debris problem. The "Conference on Disarmament" (CD) in Geneva has addressed space weapon issues repeatedly, but has been blocked since 1998. While China wants to start negotiations for PAROS, the U.S. links such negotiations with the start of talks on a Fissile Material Cut-off Treaty (FMCT), two fields which are not related technically but politically. Other attempts to introduce expert groups or subcommittees to elaborate best practices have failed. Only informal discussions under the sponsorship of United Nations Institute for Disarmament Research (UNIDIR) are taking place in Geneva without any concrete results. US representatives maintain the position that there is no arms race in space. The new U.S. National Space Policy released in 2006 clearly states: "The United States will oppose the development of new legal regimes or other restrictions that seek to prohibit or limit U.S. access to or use of space. Proposed arms control agreements or restrictions must not impair the rights of the United States to conduct research, development, testing, and operations or other activities in space for U.S. national interests".[374] The Chinese ASAT test clearly increases the pressure on the U.S. government to reconsider its position. President Putin stated at the Munich Conference on Security Policy on 10 February 2007: "In Russia's opinion, the militarisation of outer space could have unpredictable consequences for the international community, and provoke nothing less than the beginning of a nuclear era. And we have come forward more than once with initiatives designed to prevent the use of weapons in outer space".[375] Commentators from Asia are also warning about the beginning of an space competition especially in Asia: "But if Washington isn't careful, its future space activities could set a negative precedent, leading to an arms race in space between many different nations if Asia's race to the moon is any indication".[376]

Since decades, a number of proposal by states and NGO's have been made to restrict or prohibit ASAT activities in space. Other countries contributed by exploring verification procedures and the establishment of an international monitoring and space agency. In 2005 and 2006, the UNGA adopted a resolution first introduced by Russia with the title "Transparency and Confidence-building Measures in Outer Space Activities" inviting Member States to participate in proposing concrete measures on "Transparency and Confidence-building Measures" (TCBM).[377] For example, a multilateral binding "Code of Conduct" could prevent dangerous practices in space. Additionally, cooperative measures such as international space surveillance and orbital tracking can help to increase "space situational awareness". Internationally operated monitoring satellites and joint data exchange centres can help to improve space security. Unilateral declarations by states not to deploy space weapons in outer space could serve to significantly reinforce trust and confidence.

The European Union, which has become an important space actor due to its ambitious space plans and its cooperation with the United States, Russia, China, Canada, Brazil etc., should not cede the initiative in the field of future space security to other states. On 22 May 2007, ESA and the EU's Space Council adopted the joint "European Space Policy", a document which combines the EU's space activities and the "European Security and Defence Policy" (ESDP). The EU should become active independently in the range of space arms control. It is increasingly investing more, both economically and politically, into its space programme and also cooperates with Russia, China and India in the fields of launch technologies, Galileo and scientific space projects. It must be the task of its security and foreign policy to ensure that any potential future European military space programme precludes space weaponisation and that other space nations contribute to the weapon-free status of space. The same interest should be shown by the space-faring nations such as Russia, China, India, Japan and Brazil. The European Commission should establish relations with these states in this context to sign a joint declaration on renouncing space weapon deployment. Simultaneously, a move should be made to build confidence in space matters, and negotiations should be taken up to establish an adequate weapon ban regime.

---

[336] Space Security Index 2007. Waterloo/Ontario: Space Security. Org, 2007: 6.

[337] Covault, Craig. "Chinese Test Anti-Satellite Weapon". Aviation Week and Space Technology (17 Jan. 2007). www.aviationweek.com/aw/generic/story_channel.jsp?channel=space&id=news/CHI01177.xml.

[338] Fisher, Richard D. "Space to Manoeuvre". Jane's Intelligence Review (Mar. 2007): 62–63.

[339] Forden, Geoffrey. "A Preliminary Analysis of the Chinese ASAT Test". Massachusetts Institute of Technology, 2007. http://web.mit.edu/stgs/pdfs/A Preliminary Analysis of the ChineseASATTest handout.pdf.

[340] Gordon, Michael R. and Cloud, David S. "U.S. Knew of China's Test, but Kept Silence". New York Times, 23 Apr. 2007.

[341] Wright, David. "Space Debris". Physics Today (Oct. 2007): 35–40. http://www.ucsusa.org/global_security/space_weapons/debris-from-chinas-asat-test.html.

[342] Press Statement Institut für Luft- und Raumfahrtsysteme, TU Braunschweig. 16 Mar. 2007.

[343] Krag, Holger, et al. "Analysing the Risk increase in LEO due to Recent Major Fragmentation Events". Proceedings of the 2nd IAASS (International Association for the Advancement of Space Safety) Conference. 14–16 May 2007. Chicago, 2007.

[344] Taiwan launches it's E/O FormoSat-2 in May 2005, Japan has launched three surveillance satellites since 2003 and India is maintaining 3 E/O imaging satellites. It launched CARTOSAT-2 one day before China's ASAT test.

[345] Saunders, Phillip C. and Lutes, Charles D. "China's ASAT Test. Motivations and Implications". Institute for National Strategic Studies. National Defense University (INSS Special Report), June 2007: 2. http://www.ndu.edu/inss/Research/SRjun07.pdf.

[346] Morring, Frank. "Worst Ever. Chinese anti-satellite test boosted space debris population by 10% in an instant." Aviation Week and Space Technology (12 Feb. 2007): 20–21.

[347] Wright, David. "Space Debris". Physics Today (Oct. 2007): 39.

[348] For the space debris problem and space traffic management: Contant-Jorgenson, Corinne, Peter Lála, Kai-Uwe Schrogl, *eds*. "Cosmic Study on Space Traffic Management". Paris: International Academy of Astronautics (2006).

[349] UN Committee on the Peaceful Uses of Outer Space. http://www.unoosa.org/oosa/COPUOS/copuos.html.

[350] WMD Insights. "Chinese Anti-Satellite Weapon Test – The Shot heard Round the World". Issues and Viewpoints in the International Media. Special Report Mar 2007. www.wmdinsights.com/113.

[351] Ibid.

[352] Broad, J. William and Sanger, David E. "Flexing Muscle, China Destroys Satellite in Test." New York Times, 19 Jan. 2007.

[353] Ibid.

[354] O'Hanlon, Michael. "A Space Weapons Race is Not the Answer for America." Financial Times (22 Jan. 2007). http://www.brookings.edu/views/op-ed/ohanlon/20070122.htm.

[355] Kahn, Joseph. "China Confirms Anti-Satellite Test." New York Times, 23 Jan. 2007.

[356] Hagt, Eric. "China's ASAT Test: Strategic Response." China Security Winter (2007): 31–51.
[357] Johnson-Freese, Joan. China's Space Ambitions. Paris: IFRI, 2007: 20.
[358] Agence France Presse. "La Chine ne fera plus d'essai d'arme antisatellite". (12 Feb. 2007).
[359] Mulvenon, James. "Rogue Warriors? A Puzzled Look at the Chinese ASAT Test." China Leadership Monitor Winter (2007): 20. http://media.hoover.org/documents/clm20jm.pdf.
[360] Halpin, Edward, et al. eds. Cyberwar, Netwar and the Revolution in Military Affairs. Houndsmill/New York: Palgrave MacMillan, 2006: 173–195. This includes Information Warfare such as computer network attacks.
[361] Fischer, Richard D. quoting the former US attaché in Beijing, who believed that China's quest for ASAT began in the 1960s and was abandoned in 1980. ASAT Research was carried out since the 1980s ("863-program"). "Space to Manoeuvre." Jane's Intelligence Review (Mar. 2007): 62–63.
[362] Butt, Yousaf. "Satellite Laser Ranging in China." Union of Concerned Scientists Technical Working Paper (8 Jan. 2007). http://www.ucsusa.org/global_security/space_weapons/chinese-lasers-and-us-satellites.htm.
[363] Bruno, Michael, Fulghum, David A. and Covault, Craig. "Aftermath. Chinese ASAT test raises the prospects for space-control spending-eventually." Aviation Week & Space Technology (29 Jan. 2007): 28–30.
[364] Kan, Shirley. "China's Anti-Satellite Weapon Test". CRS Report for Congress (RS 22652). Washington D.C., (23 Apr. 2007): 3.
[365] Johnson-Freese, Joan. China's Space Ambitions. Paris: IFRI, 2007.
[366] Rathgeber, Wolfgang. China's Posture in Space. Implications for Europe. Vienna: ESPI. June 2007. http://www.espi.or.at/images/stories/dokumente/studies/espi_china_report_rev4-1_wf.pdf.
[367] Saunders, Phillip C. and Lutes, Charles D. "China's ASAT Test. Motivations and Implications". Institute for National Strategic Studies Special Report. Washington D.C.: National Defense Institute. June 2007: 2. http://www.ndu.edu/inss/Research/SRjun07.pdf.
[368] Forden, Geoffrey. "Evaluation of the Chinese ASAT Test". Janes Intelligence Review (Mar. 2007).
[369] The Honorable Kyl, Jon. "China's Anti-Satellite Weapons and American National Security". Heritage Lectures No. 990. Washington D.C., 1 Feb. 2007.
[370] Senator Markey Press Release. 18 Jan. 2007. http://Markey.house.gov/.
[371] Bates, Gill and Kleiber, Martin. "China's Space Odyssey: What the Antisatellite Test Reveals about Decision-Making in Beijing". Foreign Affairs (May/June 2007).
[372] The space environment is evolving rapidly in terms of actors, space applications and space traffic. Each year at least one new country accesses to space. Today ten nations have the capability of independent access to space and 28 states have the capability of "suborbital launches". Increasingly, private firms are offering commercial launch and orbital services. See Space Security Index 2007.
[373] United Nations. Prevention of an Arms Race in Outer Space. Report of the First Committee, UN-Document A/16/393. New York: Taylor, 2006.
[374] United States. The White House. US National Space Policy. 31 Aug. 2006 http://www.ostp.gov/html/US%20National%20Space%20Policy.pdf.
[375] Putin, Vladimir. Speech at the 43$^{rd}$ Munich Conference on Security Policy. 10 Feb. 2007. http://www.securityconference.de/konferenzen/rede.php?sprache=en&id=179.
[376] Shen, Dingli. "Asia's Space Race". Bulletin of the Atomic Scientists Online. 31 Oct. 2007. http://thebulletin.org/columns/dingli-shen/20071031.html.
[377] United Nations. UNGA Resolution A/RES/61/7. New York, 2006.

# 8. Basic law for space activities: A new space policy for Japan for the 21st century

Kazuto Suzuki

## 8.1. Introduction

On 20 June 2007, a bill was submitted to the Japanese's Lower House of Diet. The name of the bill is "Basic Law for Space Activities", which is a call to establish a new space policy and decision-making structure. The intention of the bill is to achieve a changeover from the old "catching-up" strategy, which focused heavily on technological development, to a more user-oriented space policy. The outcome of the Upper House elections in July 2007 and the subsequent resignation of the Prime Minister Abe makes it difficult to predict when this bill will pass the Diet, but apart from some minor opposition from the left-wing parties, there is a general consensus among wide spectrum of political activists that Japan needs a new space policy. This article discusses the reasons for creating a new legal framework for Japanese space policy, and how it would change the space activities of Japan.

## 8.2. Changes in political economy of Japan

Until the end of 1990s, there was a firm normative consensus among all actors – politicians, bureaucrats, space agencies, industry and academics – regarding Japanese space policy that it should focus on (a) "pacifist" concept based on the Diet Resolution stating "exclusively peaceful purposes" for space activities, which is interpreted as a "non-military" clause, i.e., no military involvement in the financing, development or operation of spacecraft; and (b) a "catching-up" strategy, concentrating on R&D activities to catch up with the advanced space-faring nations.[378]

However, these two normative frameworks were challenged by various political and economic factors in the late 1990s. First, severe financial constraints led to a dramatic decrease of the space budget. Nonetheless, the nominal space budget did (and still does) maintain a flat line at 180 billion yens (about 1.08 billion euros, see Figure 18). Therefore, the actual volume of spending in Japanese space programmes is on steep decline.

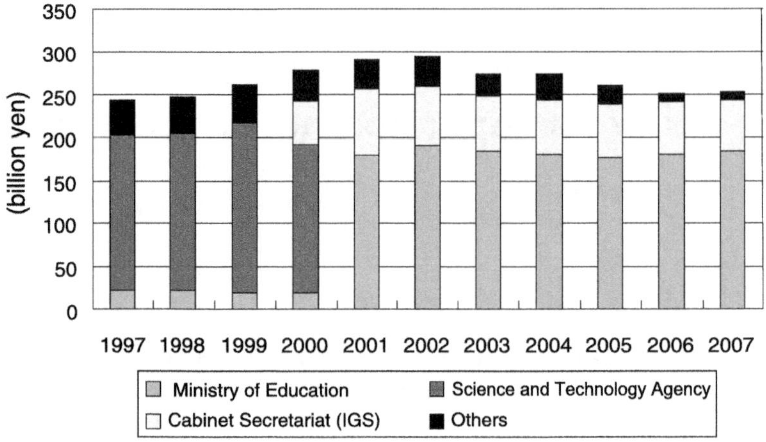


**Fig. 18:** *Space Budget of Japan (source: Society of Japan Aerospace Companies).*

This lack of funding was caused not only by budgetary constraints, but also by a lower priority and lesser political attention from the government. There were a lot of other topics with higher priority, such as the deployment of self-defence forces to Iraq, the reform of the pension system and the postal service reform. With three consecutive failures in 2003 of spacecraft (ADEOS-2, Planet-B, and H-IIA with two Information Gathering Satellite (IGS) satellites), the Japanese government began to question whether the funding for space development was being efficiently used for successful missions. Politicians had much higher expectations in the National Space Development Agency (NASDA)/Japan Aerospace Exploration Agency (JAXA), believing that the space programmes would be able to achieve greater success despite limited funding. Thus, the general trend of a shrinking budget did not change and is not expected to change for a while.

Furthermore, the administrative reform has added to the confusion regarding the Japanese space policy. The merger of the Science and Technology Agency (STA) and the Ministry of Education and Culture (MoE), the supervising ministries of NASDA and the Institute of Space and Astronautical Science (ISAS) respectively, increased the confusion of the policy-making process. The newly created Ministry for Education, Culture, Sports, Science and Technology (MEXT) has become the administrative centre for space policy, but because of the merger, the core positions for space policy are now occupied by people whose experience is mainly in the area of academic grants, exchange student affairs and elementary school education.

As regards the confusion in the area of space policy, the administrative reform has also influenced the positions at the highest level of decision-making. The Space Activities Committee (SAC), once the central inter-ministerial coordinating body that reported to the Prime Minister has became a much smaller organisation and is now only responsible for the matters concerning MEXT affairs (which covers most of the Japanese space activities) and has no inter-ministerial coordination tasks. Since SAC is now only a committee under the MEXT umbrella, any decisions by SAC are no longer final.

The final damage to space policy and implementation occurred when NASDA, ISAS and the National Aerospace Laboratory of Japan (NAL) were merged to create JAXA.[379] Many people were concerned about the differences in the organisational cultures, the long history of the institutional rivalry between ISAS and NASDA, and the difference in the administrative system (such as salaries, pensions etc.).

With these changes at the political, fiscal and administrative levels, the current Japanese space policy-making process can be described as a "paradigm lost". The larger political-economic cause has dictated the policy-making process, but no one is in a position of presenting a clear vision of what the strategic objective of Japanese space policy could be.

## 8.3. Industry's confusion

Due to the decrease in the space budget and the subsequent failures, Japanese industry has lost its confidence and motivation. The traditional contract arrangements through JAXA – a rotating prime contractors system and equal distribution of subcontracts – are no longer affordable or effective. Thus, many space companies in Japan have shrunk the size of operations and several of them have exited the market (Toshiba, one of the three major satellite manufacturers, sold its share of NT Space Corp to NEC and retreated from space business in 2007, see Figures 19 and 20).

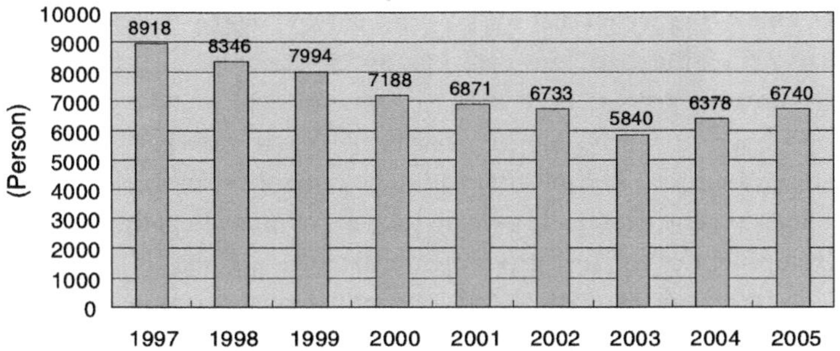


**Fig. 19:** *Number of Employees in the Space Industry of Japan (source: Society of Japan Aerospace Companies).*

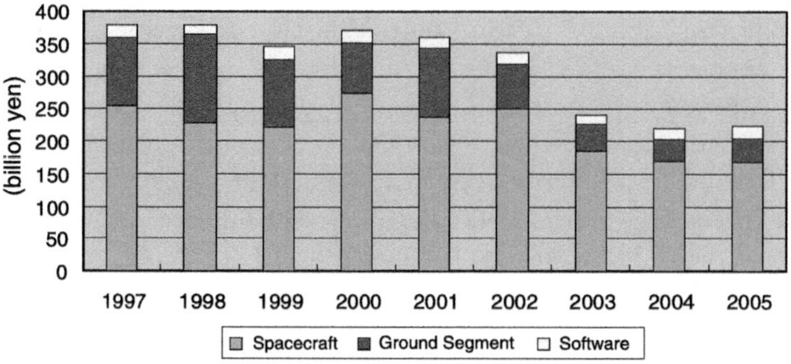


**Fig. 20:** *Turnover of the Space Industry (source: Society of Japan Aerospace Companies).*

The Japanese space industry is particularly vulnerable to the change in public spending on space because of the 1990 Accord. This accord is a by-product of the U.S.-Japan trade friction in the 1980s when the U.S. government suffered from excessive imports from Japan, and demanded the opening of Japanese public procurement. Under heavy political pressure, the Japanese government decided to sacrifice satellite procurement for non-Research and Development (R&D) purposes.[380] Since then, all application satellites except MTSAT-2 and Superbird 7 have been procured from U.S. companies. It is a widely shared view among politicians, industry and space agency that this accord is an obstacle to the strengthening of industrial competitiveness.

In this context, some Japanese companies took the initiative to launch new programmes driven by industry. The Quasi-Zenith Satellite System (QZSS) and Galaxy Express launcher (GX) programmes were launched by Mitsubishi Electric (Melco) and Ishikawajima-Harima Heavy Industry (IHI) respectively.[381] QZSS was initially designed as a Public-Private Partnership (PPP) project in which the private industry invests in the development, launch and operation of at least three satellites (the first one will be developed as an experimental satellite developed by JAXA), and it is expected that the government will be the "anchor tenant" to pay the fee to recover the costs. However, because of the confusion prevailing in policy-making, no ministry has been able to guarantee the user fee (also due to the shrinking budget) and no ministry has been able to take responsibility for this programme's success. The industry had to give up the business plan for depending on public users, and eventually gave up the programme altogether (currently, only the development of the first experimental satellite is going on).

Due to its experience with QZSS, the industrial sector is now desperate for a new policy-making structure, where the government can provide stable policy line and commitment to the PPP projects. It is also desperate for new funding for

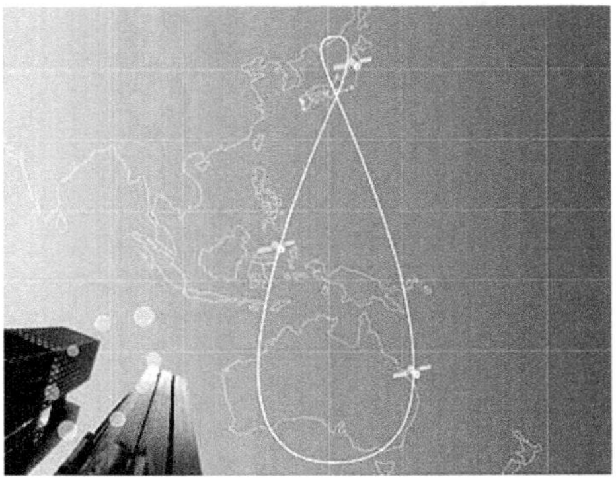

**Fig. 21:** *Orbital ground track of the Quasi-Zenith Satellite System (QZSS) (source JAXA).*

programmes which are not bound by the 1990 Accord – operational satellite programmes – for improving international competitiveness. The only possibility for the public procurement of operational satellites is through the defence programmes, which are rigidly restricted by the 1969 Resolution.

## 8.4. The Taepodong shock

While one of the normative frameworks of Japanese space policy, the "catching-up" strategy, is already facing a significant challenge, the other framework, the "pacifist" concept, is also being challenged by the changes in the international setting. The imminent threat of North Korea became visible when a Taepodong missile flew over Japan in 1998. It led to a big change in the policy paradigm. This incident put the Japanese public as well as the policy-making community in panic mode. There was strong demand to do something to prevent North Korea from launching missiles towards Japan and to protect the homeland. Thus, immediately after the Taepodong launch, the government made a decision to start a new satellite programme called IGS.

Nonetheless, the programme was seriously constrained by the legal interpretation of the 1969 Diet Resolution. Because of the "non-military" interpretation of the "exclusively peaceful purpose" Resolution, the military authority (then Japanese Defence Agency or JDA) was not allowed to participate in the development and operation of the IGS programme. Thus, although it was clear that the purpose of IGS was to monitor military activities for possible threats such as from North Korea, it was disguised as a "multi-purpose" satellite (note: it was even difficult to mention "dual-use" because this implied the possibility of the participation of JDA), which also serves civilian purposes.

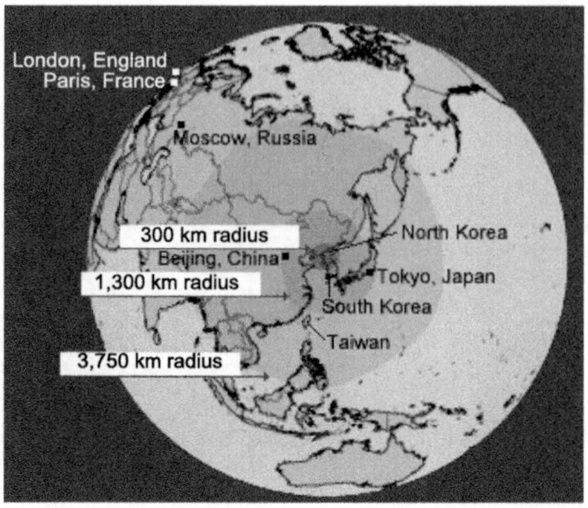


**Fig. 22:** *The Taepodong-2 has an estimated reach from 3750 km to 15 000 km (source Global Security).*

However, because of the 1990 Accord with the United States for satellite procurement, the IGS as a civilian non-R&D satellite should have been be placed under an open procurement procedure. This put the Japanese government in a serious dilemma. If the government wanted to develop the IGS as a multi-purpose satellite, the specifications of the satellite had to be open to public. The solution to this dilemma came from a careful legal interpretation. The government placed the control of the satellite not under the JDA, but rather, under the Cabinet Secretariat, a small office with national intelligence gathering mission and crisis management functions. The IGS was thus formally designed as a "crisis management satellite" with both civilian and military purposes.[382]

This incident created a wide-ranging understanding among politicians that the legal constraints of the "exclusively peaceful purpose" Resolution was too strict to provide room for manoeuvring, and that under the changing security environment in the post-Cold War period, it was nonsense to maintain such rigid pacifist rules.

## 8.5. Political action towards the changes

Although there has been increasing demand for changing the interpretation of the "exclusively peaceful purpose" Resolution, and mounting financial pressure for administrative change and reducing the space budget, no serious action has been taken by the government or politicians. The change was started in early 2005 by Takeo Kawamura, a Liberal Democratic Party (LDP) politician who just left from the seat of Minister of MEXT. During his period of office, he witnessed the failure of the sixth launch of H-IIA rocket carrying two IGS satellites. Although he was only responsible for the MEXT competence, which was the launch operation of H-IIA, the public as well as the government accused him for not properly supervising a nationally important satellite project. Because of the confusion in the area of policy-making, Kawamura thought that this was a critical failure in the implementation of the national strategy, and that something had to be done.

As soon as he stepped down from the office as Minister of MEXT, he formed a informal study group called "Consultation Group for National Strategy for Space" (also known as the Kawamura Consultation Group) composed of members of LDP working as Vice-Ministers in various ministries including MEXT, Ministry of Economy, Trade and Industry (METI), JDA and the Ministry of Foreign Affairs (MFA).

The report of the Kawamura Consultation Group proposed three new settings to establish a law to define new objectives for space activities and to set up an institutional framework to structure a more coherent space policy-making process.[383] First, the report proposed that the government should create a ministerial position with a portfolio for space affairs. This new ministerial post was to be the centre for strategic thinking and planning. Second, the report also called on the government to establish a new forum for space user ministries. This forum, under the chairmanship of the new Minister for Space, would be composed of ministers or vice-ministers from the space user ministries. This idea came from the experience of the failure of QZSS, where four ministries – Ministry of Transportation (MLIT), Telecommunications (MIC), METI and MEXT – were not able to make a compromise for sharing the financial burden of the programme. Third, this report suggested that the political community, including the members of the Consultation Group, should initiate a new interpretation of the "exclusively peaceful purpose" resolution. Because this resolution was taken by the legislative body, the Diet, it binds the action of the executive branch of the government. So, the decision to change the interpretation of the resolution has to come from the Diet members.

Furthermore, the report urged the government to seriously consider using space for achieving diplomatic objectives. Among the members of the Consultation Group, there was a strong concern about the development of the Chinese space programme. Of course, the members were impressed by the successful Chinese manned-space programme, but their concern was not about competition with respect to manned-space capability nor the space race for the Moon. Instead, their attention was devoted to the recent development of China' behaviour towards other Asian countries. In 2005, the Chinese government signed the founding agreement for Asia-Pacific Space Cooperation Organization (APSCO). It had already established the Asia-Pacific Multilateral Cooperation in Space Technology and Applications (AP-MCSTA) with 13 Member States. For many years, Japan was the leading country in this region, and JAXA and MEXT were proud to initiate the Asia-Pacific Regional Space Agency Forum (APR-SAF), which coordinates the space and promotes cooperation among the space agencies in this region. However, this organization focused only on the technical aspects of the space programmes of the various space agencies, and there was no coordination of strategy or policy. There was a wide dissatisfaction from the members that Japan was not supporting the needs of developing countries, specifically the transfer of technology and collaborative projects for space hardware, but JAXA was not able to appropriately respond to these demands.

In addition to the emerging Chinese role in the Asian region, the members of the Consultation Group paid close attention to the Chinese endeavour for using space as a diplomatic tool for bilateral relationships. Under the severe increase of the oil price, China offered its satellite technology to Nigeria and Venezuela, both oil-producing countries, for strengthening bilateral relations.[384] Since Japan is a huge oil-importing country, the Chinese activities in relations with these countries seemed to be threatening the secure supply of oil. The members of the Consultation Group wondered why Japan had not been able to do the same thing before China.

With these concerns in mind, the report of the Consultation Group was well received by the members of LDP and the government. Kawamura's initiative paved the way for Japan to transform its space policy-making process.

## 8.6. Basic law for space activities

Kawamura found that it would be appropriate to discuss his idea of reforming space policy in the LDP and consequently established the "Special Committee on Space Development (SCSD)" and became the leader of this Committee.

With the large number of Diet members in the Committee, SCSD attracted a lot of media attention and gradually the number of participants of the meeting increased. By placing the space matter on the LDP policy priority list, many Diet members began to realize the importance of space activities for national strategy, and through the media coverage the public also began to understand Kawamura's intention. In July 2006, the second North Korean missile launch campaign gave an extra boost to SCSD, because the public opinion shifted fast from safeguarding pacifist principles to a more flexible interpretation of "exclusively peaceful purpose" clause. It was in this environment that SCSD submitted the bill to pass the "Basic Law on Space Activities" in June 2007. This new bill has the following main features.

### 8.6.1. New institutional framework

The first point of this Basic Law is to set up a new Minister for Space and Space Development Strategy Headquarters (a forum of user ministries with strong authority). The Minister for Space would be a "specially designated" minister who will not be in charge of the management of the ministry, but will reside in the Cabinet Office for coordinating the policies of the different ministries. The Headquarters will include all the ministers and some appointed members from academia and industry. Although this is an ambitious challenge, there are hopes that these new institutions would provide a positive force for more political attention and dynamic in space activities. However, one concern remains about the fiscal competency of this new Minister for Space. The Basic Law avoided the issue of who will draft the budget proposal and whether the Headquarters would have the power to formulate a budget. Currently, the space budget is defined by the proposals from different ministries. The majority of the budget goes to MEXT/JAXA, one portion goes to the Cabinet Secretariat for IGS, and the rest is split into different ministries for the utilisation of space assets (see Figure 18). As long as this budgetary structure remains, it would be difficult for the Minister for Space and Headquarters to take initiatives for user-driven programmes, because the user ministries would be reluctant to spare their limited budget, and MEXT would refuse to reallocate its own budget. However, one member of SCSD suggested that the Headquarters should be the final decision-making body for the allocation of budget by bundling all budget requests from the various ministries and negotiate with the Ministry of Finance on behalf of those ministries. This would give a significant leverage to the Headquarters because any budget request would have to go through the Headquarters, and the ministries would lose their competence for defining a space programme without the consent of the Headquarters and the Minister for Space.

### 8.6.2. New interpretation of "Exclusively Peaceful Purpose"

The second point of the Basic Law is the question of security. As discussed above, the "exclusively peaceful purpose" Resolution was under pressure in the changing security environment for Japan in the post-Cold War period. Article 2 of the bill states that "Our space development shall observe the Outer Space Treaty and other international agreements, and shall be conducted in accordance with the principle of pacifism upheld in the Constitution". In other words, the traditional interpretation of "exclusively peaceful purposes" as "non-military" would no longer apply.

Instead, the policy would adopt the international standard interpretation of the "peaceful use" of space as the "non-aggressive" or "non-offensive" use of space. The new bill would accordingly enable the Japanese defence authority to become involved in the development, procurement, and operation of space systems.

In addition, Article 14 states that "the government shall take the necessary measures to promote space development that will contribute to international peace and security and also to our nation's security." Because this statement is so general, Article 14 could be interpreted as allowing the government to use space systems for aggressive purposes. But because Article 2 stipulates that the use of space systems for national and international security comply with both the framework of international agreements and a pacifist Constitution, it implies that Japan may use its space assets for crisis management and disaster monitoring in Asia and for peacekeeping missions outside its territory. Article 2 also suggests that Japan can use early-warning satellites for its missile defence, as this falls into the category of self-defence.

Thus, the Basic Law is designed to strengthen the Japanese role in the dispute settlement and crisis management with a peaceful means. It only tries to change the interpretation of the Diet resolution which prevented any use of space assets by military authority.

### 8.6.3. Industrial implications

The third point of the Basic Law is about the "industrialization" of the space industry. Since 1990, when the Accord with the United States for satellite procurement entered into force, the Japanese satellite industry lost its opportunity to improve international competitiveness through government programmes. During the period of commercialization in the late-1990s, the domestic industry was not able to enter the booming market (in retrospect, this might have limited the damage from the downturn of the market though). Nonetheless, a long history of concentrating on R&D and technological development had made Japanese industry entirely reliant on the government R&D funding, which is now decreasing due to the fiscal constraints. Furthermore, because of the nature of government-funded R&D projects, Japanese industry failed to concentrate on improving international competitiveness, i.e., improving reliability, reducing costs and meeting with delivery deadlines. If the industry continues to depend on government procurement, Japanese industrial and technological capability will inevitably face a cul-de-sac.

Thus, the Basic Law urged the government as well as the industry to shore up their effort of "industrialization", i.e., strengthening industrial capability and autonomous business competence. For the first time in space-related legislation in

Japan, the concept of "competitiveness" appears in Article 4. For a long time, there was no legal framework to encourage the Japanese space agency and industry to promote international competitiveness, which was due to the idea of "catching-up". Article 4 is therefore a manifesto for leaving the past far behind.

Moreover, Article 16 defines that "the government shall take into account the constant and regular procurement of items and services from private entities and use private entities' capability to advance private space business activities (including R&D). The government shall also take the necessary measures to improve launch sites, test and research equipment, other installations and facilities, technology transfer, the business incorporation of R&D results, and tax and financial measures for facilitating investment by private entities". This statement is designed to provide a legal foundation for the government to take the "anchor tenancy" approach, which is borrowed from the idea of PPP seen in Europe, particularly, the experience of Private Finance Initiative (PFI) scheme of Skynet 5, the British military communication satellite. Through "constant and regular procurement", the government would be able to make multi-annual plans for the procurement of services from the private sector. This scheme may be able to avoid infringement with the 1990 Accord since it is not a pure "public procurement". The procuring body is not the government agency, but a private enterprise, and the government only plays the role of principal customer. So the SCSD believed that if the Minister for Space takes the initiative and the Headquarters functions properly, the government should be able to offer stable "anchor tenancy" to the private enterprise, and the "industrialization" would be successful.

### 8.6.4. Promoting space science

Although the Basic Law focuses extensively on the user-side of space projects, it does not neglect the importance of R&D. However, its interest in R&D is not

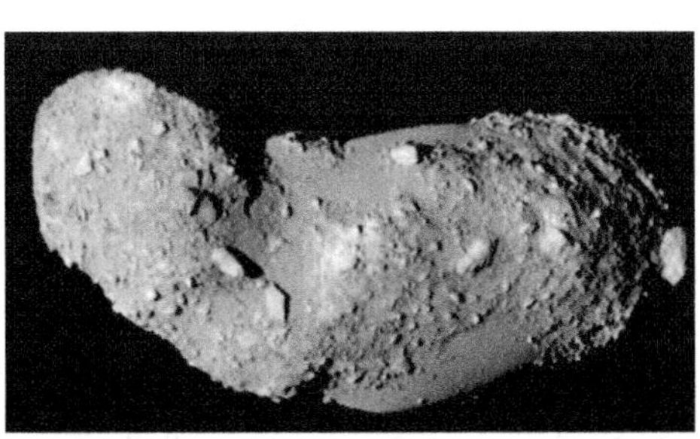

**Fig. 23:** *Hayabusa's picture of the asteroid Itokawa.*

based on the concept of technological "catching-up", but on originality and on national prestige. The experience of the asteroid mission "Hayabusa" was a good example of the new concept of R&D. For many years, the Japanese space R&D focused on reproducing the technology available elsewhere. But the Basic Law states that copying someone else's technology is not the way to improve industrial competitiveness or the strategic importance of space. Instead, it argues that developing original planning and technology, which would fascinate the public and the international space community, would serve Japan better.

## 8.7. Implications for the future of Japan's space policy and its role in the world

The Basic Law would inevitably change the practice of Japanese space policy. For a long time, NASDA/JAXA has been the central core of policy-making with a strong focus on R&D without much strategic implications. But the installation of a Minister for Space and Space Development Strategy Headquarters would become the main body to formulate long-term strategy and programmatic affairs.

It is predicted that JAXA's autonomy of policy-making would be largely undermined. The Minister of Space and the Headquarters will have the control of the entire space budget and programme planning. Thus, JAXA would become an executing strategic agency, while it would have certain autonomy for making decisions on basic R&D programmes. Furthermore, it is likely that the former ISAS agency would be separated from JAXA to grant more autonomy for space science. The new institution is designed to change the "bottom-up" approach into a "top-down" culture in Japanese space policy-making so that it would not be appropriate to conduct scientific programmes where scientists need academic freedom.

This change to JAXA's role would have further implications on the international relations of Japanese space policy. As it was discussed within the Kawamura Consultation Group, the purpose of the Basic Law is to strengthen Japanese diplomatic relationships through space cooperation. The members of the Kawamura Consultation Group felt that the international cooperation programmes led by JAXA were not contributing to the foreign policy of Japan. Thus, the Minister for Space shall take initiatives for promoting international cooperation in association with Japanese foreign policy objectives. Currently, it is believed that cooperation with Asian countries would be the main subject since the new Prime Minister Fukuda has a strong interest in expanding cooperation in this region. Also, Japanese traditional foreign policy objectives – cooperation with the United States – would be a high priority. Nonetheless, this does not mean that Japan would automatically choose to participate in the U.S.-led exploration programmes.

However, over the long run, cooperation with Europe would satisfy the strategic needs of Japan. There is a lot in common between Japanese and European space policies. First, the similar size of the budget and the levels of technology would complement each other. One good example is the alliance with Arianespace and Mitsubishi Heavy Industry (MHI) for launch services. Japanese industrial capability would be an ideal partner for Europe to pursue the "International Traffic in Arms Regulations (ITAR)-free" projects. Second, space cooperation would strengthen the political and strategic relationships between Japan and Europe. Given the increasing demand for participating in international peacekeeping operations, Japan shares the similar political and strategic objectives in international politics. If Japan and Europe increase opportunities for joint peacekeeping operations, space cooperation would make more sense. Furthermore, Japan and Europe share common objectives in the environmental domain where space cooperation and data exchange would make an important contribution. Third, the principle of the Basic Law – user-oriented space policy – would coincide with European space policy objectives. There is a lot to learn from each other for the use of the space system for Intelligent Transportation System (ITS) or for monitoring agriculture and land use. The Basic Law would thus provide these opportunities to strengthen cooperation between Japan and Europe.

## 8.8. Conclusion

The Basic Law is a legal document, but it has the power to define the course of Japanese space policy for the foreseeable future. In short, the Basic Law provides a new model for Japanese space policy that would meet the challenges of 21st Century. The Kawamura Consultation Group and LDP's SCSD were clearly aware of the domestic and international political-economic shifts, and the industry was happy about this initiative. However, because of the Upper House of Diet election in July 2007, the LDP is in a very difficult position to manage the parliamentary strategy and the variety of governance issues, such as budgetary priorities. The Democratic Party of Japan (DPJ), the largest party in the Upper House, is generally in favour of the Basic Law and the new space policy, but unpredictability is much higher than before. Nonetheless, the new Basic Law is considered the only way for revitalising the ailing Japanese space industry and maintaining Japanese leadership in Asia. Irrespective of the outcome of this political turmoil, what is certain is that Japanese space policy will no longer be what it once was.

[378] Kazuto, Suzuki. "Administrative Reforms and Policy Logics of Japanese Space Policy". Space Policy 22 (2005): 11–19.
[379] See Godai, Tomifumi and Masahiko Sato. "Reorganization of the Space Development Structure in Japan". Space Policy 19 (2003): 101–109.
[380] Sato, Masahiko, Toshio Kosuge, and Peter van Fenema. "Legal Implications on Satellite Procurement and Trade Issues between Japan and the United States". Proceedings of the Institute of International Space Law Conference, 1999.
[381] Suzuki, Kazuto. "Adopting the European Model: Japanese Experience in Implementing Public-Private-Partnership in Space Program". Proceedings of Council for European Studies Fifteenth Biennial International Conference. Chicago: The Drake, 31 Mar. 2006.
[382] Sunohara, Tsuyoshi. Tanjo Kokusan Spai Eisei (The Birth of National Spy Satellite), Tokyo: Nikkei Publishers, 2005.
[383] The report of the Consultation Group for National Strategy for Space: Towards a construction of new institutions for space development and utilization, Tokyo: Mimeo. Oct. 2005.
[384] "China Prepares To Export First Satellite". Space Daily. 3 July 2005. http://www.spacedaily.com/news/china-05zzzu.html; "China, Venezuela sign satellite launch agreement". China Daily 12 Nov. 2005. http://news.xinhuanet.com/english/2005-11/02/content_3718959.htm.

# 9. The IPCC report – In need of Earth observations

Jean-Louis Fellous

## 9.1. Introduction

The results of modern science of climate change are persuasive, in that they include a large body of observational evidence characterizing changes that have already occurred. This science has developed to the point where we can connect the observed changes with the causes that are responsible for them. The resulting picture demonstrates that human activities dominate in causing the recently observed climate change.

On 2 February 2007 in Paris, the Intergovernmental Panel on Climate Change (IPCC) released a Summary for Policymakers (SPM) of its Fourth Assessment Report (AR4). This SPM[385] largely reaffirmed, with in many cases a diminished uncertainty, the conclusions of previous IPCC assessments[386–389]. Among the main conclusions were the following:

- Warming of the climate system is *unequivocal*, based on several different kinds of observations (Figure 24).
- Most of the observed increase in globally averaged temperatures since the mid-20$^{th}$ century is *very likely*[390] due to the observed increase in human-caused greenhouse gas concentrations in the atmosphere.

Overall, this document represents a more extensive and more certain assessment of climate change science than ever before, and it greatly strengthens the scientific case for the reality and seriousness of human-induced climate change. However, the issue remains as to whether the current global observation system is adequate to monitor climate, and to detect (and attribute) climate change. This issue has long been debated, and important developments have occurred in the past year, which are presented in some detail in this article.[391]

## 9.2. The Intergovernmental Panel on Climate Change (IPCC)

The IPCC was established in 1988 to provide an authoritative assessment of results from climate science as an input for policymakers. IPCC reports are

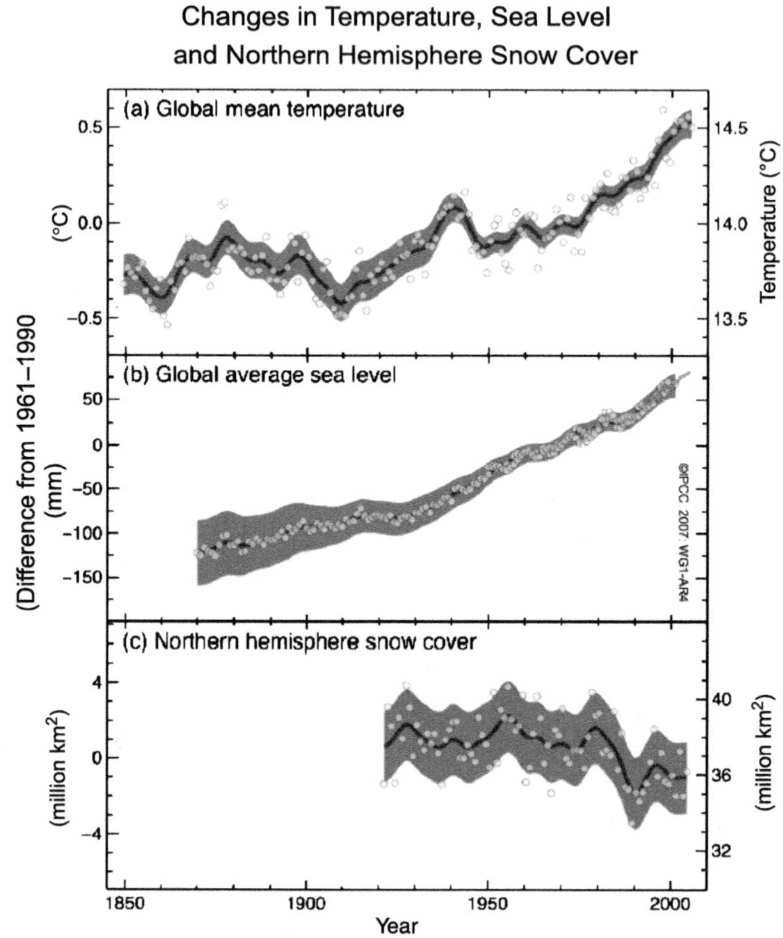


**Fig. 24:** *Observed changes in global mean temperature, sea level and Northern Hemisphere snow cover. Satellites represent a significant data source since the 1970s (source: IPCC AR4).*

very influential, both because of the quality of the scientists involved in writing them and of the rigorous editorial process which they undergo. In a series of reports beginning in 1990, culminating in the AR4, the IPCC has expressed increasing certainty that human activity contributes significantly to global climate change, and that this contribution will increase in the foreseeable future. A sense of the increasing levels of confidence and certainty in successive IPCC assessment reports can be gained by comparing the following headline statements:

- "The balance of evidence suggests a discernible human influence on global climate."–IPCC Second Assessment Report (1995).
- "There is new and stronger evidence that most of the warming observed over the last 50 years is attributable to human activities."–IPCC Third Assessment Report TAR (2001).
- "Warming of the climate system is unequivocal, as is now evident from observations of increases in global average air and ocean temperatures, widespread melting of snow and ice, and rising global average sea level."–IPCC AR4 (2007).
- "Most of the observed increase in globally averaged temperatures since the mid-20$^{th}$ century is **very likely** due to the observed increase in anthropogenic greenhouse gas concentrations."–IPCC AR4 (2007).

## 9.3. Deficiencies in IPCC assessments

While IPCC Assessment Reports represent the 'best available knowledge' at a given time, they are heavily dependant upon new research conducted and results achieved between reports. After the publication of the TAR considerable efforts were devoted to correct for gaps and deficiencies in preparation of the AR4. It is interesting to look at the recommendations that were the outcome of one of the many workshops[392] held in the period 2001–2006, devoted here to the study of *extreme events*, as one important indicator of climate change. Data availability (due to the format of archived data, and cost), declining numbers of observing stations, data quality (homogenisation), and short duration of datasets were amongst the major concerns expressed. As stated by the scientists attending this workshop "*It goes without saying that above all else, high quality data is the indispensable resource that is required to quantitatively assess changes in [temperature] extremes, and to assess whether human activity is changing the intensity, frequency and duration of extremes.*" Other serious concerns related to the decline of *in situ* networks, and the many technical (or political) obstacles to a free and open access to, and exchange of, long-term, high quality, high frequency data for temperature, precipitation, winds and waves, etc. It was further noted that important regional effects were poorly observed, particularly in the Southern Hemisphere.

Despite all efforts made, the lack of adequate data, resulting from the lack of an adequate observing system, still hampers some of the conclusions of the IPCC. Obviously a comprehensive and sustained global observing system for climate would improve understanding of the Earth's climate and enhance the ability to monitor and predict climate change. Earth observation satellites are a key element in this regard.

## 9.4. Observing climate and climate change

Sitting somewhere between applied research and an academic exercise of fluid dynamics, climatology became a major research field at the same time our perception grew that human influence on the planetary environment had reached an unacceptable level. The raised consciousness that mankind is altering the Earth's climate has changed climatology. Demand grows for a deeper understanding of climate drivers and outcomes. At what rate is climate changing? Is climate change accelerating? What are the respective extents of human-induced and 'natural' causes? How will the climate be like in fifty years from now? What are the long-term effects of human influence on the planetary environment? How can we best prepare for, and even adapt to, changes in local and regional climate? These questions do not come only from the science community, but are increasingly posed by policy-makers and society at large. How these questions are addressed can have major economic consequences for both the developed and developing world.

### 9.4.1. Climate modelling and data assimilation

In climate as in every science, measurements and observations are required to validate (or contradict) assumptions and theories. But the Earth's climate is not controlled as in a laboratory experiment. Natural variability and turbulence characterise the behaviour of the fluids that envelop the Earth, making it difficult to forecast their long-term evolution. Meteorologists use numerical models simulating three-dimensional processes to forecast the weather. Nowadays observational data are continuously assimilated into models, so as to minimize the differences between their predictions and what actually happens in the real world.

The requirements for an adequate climate observing system are far more stringent than those being used by the meteorological community. Climate analysis and prediction pose different problems, as measuring the small changes associated with the postulated climate change is far from being a simple task, especially as the time horizon for prediction is from decades to a century and more.

Building new measurement systems using modern technology also presents a daunting challenge when it comes to detecting surface temperature trends of one-tenth of a degree over a decade, or variations in the solar constant as small as one-thousandth over the same period of time, or a sea level rise of a few millimetres per year.

### 9.4.2. The climate quality challenge

Natural climate variability occurs on time scales ranging from years to hundreds of millennia. The human-caused (anthropogenic) climate perturbation is superimposed on this natural variability, or may simply modify it in ways that are still to be fully understood. Climate change signals can be detected only if they are greater than the background natural variability. Assessment of small climate changes through observations over long time periods implies the availability of time series that are both *accurate* and *stable*. Here *accuracy* is measured by the systematic error of the data with respect to the truth, once random errors have been averaged out. *Stability* refers to the long-term evolution of the measurement accuracy, say over a decade. Climate change is expected to be 'small' over global scales, while regional changes may be quite large, such as retreating glacier edges, desertification or coastal zones submersion. In fact one meter rise in sea level is not quite so small, nor is a 5 °C increase in the global mean surface temperature over the next 100 years (the difference between today's temperature and that of the 18,000-year-age when the Earth was in full glaciation). Nevertheless detecting global climate change is much more demanding than monitoring regional impacts. Moreover, detection alone is not enough: observations should be accurate and stable enough to prove beyond any doubt that climate change is occurring, and to allow the forcings and feedbacks involved to be fully evaluated.

This has many implications for the characteristics of an observing system able to deliver climate-quality data. Measurements should be taken with accurate, calibrated instruments, converted into geophysical data, quality-controlled and stored in standard format. Data sets should be precise enough for the early detection of trends over the next decade, homogeneous in location, time and method, uninterrupted and long enough to resolve decadal trends, and with sufficient coverage and resolution to permit a description of spatial and temporal patterns of change. Needless to say, this is not an easy set of requirements, and the answer to the question of the full adequacy of current climate observing systems to facing this challenge is clearly *negative*.

### 9.4.3. Advances in observational techniques

Most significant advances in technology for Earth sciences have affected the observing systems, notably the advent of satellites, but also a variety of new sensors able to provide accurate measurements in a hostile environment, such as long-lived profiling floats in the ocean. Advances in computing power, in turn, have allowed the development of more realistic models of the Earth system and its atmospheric, oceanic and terrestrial compartments. In addition, new mathematical techniques have allowed the timely use of data to improve the quality of forecasts. Among others, progress in electronics miniaturisation, antenna design, communication rates, microwave radar techniques, and stability of oscillators have made it possible to fly active sensors on board remote sensing satellites capable of acquiring all-weather, day and night observations and to transmit to the ground huge amounts of data with ever-increasing resolution in space and time. Progress in detector sensitivity has allowed atmospheric sounding of moisture and temperature profiles and the detection and monitoring of minor atmospheric constituents.

### 9.4.4. Satellite-based climate observations

Meteorologists benefit from an operational suite of highly optimized weather satellites, owned by the United States, Europe, Russia, Japan, China, and India, comprising two series of several platforms each, respectively circling the planet in geostationary and pole-to-pole ('polar') orbits. In parallel with their effort devoted to meteorological observations, space agencies initiated a number of experimental missions, with a view to extend the capabilities offered by remote sensing to ocean and terrestrial observation, as well as to atmospheric chemistry. For example, the European Space Agency's ERS-1 and 2, launched in 1991 and 1995, were pioneering satellites for Earth observation. Over the last twenty years, a wealth of new techniques has flourished.

An incomplete list of the climate variables accessible to space-based passive measurements (and sometimes only to those measurements) includes solar irradiance (total and spectral), Earth radiation budget components (net incoming solar radiation, outgoing long wave radiation, cloudiness), atmospheric temperature, water vapour, ozone, aerosols, precipitation, carbon dioxide, methane, vegetation and forest cover, snow cover, sea ice, sea surface temperature, ocean colour (related to phytoplankton surface concentration), lakes areas, glacier outlines, etc. Active sensors give access to ocean currents, sea and lakes level, ocean surface wind vectors, sea state, soil moisture, cloud top heights, precipitation rates, ice sheets topography, etc.

Of course satellite observations have limitations: electromagnetic waves do not penetrate far below the ocean or ground surface; also, some lower atmosphere parameters are practically out of reach from space observation, due to atmospheric opacity; satellite lifetimes are rarely longer than a few years; instruments degrade in the harsh orbital environment; calibration of sensors is often difficult and unstable. No space observation is self-sufficient, and satellite data must be complemented and validated by *in situ* measurements.

## 9.5. Challenges and opportunities in building a global climate observing system

### 9.5.1. The financial and geographic challenges

Since the Earth is a complex, integrated system, one must measure all climate-critical variables across all climate-relevant Earth subsystems. This is an enormous undertaking – one that fully exceeds not only the financial abilities of any individual country, but also their geographic access. And while one country might fly environmental satellites to routinely collect global surface and near-surface data from space, these data alone are not sufficient to fulfil all of the data needs. Surface and sub-surface measurements continue to require local (*in situ*) measurement in order to meet climate modelling requirements.

### 9.5.2. The compatibility challenge

There are extensive challenges arising from the varying, original purposes of specific observing systems that also supply data for climate applications. Differing requirements, priorities, and budgets drive the capabilities of *in situ* and remote sensing systems, none of which is fully dedicated to climate. Short-duration research projects; long-term regional and national weather prediction; wildland fire detection for fire fighting or biomass change detection; sea ice monitoring for safety of navigation or local weather prediction; these are among the many and varied purposes of today's diverse observing systems. More complexity results from the differing data collection, processing, and storage schemes for those observing systems and from their diverse dissemination mechanisms and paths. Similarly, differing data policies add to the challenge.

In order to gain financial and political support for the development of new observing systems, some countries are required to include elements of cost recovery into the business case for their system. In some other instances, system providers may include free (or limited cost) access to data – although perhaps on a delayed or lower resolution basis.

### 9.5.3. The knowledge challenge

Building on recent advances in satellite systems described above, promising new techniques are emerging, using new generation sensors (e.g., lasers, interferometers, imaging altimeters) or formation-flying satellites (e.g., the A-Train) and giving access to new parameters (e.g., sea surface salinity, river flow, ice mass balance). All these exciting new concepts have been developed by space research agencies, based on selected proposals received in response to announcements of opportunity. Even though most public research budgets devoted to Earth sciences are declining, in all countries there has been a steady flow of new ideas and new missions. Most of those prove highly successful, with performances much above expectations. Some can really be considered as 'one-shot' experiments, while many have the potential of becoming operational sources of climate data, once demonstrated in flight – which raises the delicate issue of transitioning observing systems from research to operational status.

### 9.5.4. The continuity challenge

Although recognition of this transition requirement is growing, building the migration path in the planning for current and future systems remains difficult. Advanced techniques rarely flow quickly or smoothly from research into operations. The funding needed to 'bridge' the results of a research programme from one organisation into the next-generation systems in an operational programme of another organisation is typically not within the budget of either institution. And in fact some of the more promising tools and techniques may prove entirely too expensive for long-term replication and operation. The challenge of bridging the gap between research and operations has been compared to "crossing the Valley of Death." Fortunately, organisations and governments are beginning to address this challenge.

## 9.6. An optimistic view

Clearly, a global observing system for climate will result from the combination of diverse systems, both space-based and *in situ*, used in conjunction with models, networks of data distribution, and ready access to the holdings of a wide array of environmental data archiving centres. Given this reality, one can readily see the value of establishing broad agreements for international cooperation and collaboration across the many facets of realizing a global climate observing system capability.

Today, despite all difficulties, international cooperation and collaboration in observing systems take place on many levels and across a number of application domains. One of the longest duration and most extensive of these endeavours is the World Weather Watch (WWW) under the World Meteorological Organization (WMO). The WWW combines observing systems (Figure 25), telecommunication facilities, and data processing and forecasting centres – operated by WMO members – to make available meteorological and related geophysical information needed to provide efficient services in all countries. While the WMO does not itself own or operate these capabilities, member countries have long recognized the mutual benefits possible from working together throughout the design, deploy-

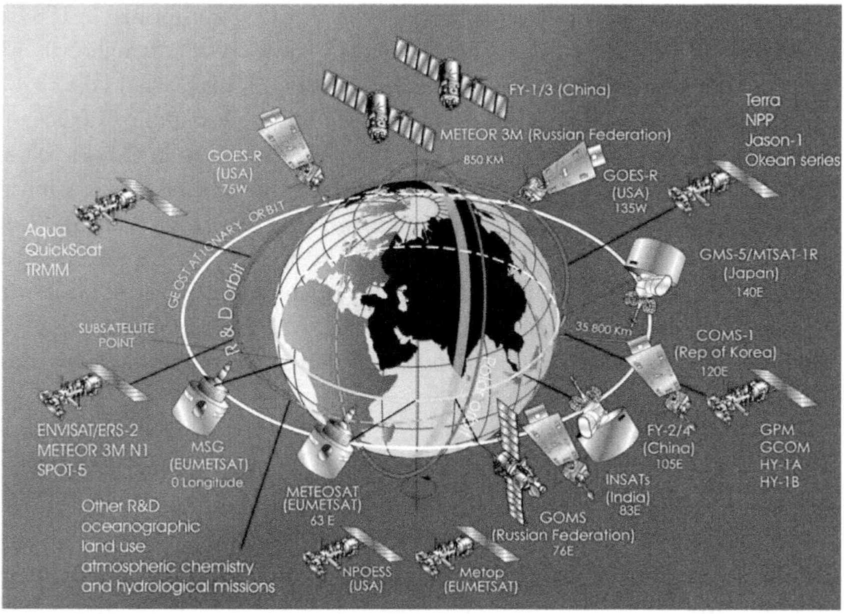


**Fig. 25:** *The space-based component of the World Weather Watch in 2006 (source WMO).*

ment, operation and utilization of data from this network of observing systems. In many ways the WWW serves as a role model for the results that are sought in the area of climate observations.

### 9.6.1. From GCOS...

The GCOS (Global Climate Observing System) programme was established in 1992 to ensure that the observations and information needed to address climate-related issues are obtained and made available to potential users. While the scientific community has been strongly supportive of the GCOS activity in principle, progress remains frustratingly slow. Like WWW, GCOS itself neither makes observations nor generates data products. In 2004, GCOS published its Second Adequacy Report,[393] where the gaps and deficiencies that affect the present climate observing system were analysed, and an Implementation Plan[394] providing a ten-year strategy for their correction was further released.

### 9.6.2. ...and CEOS...

The Committee on Earth Observation Satellites (CEOS) was formed in 1984 as an international mechanism to facilitate the coordination of international civil space-based missions designed to observe and study the planet Earth. In the 1990s, CEOS member agencies and associated organisations, including GCOS and others, recognized that in order to increase the effective utilization of space-based Earth observation data, a more thorough understanding was needed of major user requirements and how those requirements might be better supported by Earth observation data. Through a series of prototype projects the concept emerged of establishing a broader collaboration among organizations involved with the planning and coordination of observing systems. And most importantly, it was recognized that for useful coordination to take place, users of these systems must also take part in the planning activities.

### 9.6.3. ...through IGOS ...

This led to the formation in 1998 of the Integrated Global Observing Strategy (IGOS) Partnership, which brought together activities that coordinate and promote the development of major Earth and space-based systems for global environmental observations of the atmosphere, oceans and land. The IGOS effort is widely recognized for its successful development of comprehensive, community-consensus 'theme' reports, which identify the Earth observation requirements of selected, major application domains (e.g., ocean observations, geo-hazards, atmospheric chemistry, water cycle, carbon cycle) and present clear and compelling recommendations for action across sensor capabilities, system operational features, model enhancements, and data dissemination strategies.

Yet in spite of such laudable efforts and the growing recognition of the vital role of Earth systems monitoring, much work remains to realize a comprehensive and sustainable climate observing system. In recent years, several high-level events have for the first time brought political level attention to the need for progress in Earth observing systems. At the 2002 World Summit on Sustainable Development (WSSD) at Johannesburg, South Africa, governments formally recognised the urgent need for coordinated observations relating to the state of the Earth. Later, at a meeting of the Group of 8 Industrialized Countries (G8) Summit in June 2003 in Evian, France, the Heads of State affirmed the importance of Earth observation as a priority activity.

### 9.6.4. ... to GEO and GEOSS

In July 2003, the first Earth Observation Summit was convened in Washington, D.C., USA. There, governments adopted a declaration signifying *a political commitment* to move towards the development of a comprehensive, coordinated, and sustained Earth observation system of systems. The Summit established an *ad hoc* intergovernmental Group on Earth Observations (GEO), tasked with the development of an initial 10-Year Implementation Plan. In 2005, a standing GEO Secretariat was established with an action plan to realise a Global Earth Observation System of Systems (GEOSS).

The vision for GEOSS is grand – to realize a future in which decisions and actions for the benefit of humankind are informed by coordinated, comprehensive and sustained Earth observations and information. GEOSS aspires to involve all countries of the world, and to cover *in situ* observations as well as airborne and space-based observations. The established Earth observation systems provide essential building blocks for GEOSS. They all retain their existing mandates and governance arrangements, supplemented by their involvement in GEOSS. Probably the most significant promise is that through GEOSS, participating systems will share observations and products with the system as a whole and take necessary steps to ensure that the shared observations and products are accessible, comparable, and understandable. In the climate domain, important goals for GEOSS include ensuring the sustained provision of data relevant to climate studies, promoting the enhancement of climate observing systems (especially in the terrestrial and ocean domains), improving access to quality-assured climate data, and assisting international coordination of climate observations. Rather than attempting to develop yet another plan for realizing a global climate observing system, GEOSS recognizes and supports the GCOS Implementation Plan.

## 9.7. A less optimistic view

GEO provides the long-awaited political framework through which the demand to establish and maintain Earth observing systems comes from high levels within governments. However, without a firm commitment and concerted action there is still a risk that existing systems will degrade in the coming years. Much has been achieved by the current systems, but the failure to take the opportunity afforded by GEOSS to rectify identified observation system deficiencies will mean that the potential to obtain substantial added value from the global observational network will be lost for the foreseeable future. In certain important aspects (e.g., in surface climate, upper atmosphere, hydrological observations) the already insufficient observational capacity is likely to continue the decline that has been evident for several decades unless a decisive intervention is made. This is also true for satellites, which require long development times and involve substantial funding. Examples of forthcoming crucial

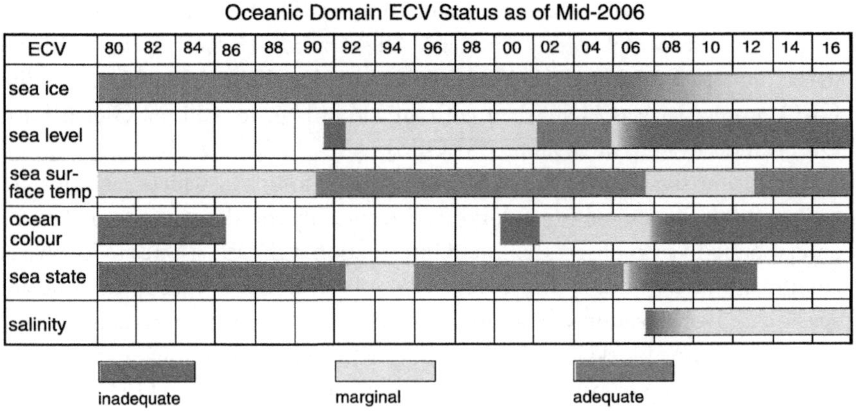


**Fig. 26:** *Status of space observations of ocean Essential Climate Variables. After a "golden age" in the early 2000s, this bar chart shows a risk of quick degradation beyond 2007, when most current ocean satellite missions come to end of life.*

data gaps or interruptions in the so-called 'Essential Climate Variables' are unfortunately numerous (Figure 26). Practically all ongoing experimental missions, however promising their results, remain without any plan for continuity.

The U.S. National Polar-orbiting Operational Environment Satellite System (NPOESS) was de-scoped last year due to cost overruns. There was a severe reduction of the number of NPOESS satellites from six to four, and the "de-manifesting" of all five NPOESS climate science instruments: the Ozone Mapping and Profiler Suite (OMPS); the Total Solar Irradiance Sensor (TSIS); the Earth Radiation Budget Sensor (ERBS); the Radar Altimeter (ALT); the Aerosol Polarimeter Sensor (APS). The Conical Scanning Microwave Imager (CMIS) was also substantially reconfigured. In early 2007, the U.S. National Research Council Decadal Survey[395] formulated recommendations to restore these capabilities.

## 9.8. Some encouraging prospects

Still, some positive signs should be noted, in the context of GEOSS. Prompted by a request of the United Nations Framework Convention on Climate Change (UNFCCC) Conference of the Parties and as part of its contribution to GEOSS climate needs, CEOS developed a plan in 2006 to respond to GCOS requirements for space-based observations[396,397]. The CEOS response is a comprehensive collection of 59 actions aimed at fulfilling climate observing needs in the atmosphere, ocean and terrestrial domains, as well as a number of cross-cutting requirements. Moreover, in order to harmonise efforts between space agencies to deploy Earth observation missions and with the aim to close emerging data gaps, CEOS has established the concept of Virtual Constellations for GEO, whereby a number of satellites or instruments and their observations, when coordinated in their operation and exploitation, have the potential for integration/merging of data and derived information to contribute to an enhanced (quantitative) analysis/measurement goal. Four CEOS prototype Constellations are currently considered, all significantly contributing to the global climate observing system, with focus on Precipitation, Land-Surface Imaging, Ocean Surface Topography, and Atmospheric Composition.

As part of its recurring process of reviewing its Global Observing System (the space-based WWW), the WMO has recently set 2025 as a new horizon, with widened requirements to address the needs for satellite observations for climate. Expert teams have identified priority thematic issues (passive radiometric sounding, radio-occultation sounding, ocean altimetry, Earth radiation budget, atmospheric chemistry and surface wind vectors) that require resolution.

In Europe, the Global Monitoring for Environment and Security (GMES) initiative launched in 1998 by the European Commission (EC) and European Space Agency (ESA) has moved from concept to reality, with a series of major funding initiatives since 2001. After an EC Communication affirming the intent to devote significant funding to the establishment and maintenance of an infrastructure for GMES as part of its contribution to GEOSS, the Ministerial Council of ESA held in Berlin, Germany in December 2005 took the decision to build a series of preoperational satellite missions, called Sentinels. Initial funding by ESA Member States will cover development costs, while the EC should support the continuation of the series. If confirmed the Sentinels will ensure the continuing acquisition of many useful climate data sets. However, a critical element in the success of GMES is the establishment as of 2008 of operational GMES Services currently based on data from satellites over their nominal lifetime, while the first Sentinels are planned to be launched in 2011–2012 at the earliest.

The U.S. has also initiated a coordinated, national effort to achieve a U.S. Integrated Earth Observation System (IEOS). The IEOS plan recognizes that climate observations need to be taken in ways that satisfy the climate monitoring principles (established by GCOS and formally endorsed by the UNFCCC and CEOS), ensuring long-term continuity and supporting the ability to detect small but 'persistent' signals. Various options to recover the lost NPOESS climate capabilities are being studied at present by NASA and NOAA. Deliberations are ongoing and should be resolved on time for the preparation of their Fiscal Year 2009 budgets. Discussions are focusing on a possible prioritisation of the de-manifested instruments, and then on the identification of flight opportunities, including how international collaboration might assist. A first positive outcome of this process has been the announcement in April 2007 that the OMPS Limb instrument, an ozone sensor, was reinstated on the NPP (NPOESS Preparatory Project) platform, to be launched in 2009, directly addressing one of the recommendations of the National Research Council (NRC) Decadal Survey.

In the past 20 to 30 years, the United States and Europe have provided the largest part of the climate observing systems. But many new contributions to the climate observing networks and satellites are now becoming available, including Earth observation satellites from Japan, China, India, Brazil, Argentina, South Africa, Thailand, South Korea, and others. This raises new issues, including that of data distribution policies and sensor inter-calibration, and these are now being addressed.

The U.S. and European efforts are promising, and the newcomers' contribution is quite encouraging. A global climate observing system seems at hand, but the opportunity has to be seized now. The European GMES and the international GEOSS endeavours have to be success stories.

385 IPCC Working Group I. IPCC Fourth Assessment Report. The Physical Science Basis – Summary for Policymakers (SPM). Cambridge: Cambridge University Press, 2007. http://ipcc-wg1.ucar.edu/wg1/wg1-report.html.

386 The most recent IPCC reports are available online at http://www.ipcc.ch and are published in print form by Cambridge University Press (see reference list).

387 Jenkins, Geoffrey J. et al., eds. IPCC First Assessment Report: Scientific Assessment of Climate change. Cambridge: Cambridge University Press, 1990: 365.

388 Houghton, John T, et al., eds. "IPCC Second Assessment Report: The Science of Climate Change. Contribution of Working Group I to the Second Assessment of the Intergovernmental Panel on Climate Change". Cambridge: Cambridge University Press, 1995: 572.

389 Houghton, John T, et al. eds. "IPCC Third Assessment Report: The Scientific Basis". Contribution of Working Group I to the Third Assessment Report of the Intergovernmental Panel on Climate Change (IPCC)". Cambridge: Cambridge University Press, 2001: 944.

390 "*Very likely*" in the IPCC context means an assessed likelihood of at least 90%, while "*likely*" means at least 66%.

391 The author acknowledges that this paper reuses some content of two chapters in a multi-authored book edited by Prof. Catherine Gautier and him, to be published under the title "*Facing climate change together*" in 2008 by Cambridge University Press. A French version of this same book, under his and Prof. Gautier's direction, has been published in August 2007 by Odile Jacob Sciences under the title "*Comprendre le changement climatique*". The author is particularly indebted to Richard Somerville and Jean Jouzel, for their permission to use elements from their chapter on "The global consensus and the IPCC", and to Helen Wood, whose chapter (co-authored with him) on the "Observing system for climate" has also provided quite relevant resources.

392 IPCC Workshop on Changes in Extreme Weather and Climate Events report. Beijing, China, 11–13 June 2002. Geneva IPCC. http://.isse.ucar.edu/extremevalues/ipcc.pdf.

393 "GCOS Second Adequacy Report: The second report on the adequacy of the global observing systems for climate in support of the UNFCCC. GCOS-82". WMO/TD n°1143. 2003: 85. http://www.wmo.ch/web/gcos/gcoshome.html.

394 GCOS Implementation Plan: Implementation plan for the global observing system for climate in support of the UNFCCC. GCOS-92. WMO/TD n°1219. 2004: 153. http://www.wmo.ch/web/gcos/gcoshome.html.

395 National Research Council. NRC Decadal Survey: Earth Science Applications from Space: National imperatives for the next decade and beyond. Washington D.C.: NRC 2006: 400 http://www.nap.edu/execsumm_pdf/11820.pdf.

396 GCOS Satellite Supplement. Systematic observation requirements for satellite-based products for climate. GCOS-107. WMO/TD n°1338. 2006: 103 http://www.wmo.ch/web/gcos/gcoshome.html.

397 CEOS Response to GCOS Requirements: Satellite observation of the climate system. 2006: 54 http://www.ceos.org/pages/pub.html.

# 10. Space entrepreneurship – Status & prospects

Joerg Kreisel & Burton H. Lee

## 10.1. Introduction

Space entrepreneurship is a frequent topic of discussion today within the European and United States space communities. To some observers, space and entrepreneurial activities may not seem complementary at first glance, particularly when considered in the context of a sector which seems to prefer costly large-scale projects under government leadership or major multinational companies building highly complex space systems. Traditional space industry typically views the entrepreneurial space sector with a combination of hope and interest in new approaches to technology development and risk management, as well as with misunderstanding, suspicion and unmet expectations. While some space startup firms have experienced market and financial success, outside of the satellite industry, most to date have not brought major changes to the space industry and the broader economy as expected.

Over the decades, however, a broad and robust spectrum of commercial space activities have evolved that account for a substantial portion of today's space business. As in the Information Technology sector, space entrepreneurs have

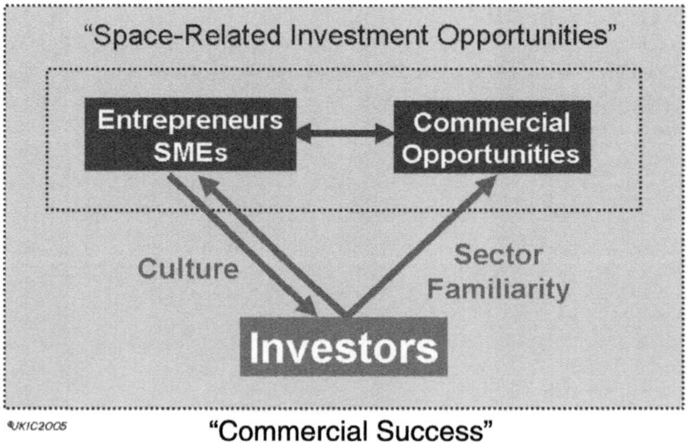


**Fig. 27:** *"Bridging the Gap" between Space Ventures & Investors.*

demonstrated their ability to contribute important technical innovations, capital and cost efficiencies, new business models, an ability to effectively serve niche markets, and faster times-to-market than traditional aerospace corporations. Entrepreneurial space, however, does not exist in a separate universe, but remains closely linked to broader market developments and the availability of seed- and early-stage finance within the wider innovation environment and policy context. Seed- and early-stage space ventures in both regions face significant financing problems today due to the continuing mismatch between investor expectations and space startup business models and risk profiles.

The authors review international developments in space entrepreneurship and venture finance that are influencing the conduct and scope of traditional space business. Practitioners' insights are employed to clarify fundamental concepts and illustrate important issues and trends in the U.S. and Europe. Finally, policy recommendations aimed at encouraging sound and sustainable space entrepreneurship are provided. The authors in particular note that today's space policy community does not integrate well with science and technology policy organizations that focus on enhancing national, regional or sector-specific innovation pipelines, or with the investment community. Despite the potential positive benefits associated with developing a vibrant space entrepreneurial community, policy makers in Europe and elsewhere continue to pay insufficient attention to this important innovation sector.

## 10.2. "Commercial opportunities" in space?

Space business has evolved from its early non-commercial market foundations, dominated by government procurement, into a major commercial industry based on a mix of industrial and consumer markets. Driven largely by developments in space telecommunications, commercial space business today accounts for more than half of the space industry's consolidated global revenues. Given the complex market structures and financing mechanisms involved, however, it remains difficult to clearly distinguish between the major types of commercial space opportunities.

The global space sector has been traditionally organized along technology and programmatic lines which, from a financial and commercialization perspective, is no longer an appropriate approach. Commercial space activities should instead be assessed in terms of the following generic characteristics and differentiators:

- Business nature and model
- Risks

- Market characteristics
- Timeframe
- Financing needs, and
- System and technology

The nature of a business is very important to be understood, because within the same category (see below) business may be executed in very different ways (example: lease of satellite capacity vs. operation and ownership of satellites). Risks include both business and technical uncertainties. The underlying financial structure and investment mechanisms of a business are also an important differentiator.

*Classification of commercial space opportunities and markets*

Applying the objective criteria described above leads to the following proposed alternative classification of space-related business which we distinguish primarily by 1) nature of business, and 2) the level of total risk:

- **"CORE" (Space):** space business as traditionally perceived and performed. This includes manufacturing of space hardware, space transportation services and the operation of foundational space infrastructure elements and associated science or R&D. Customers here are predominantly government and major industrial organizations, but also encompass a growing number of private entities, e.g. satellite operators or privately operated space activities as promoted in recent years.
- **(Space) "TOURISM":** includes the various facets of next generation space tourism, e.g. sub-orbital and orbital spaceflights, which are currently being promoted. Customers here are primarily high-end consumers.
- **"UTILIZATION":** covers the use of space-borne assets and their data transmission capabilities for terrestrial applications. This includes satellite downstream value-added services (DVAS), which combine and increasingly integrate generic satellite applications – telecommunications, navigation and Earth Observation – around multiple payloads and integrated terrestrial systems, into e.g. location-based services. Customers in these markets comprise a mix of government and industrial organizations, together with mass-market consumer end users.
- **"DIFFUSION":** refers to traditional "technology transfer" (*from* space) by which technologies and processes developed in the course of space activities are transferred to – or diffused into – non-space industries (also referred to as "Spin-Offs" from a space-centric perspective). Spin-offs take place via either licensing or new business creation. A number of space spin-offs have been realized in recent decades and are often used for political lobbying efforts, and for justifying

government space agency missions and budgets. In most cases, companies found in this category are small and medium sized enterprises (SMEs). Customers include a wide range of industries, from large to small around the world.

- **"INFUSION":** is the reverse form of technology transfer (*to* space), by which mature technologies and processes developed and applied in other industries are transferred to – or infused into – space systems and ventures (also referred to as "Spin-Ins" from a space-centric perspective). Spin-ins are a relatively recent development which is driven primarily by efficiency and marketplace competitive advantage issues; these typically occur via corporate licensing or acquisition activities. Infusion customers are primarily prime contractors or

| COMMERCIAL SPACE OPPORTNITIES & MARKETS | | | |
|---|---|---|---|
| **Category** | **Areas Covered** (Examples) | **Risk Level** | **Total Annual Market Size** (# End Customers, 2010-2015) |
| **CORE** (B2B, B2G) | o **Manufacturing** (Launchers, Satellites, Smallsats, Probes, Ground Segment, etc.)<br>o **Launch Service Providers** (All Types of Space Transportation)<br>o **Satellite Platform Operations** (Upstream Segment)<br>o **OOS** (On-Orbit Servicing: A&R Tasks, Life Extension, Assembly, etc.)<br>o **Space Stations** (Traditional)<br>o **Exploration & Exploitation** (e.g. Resources) | **High** | **10's** |
| **TOURISM** (B2C) | o **Orbital** (Space Flights & Walks, Hotels, etc.)<br>o **Sub-Orbital** (Space Flights) | **High** | **100's – 1.000's** |
| **UTILIZATION** (B2C, B2B, B2G) | o **Satellite Applications** (Downstream Segment)<br>▪ Telecommunications<br>▪ Navigation/Location-Based Services<br>▪ Earth Observation/Remote Sensing<br>▪ Combinations (convergence) | **Medium** | **1.000.000's– 100.000.000's** |
| **DIFFUSION** (B2B, B2C) | o **Spin-Offs** (Technology Transfer) :<br>▪ Medical Equipment<br>▪ New materials<br>▪ Sensors<br>▪ Processes<br>▪ Etc. (basically all sectors) | **Low-Medium** | **1.000's – 10.000's** |
| **INFUSION** (B2B, B2G) | o **Spin-Ins** (reverse Technology Transfer):<br>▪ Selected key technologies & materials<br>▪ Selected processes & manufacturing<br>▪ Platforms | **Low-Medium** | **1.000's – 10.000's** |
| **EMOTION** (B2C, B2B) | o **Consumer Merchandizing**<br>o **Mass Media**<br>o **Terrestrial Space-themed Entertainment** (Training, Site Visits, resorts, etc.)<br>o **Events** (space-themed)<br>o **Fashion**<br>o **Education & Outreach** | **Low** | **1.000.000's – 100.000.000's** |

*JKIC & B.H. Lee 2007 - Not considered: "Supply Chain" and business opportunities in specialized support service sectors such as e.g. legal, finance and insurance.*
*(B2C: business-to-(end)customer; B2B: business-to-business; B2G: business-to-government)*

**Fig. 28:** *Commercial Space Opportunities & Markets: Definition & Overview.*

public entities, but in some cases also include SMEs found operating in global supply chains.

- "**EMOTION**": comprises all businesses that rely on the image and flair of space – its emotional power and perception – with the exception of "Tourism" as described above. Customers here are primary mass-market end consumers, and may also include media and travel agencies.
- "**SUPPLY CHAIN**": covers a broad range of systems, subsystems and components as well as services mostly linked with "Core", "Tourism" and "Utilization" businesses. This class is NOT investigated here in any detail!

It is useful to also consider each of these industry categories according to their respective level of risk. Risk levels assigned represent the generic risk involved – technical, market, financial, or execution – due to the nature of activity of each category. Figure 28 provides a breakout of the major commercial space opportunities and markets, along with their relative levels of risk and market size in terms of number of end customers.

Due to the complexity of space business, the above classification is not precise in all respects. This is partly due to the substantially different perceptions of risk, investment opportunities and availability of finance in Europe and the United States. Divergent cultural attitudes regarding technical innovation and adoption, and new business creation, also play an important role here. As a result, the level of entrepreneurial activity in space, and the associated supply of finance – especially small and large-scale early-stage equity – is today substantially higher in the United States. Figure 29 provides a qualitative comparison of Europe vs the United States in each of the above commercial space sectors.

| | SPACE BUSINESS LANDSCAPE | | | | | | | |
|---|---|---|---|---|---|---|---|---|
| | EUROPE | | | | UNITED STATES | | | |
| **Category** | Perceived **Risk** | Business **Potential** | Supply of **Finance** | Entrepr. **Activity** | Perceived **Risk** | Business **Potential** | Supply of **Finance** | Entrepr. **Activity** |
| **CORE** | ●●● | ● | - | ● | ●● | ●● | ●●● | ●● |
| **TOURISM** | ●●● | - | - | ● | ●● | ●● | ●● | ●● |
| **UTILIZATION** | ●● | ●● | ● | ●●● | ●● | ●●● | ●● | ●●● |
| **DIFFUSION** | ●● | ●● | - | ●● | ● | ●● | ● | ●●● |
| **INFUSION** | ● | - | - | - | - | ● | ● | ● |
| **EMOTION** | ●● | ● | ● | ● | ● | ● | ● | ● |

*JKIC & B.H. Lee 2007 - (Qualitative Level of Relative Strength: ● Low, ●● Medium, ●●● High)*

**Fig. 29:** *Space Business Landscape: Europe vs. United States.*

It is worth noting that some businesses in the space sector are very entrepreneurial in terms of culture and execution but do directly fit the above definitions.

## 10.3. Types of space entrepreneurs

It is a challenge to define "space entrepreneurs" in a manner that properly reflects the various types of space-related businesses in Europe and the United States. The above classification of major industry segments, however, permits us to group entrepreneurs into the following general categories:

- "**Super Angels**". These entrepreneurs are extremely high net worth individuals (EHNWI's) who possess the means to self-fund a space startup company independent of the international financial markets. Examples of these individuals include Jeff Bezos (Blue Origin), Robert Bigelow (Bigelow Aerospace), Richard Branson (Virgin Galactic), John Carmack (Armadillo Aerospace), and Elon Musk (SpaceX). Tom Pickens (SpaceHab), may also be considered a potential new entrant into the Super Angel "club". With the exception of SpaceHab, each of these companies initiated operations within the past four years, with a focus on developing a new generation of commercial space transportation and habitation systems aimed at space tourism and other consumer, enterprise and government markets.
- "**Core+Tourism**" **entrepreneurs.** In contrast to Super Angels, this class of space entrepreneur does not fall within the EHNWI classification and thus must rely on direct access to global seed- and early-stage financial markets to fund their companies. These firms are also typically focused on high risk "Core" and space tourism technologies and markets closely associated with commercially-based human presence and foundational infrastructure in space, but due to the scope of their envisaged business, may lack the sustained financial resources needed to weather cashflow problems caused by unforeseen technical problems, delays in market development, or investor jitters. These space entrepreneurs usually possess a substantial space background and include George French (RocketplaneKistler), Jeff Greason (XCOR Aerospace), Rick Tumlinson (Orbital Outfitters), David Gump (t/Space) and Jim Benson (Benson Space Company).[398]
- "**UtilizationPlus**" **entrepreneurs.** This category of entrepreneur also relies on external financing to fund his businesses, but is instead focused on moderate-to-low risk space technologies and application markets. As a result, their financing

needs are typically lower than those of "Core" and "Tourism" companies. This class of space entrepreneur can come from a variety of industry backgrounds, including information technology and space. Prominent European examples include Manfred Krischke and Wolfgang Biedermann (RapidEye), and Peter Grognard (Septentrio). This category of entrepreneur is less well-known within the broader space community, but represents an important if neglected group of companies, and thus deserves special attention. Most of these businesses are not "space" startups, but rather mainstream "high-tech" organizations that are transferring space-related innovations to terrestrial markets.

- **Institutional entrepreneurs**. In contrast to the above groups which stress the role of individual entrepreneurs in narrowly-focused space industry market segments, institutional entrepreneurs tend to work across all six industry categories. We can identify within this community both key individuals and the business units to which they belong, as follows:

    - **(Space) "Intrapreneurs"**. This category covers individuals, who as part of corporate incubator or in-house business unit developments pursue entrepreneurial activities in the space sector. European examples are Jörg Hermann (Infoterra), Helmut Luttmann (ISS Industrial Operator Team) and Philippe Richard (ONERA Commercialization).
    - "**Enterprise Incubators**" develop and fund space startup companies internally, or perhaps with government support, and then spin them out as independent ventures. Examples here are SpaceTech, Eighth Continent, and the Houston Technology Center (HTC) in the US. Potential future players in the US enterprise incubator arena are the many contract R&D firms and non-profit research institutes that develop new space-related technologies with Small Business Innovation Research (SBIR) program or other research funding from NASA, DARPA, the Air Force and other federal space and science and technology organizations.[399] Examples here include Orbitec, LaunchPoint, SRI International and The Aerospace Corporation.

## 10.4. Space entrepreneurs and opportunity fit

Linking the categories of commercial opportunities to the types of space entrepreneurs leads to the high-level relationship shown below (see Figure 30).

This perspective does not take into account space entrepreneurs who have adopted a strategy aimed at exploiting "Core" and "Tourism" market opportu-

| | PLAYGROUNDS FOR ENTREPRENEURS | | | |
|---|---|---|---|---|
| | SPACE ENTREPRENEURS (Categories) | | | |
| OPPORTUNITIES | Super Angels | Core+Tourism | UtilizationPlus | Institutional |
| Core | ●●● | ●●● | | ● |
| Tourism | ●●● | ●●● | | ● |
| Utilization | ● | ● | ●●● | ● |
| Diffusion | | | ●●● | ● |
| Infusion | | | ● | ● |
| Emotion | | | ● | ● |

*JKIC & B.H. Lee 2007 - (Qualitative Level of Relative Strength: ● Low, ●● Medium, ● ● ● High)*

**Fig. 30:** *Opportunities for Space Entrepreneurs.*

nities at the sub-system and component-level. A few individuals such as Jim Benson, SpaceDev founder, have been quite successful as suppliers, be it to traditional government programs or to newer space tourism ventures. The industry classification presented above, however, focuses only on OEM-level activities (or system level) to identify major commercial space sector categories.

The following diagram underlines the complex landscape of space entrepreneurship and even moreso how the above-described characteristics of various commercial opportunities impact industry structure (see Figure 31).

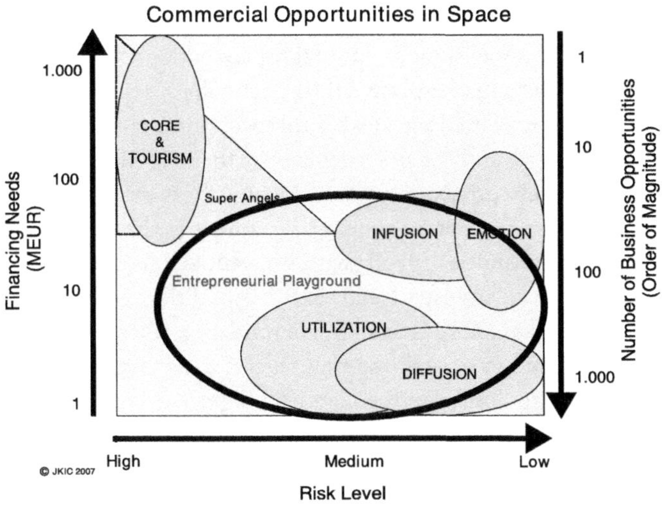


**Fig. 31:** *Commercial Opportunities in Space & Entrepreneurs.*

## 10.5. Relevant developments in European and U.S. commercial space

The 2006–2007 period witnessed numerous developments that have had a positive impact on space entrepreneurship in Europe and the United States. Selected trends, events and other activities of note are summarized below.

### 10.5.1. Industry trends

Although the number of entrepreneurial and space-related ventures in Europe has shown some growth over the past two years, observers believe that European space entrepreneurs will continue to face difficult times and relatively poor success rates in the long-term. Reasons range from their lack of business acumen and the absence of dedicated financing to sub-optimal support by the relevant organizations. Recent surveys of the state of space entrepreneurship by JKIC together with selected members of European Private Equity & Venture Capital Association (EVCA) have revealed that while most European space startup companies do not meet venture capital funding requirements, angel financing remains a viable option, even as it continues to be relatively underdeveloped in Europe in comparison to the United States.

Space startup deal flow in Europe, however, is increasing and the quality of investment opportunities appears to be slowly improving. The June 2007 announcement by EADS of its intent to pro-actively develop a commercial space tourism program creates new momentum for space entrepreneurs to also enter the emerging space tourism marketplace. OHB Technology's takeover of the German space company Kayser-Threde creates a third space prime contractor of substantial size – besides EADS and Thales-Alenia – with a significantly different and stronger corporate entrepreneurial culture. Finally, the recent decision to continue the struggling Galileo navigation system program paves the way for multiple new entrepreneurial opportunities, albeit, with substantial delays.

Seed- and early-stage industry trends in the U.S. differ markedly from those in Europe. Continuing market leadership, consolidation, diversification and innovation can be observed in several important areas. Zero Gravity Corporation, for example, continues to dominate the parabolic flight entertainment sector, while Virgin Galactic appears to be solidifying its lead over suborbital space tourism competitors. Although important progress can be observed in this emerging market, suborbital operators continue to face uncertainties in vehicle development, flight test, FAA safety certification and state-level liability protection.

Within the orbital space transportation arena, in contrast, first-round Commercial Orbital Transportation Systems (NASA COTS) winners SpaceX and RocketplaneKistler (RpK) received early attention, but after RpK failed to reach critical financing milestones in late 2007, NASA has re-opened the field to new competitors. Planetspace, a U.S.-Canadian "upstart", appears to have achieved a measure of financial and technical viability in recent months, while SpaceDev can be expected to stake out a strong position based on continued progress with its propulsion and other vehicle technologies, along with its strong management and finance team. Transformational Space Corporation (t/Space), Constellation Services (CSI) and other first-round COTS proposers have also expended considerable effort to seek out new government customers, particularly for reusable launch vehicle (RLV) technologies that can support military spacelift and operationally responsive space (ORS) concepts. Other entrepreneurs are diversifying within the NASA marketplace as suppliers to aerospace primes on major Vision for Space Exploration contracts.

Within the consumer entertainment marketplace we note two startups of particular interest – Xtreme Space and SpeedUp – that have proposed innovative technology concepts aimed at expanding space-oriented extreme sports into "space diving" and rocket-powered motorcycles.

### 10.5.2. Startup companies

A surprisingly wide variety of non-traditional and entrepreneurial space businesses can be found in Europe. Surveys of these "EuroSpaneurs"[400] (reveal a high level of activity in the area of satellite applications downstream ICT ("Utilization") and in technology transfer ("Diffusion"), with approximately 100–200 new businesses being created in recent years according to surveys on the European Space Incubator Network ESINET. Within the traditional ("Core") space sectors, in contrast, few startup companies can be identified, with the bulk of these being found in satellite manufacturing. Entrepreneurial startups in the "Tourism" category, for example, do not really exist in Europe today, with the exception of recently formed Galactic Suite. The following selected cases give a feel for the dimension, scope and potential of space entrepreneurship in Europe: SSTL ("Core"), RapidEye ("Core" & "Utilization"), OSSL ("Core"), Virgin Galactic ("Tourism"), MARSS ("Utilization"), iOpener ("Utilization"), EuroHeatPipes ("Diffusion"), Ondas Media SA ("Utilization").

American space entrepreneurs, in contrast, have demonstrated substantially greater ability to enter "Core" markets due to the greater presence of Super Angels and directed incentive prizes. During 2006 and 2007 several startup firms continued their progress in achieving a number of critical technical and business milestones: Armadillo Aerospace ("Core & Tourism"), Bigelow Aerospace ("Core & Tourism"), Blue Origin ("Core & Tourism"), Scaled Composites ("Core" and "Tourism"), SpaceX ("Core"). Most of these startups remain on schedule for initial commercial flights in the 2009–2011 timeframe. The great majority of American space startups, however, can be considered seed- or early-stage ventures that are active in the "Tourism", "Utilization", "Diffusion", "Infusion" and "Supply Chain" sectors. The bulk of their equity funding is generally raised through personal networks within the US angel investor community, and may in some cases be supplemented by federal SBIR grant funds.

### 10.5.3. Entrepreneurial space finance

The European seed- and early-stage "space" finance sector remains relatively under-developed in comparison to the U.S. The U.S. entrepreneurial space finance sector, in contrast, has witnessed important advances across the startup funding pipeline during the past two years.

### 10.5.4. Angel finance

Sustained weakness in angel finance for seed-stage European space entrepreneurs reflects a broader situation across the continent. With the exception of the United Kingdom, large organized angel investor networks rarely exist within Europe; "Super Angels" are likewise not present in the space startup sector, with Richard Branson being the sole exception. In contrast, during 2006–2007 the U.S. witnessed serious efforts to better organize, educate and connect the U.S. angel investor community with seed- and early-stage commercial space entrepreneurs. Two developments here are noteworthy:

- **Space Angels Network**[401] is spearheading progress here with an online platform where approved entrepreneurs seeking equity funding in the 50 000 U.S. dollars to 5 million U.S. dollars range can post their business plans for review by accredited angel investors. Space Angels Network is expected to have a significant impact on improving the quality and sustainability of space-related startup companies, and in reducing today's bottleneck in seed-stage funding.
- **Boston Harbor Angels** agreed in June 2007 to a significant investment in XCOR Aerospace. This is the first entry of mainstream U.S. angel investor networks into early-stage commercial space finance.

### 10.5.5. Venture capital

European VCs have not looked favorably upon space-related ventures since the early 1990s for a variety of reasons; some selected funds, however, do consider deals today on an occasional basis. Worthwhile mentioning is, that ESA entered new territory with a VC fund project intended to support new businesses focused on "Diffusion" and "Utilization". In Summer 2006, ESA selected the UK-based VC firm E-Synergy to raise and manage the "European Special Applications Fund" (target 40 million euros) sponsored by ESA with 5 million euros.

In the United States, Northrop Grumman's acquisition of Scaled Composites in June 2007 attracted some notice from VCs who have frequently pointed to the absence of successful exits as an ongoing barrier to investment. Nevertheless, today there exist few space-focused venture funds, as we describe below:

- **Red Planet Capital** (RPC) was created by NASA in 2006 as a 75 million U.S. dollars analogue fund to the CIA's In-Q-Tel. RPC's core "Infusion"-oriented mission was to identify startups with key private sector technologies which could support NASA's exploration mission. NASA's subsequent pullback from Red Planet led to its reincorporation as Astrolabe Ventures with a broader investment focus which continues to include aerospace. Astrolabe is currently raising two private funds in the U.S. and Europe with a combined size of 400 million euros, to invest from 250 000 U.S. dollars to 5 million U.S. dollars per company over multiple rounds of financing.
- **SpaceTech** was formed by Tom Pickens in mid-2007 as a venture fund and incubator division within SpaceHab as part of the company's goal of diversifying its business base.
- **SpaceVest**, which pioneered space venture capital during the 1990s, changed its name to Redshift Ventures and has, in effect, retired as a space-focused VC firm.

### 10.5.6. State equity funds and infrastructure investment

States are now entering as major players in aerospace venture and infrastructure finance. Two initiatives are worth noting here:

- **Space Florida**, the new state economic development agency dedicated to keeping and attracting space startups and major aerospace corporations, has received authority from the state legislature to support both equity and debt financing in support of small and large aerospace companies. Florida has announced that it plans to raise a 150 million U.S. dollars state-private sector equity fund aimed at aerospace, alternative energy and technology startups.

- **New Mexico** has committed state and county funds exceeding 125 million U.S. dollars in support of the construction of Spaceport America which will form the cornerstone of Virgin Galactic's U.S. operations. The New Mexico Spaceport Authority is authorized to issue revenue bonds in support of its effort to attract new tenants to the spaceport. The State Investment Council announced in 2006 that it would invest up to 20 million U.S. dollars in equity in a qualifying company receiving a NASA COTS first-round award, and that agreed to locate its primary operations in New Mexico.

### 10.5.7. Directed incentive prizes

Space prizes have emerged in the U.S. as an important and growing instrument for supporting innovation-focused incentive finance. Following the successful Ansari X-Prize flight in 2004, NASA and the X-Prize Foundation (XPF) have embraced funded prizes as a new tool for incentivizing and financing global entrepreneurial space companies in key technology domains. NASA's efforts in this area are part of a broader policy initiative by Congress and federal agencies to employ prizes as a new financial instrument aimed at accelerating entrepreneurial innovation in strategic areas of U.S. national interest. Major developments here include the following:

- **Lunar Lander Challenge (LLC)** is a 2.5 million U.S. dollars prize funded by NASA and administered by the XPF. The first and second rounds were held in October 2006/2007 at the X-Prize Cup in New Mexico. A third competition round is planned for 2008.
- **Other NASA-sponsored Prizes**. Other major prizes competed by NASA and partner foundations during this timeframe include the 2006 Tether Challenge (200 000 U.S. dollars, no winners), the 2006 Beam Power Challenge (200 000 U.S. dollars, no winners), the 2007 Astronaut Glove Challenge (200 000 U.S. dollars, one winner) and the 2007 Regolith Excavation Challenge (250 000 U.S. dollars, no winners). A new round of prizes has been announced for 2008.
- **Google Lunar X-Prize.** The XPF and Google announced in September 2007 the Google Lunar X-Prize, a robotic race to the moon with prizes valued at 30 million U.S. dollars.

### 10.5.8. Incubators and other activities

ESA's Technology Transfer Office runs three ESA-in-house incubators at major ESA centers and the European Space Incubators network (ESINET, approximately 50 space-related incubators throughout Europe) managed by the European Business & Innovation Centre Network (EBN, Brussels). Other ESA-led activities include technology brokerage and in-house grant-type activities like the ARTES program to support commercial applications in space telecommunications, the Earth Observation Market Development program (EOMD) and the Innovation Triangle Initiative to identify new ideas for products, processes or services. These latter programs are of benefit in special cases but do not especially address space entrepreneurs.

The European Commission (EC) launched three space-related activities under its EuropeInnova Initiative, which are focused at linking investors with space-related business ventures (INVESAT: "Utilization", FINACESPACE: "Diffusion" and CASTLE: related networks). ESA is part of all those projects as ESA and the EC work closer together these days.

In the United States, the Colorado School of Mines announced in 2007 the formation of its "Eighth Continent Project", which will include a space startup incubator presently under development.

### 10.5.9. Important events

Several events dedicated to space entrepreneurship, venture finance and space tourism generated growing momentum for innovative financing and technical approaches to space commerce. In April 2007 ESA hosted the 1$^{st}$ ESA Investment Forum, which was the first event of its kind in Europe ; the Forum was considered a success since investors were impressed by the quality of the 22 companies and followed up with some of them. A second very important and high-quality milestone conference was the Space Venture Finance Symposium (Dallas, Texas; May 2007) organized by Innovarium Ventures and the National Space Society. This event covered the complete space venture equity finance pipeline with discussions by leading experts from around the globe, and resulted in financing deals for two of the 17 startups presented. Additional significant events included the following:

- Space Tourism Conference (London, UK; June 2006).
- X-Prize Cup and the International Symposium on Personal Spaceflight (Las Cruces, New Mexico; October 2006/2007).

- Space Investment Summits I and III (San Jose, CA; New York, NY): hosted by the (traditional) aerospace industry-sponsored "Space Commerce Roundtable" group.
- Esther Dyson's Flight School (Aspen, Colorado; July 2006/2007): aviation and space entrepreneurs convocation.

## 10.6. Common themes and findings

Viewed within the current and historical context, it is evident that space entrepreneurship positively impacts the broader space sector and contributes to macro- and regional economic development and growth. Space entrepreneurs offer several important capabilities and advantages which can be difficult to create and sustain in larger aerospace corporate cultures: 1) an ability to build fast-moving and efficient teams and organizations geared around a core mission and competency, 2) the capacity for innovative "out-of-the box" thinking with the potential to generate substantial cost reductions over traditional space industry approaches, and 3) a culture and financial structure which accepts higher levels of technical and market risk.

Entrepreneurial space ventures in both regions, however, face significant financing problems due to the large gap in understanding between the space and investment communities. This problem is particularly acute in Europe and Asia, which lag behind the U.S. in their ability to provide space entrepreneurs with the market, innovation and financial environment needed to support sustained risk-taking. As a result, new space ventures in Europe are mostly geared towards lower risk opportunities at the sub-system or component level, or in the area of satellite services and space technology spin-offs.

The following interpretations of the broader success and impacts of space entrepreneurship in the U.S. also offer relevant insights for Europe:

- Space startups possess a highly successful track record within certain segments of the U.S. space industry, and particularly within the "Utilization" sector, if one considers that many of today's major telecom firms (XM Satellite Radio, Sirius, Echostar, etc), were startups only a decade or so ago.
- Current efforts to extend space entrepreneurship into "Core" markets, and in particular to transportation technologies and services, should thus be viewed as a logical extension of these historical successes into new areas.
- The U.S. small satellite sector, which began in the early 1990s as a result of entrepreneurial initiatives in the "Core" space arena, is now beginning to

reach maturity in the eyes of investors and the mainstream aerospace industry, as evidenced by the August 2007 acquisition of AeroAstro by Radyne Corporation.

- Early signs of growing convergence between the aviation and space industries are evident today. This is evident in the blurring of boundaries between aircraft/ spacecraft technologies (SpaceShipTwo, Rocketplane XP, X-Racer); between air and space markets (Zero Gravity Corporation, Virgin Galactic); in the emergence of hybrid spaceports/airports; in continuing movement towards an integrated Space-Air Traffic Management System by the FAA; and in the regulatory, legal and policy regime surrounding the integration of commercial spaceflight, UAVs, and Very Light Jets. This convergence is most pronounced within the space and air entrepreneurial sectors, where, e.g. UAV company Aurora Flight Sciences recently acquired Payload Systems Inc.

It is important to identify key underlying policy issues and barriers associated with efforts to create a growing and sustainable space startup and entrepreneurial finance sector in Europe and the United States. Critical macro-level issues, findings and questions facing this portion of the aerospace innovation industry on both sides of the Atlantic include the following:

**Awareness and credibility:** Traditional industry and space agencies do not today take full advantage of the innovation capacity inherent in entrepreneurial companies. Important opportunities for synergy and collaboration are frequently overlooked because of institutional bias against small companies, lack of understanding of space seed- and early-stage venture innovation potential by technical or political decision-makers, or high barriers to entry faced by startup firms. The

| | SPACE ENTREPRENEURSHIP | | | |
|---|---|---|---|---|
| | Intensity & Impact | | | |
| **OPPORTUNITIES & MARKETS** | **Penetration Level** | **Space Sector Impact to Date** | **Innovation Output** | **Macro-Economic Effects** |
| **Core** | ● | ● | ●●● | - |
| **Tourism** | ●●● | ●●● | ●●● | - |
| **Utilization** | ●● | ●●● | ●●● | ●●● |
| **Diffusion** | ●● | ● | ●●● | ●●● |
| **Infusion** | ●● | ●● | ● | ● |
| **Emotion** | ● | - | - | ● |

*JKIC & B.H. Lee 2007 - (Qualitative Level of Relative Strength: ● Low, ●● Medium, ●●● High)*

**Fig. 32** *Intensity & Impact of Space Entrepreneurship in the US & Europe.*

table below captures the qualitative impact of space entrepreneurship with the major industry categories described above. The table summarizes the qualitative impact of space entrepreneurship on 1) the immediate space sector, and 2) the larger economy. "Penetration Level" indicates the approximate degree to which startup companies constitute a portion of the existing or historical industrial base in each sector. "Space Sector Impact to Date" shows the extent to which entrepreneurial firms have made an impact on the larger space sector, and, "Innovation Output" indicates the relative degree to which these startup companies have produced significant technical or business model innovations. Finally, "Macro-Economic Effects" describe the qualitative level of impact on the larger macro-economy of said innovations.

Although an acceleration of space entrepreneurship would not require excessive funds, policy-backed and focused actions involving appropriate expertise are important. Currently, however, in Europe programs like Galileo substantially absorb attention and budget.

- **Irrational exuberance and space entrepreneurship:** A clear positive trend to be highlighted is the growing maturity and realism to be found within the space entrepreneurship community, along with an associated reduction in the frequent "hype" and presentation of immature or infeasible space business concepts. The space sector, however, continues to see occasional "dreams" that are brought forward by startup companies and space agencies, and even by financiers, as a result of poor technical and business judgment, inadequate decision-making processes, and the absence of ongoing monitoring of entrepreneur technical and financial performance.
- **Dedicated seed- and early-stage equity finance:** Outside of the satellite telecom arena ("Utilization"), well-structured and organized seed- and early-stage equity financing and investor communities are today essentially absent on both sides of the Atlantic. Reasons for this chronic structural gap in space venture finance include lack of familiarity with the sector and associated market opportunities; suboptimality in the quantity and quality of deal flow as compared to other mainstream investment sectors; difficulties associated with performing due diligence on space startups; and the absence of visible and repeatable investor success stories and "exits". To date, the only financial successes of any substantial scope have been realized primarily by early investors in "Utilization" market startup companies, most notably satellite telecommunications operators. A potential emerging trend here is the acquisition of "Core" entrepreneurial space startups by aerospace primes (e.g. Northrop Grumman's purchase of Scaled Composites in 2007), an event which may encourage additional participation by

early-stage investors. In general, capital formation rates in the commercial space industry, particularly for seed- and early- stage ventures, are not yet adequate to sustain the creation of next generation space startup companies.

- **Broad differences in macro-level innovation environments:** European industrial and mass consumer markets for space-related products & services remain much weaker than in the U.S. Regarding attitudes towards risk, European investors tend towards far more conservative and risk-averse approaches than their American counterparts when it comes to investing in technology-based startups. Finally, in contrast to the United States, Europe has done little since 2004 to create funded incentive prize programs – or a broader prize policy framework – as a part of new efforts to promote and support entrepreneurial innovation in the space sector. The absence of this new form of early-stage finance remains an important gap in Europe 's larger innovation policy framework, which continues to widen vis-a-vis the United States.

Often industrial and research sectors are compared to derive policy recommendations based on quantitative benchmarks and qualitative indicators. It is customary in Europe to compare industrial sectors with new industries in the U.S. Analogies to space can be found in the biotech and semiconductor industries, both of which have long lead times and require massive upfront capital investment. Since biotech, however, does not represent a real success at this stage, while semiconductors have become a star performer, derived policy recommendations aimed at the entrepreneurial space sector would be misleading, especially since sound data are difficult to obtain and measures of the space-special culture and process-related issues remain under development.

## 10.7. Recommendations

The following recommendations are offered for concurrent consideration by the U.S. and European space policy and innovation and entrepreneurship policy communities. The authors first present several propositions common to both regions, and then follow with a number of specific recommendations aimed at European policy makers in particular. Proposals for action made here, however, can only reflect high-level issues, as further investigation and study is required to implement proper actions at a sector level.

- The creation of viable space sector startup companies that can build investor confidence by achieving successful "exits" should be adopted as a major policy goal.

- Space entrepreneurship policy studies and actions should be considered in combination with seed- and early-stage space venture finance policy issues, and within the broader aerospace innovation pipeline policy context.
- Dedicated incentives aimed at encouraging investment in seed- and early-stage space-related business ventures should be developed.[402]
- Space entrepreneurship initiatives and programs require improved monitoring and assessment, including the use of experts to properly review selected cases and issues.
- Improved coaching and mentoring of seed- and early-stage space venture management teams is needed to increase success rates and attract the attention of the investment community.
- An expansion of "space policy" to include "space innovation policy" is desirable. The European and U.S. space policy communities must broaden their focus to include aerospace sector innovation and entrepreneurship policy issues as areas of valid discussion and research.
- **Space Innovation Policy Conferences**. It is recommended that two space sector innovation policy conferences be organized, one in Europe and one in the US, to discuss in a systematic manner aerospace sector innovation policy issues as they affect space entrepreneurship and venture finance in both regions.

*"No Deposit – No Return!"*

## 10.8. Relevant policy actions for Europe

- The European Commission should clearly link its general innovation policy with space-specific innovation matters and gear part of its financial support schemes to space entrepreneurship. ESA, in turn, should consider focusing on technology issues associated with seed- and early-stage space business ventures. National space agencies and "space regions" need to develop best fitting sub-activities focused at centered key competencies. Gaps in policy coverage should be closed by actions and experts at the appropriate level.
- European space and other innovation incentive prizes should be created. Education and lobbying efforts in support of the creation of formal innovation incentive prize programs, funded through a combination of government, industry, private and foundation monies should be undertaken, which are complementary to those currently being offered out of the United States.
- Experts in European space entrepreneurship and venture finance need to be identified, aggregated and receive enhanced visibility within the space venture and broader technology innovation pipeline community.

- In order to elaborate relevant policy items and work plans, a "spacial" independent task force should be created.
- The broad perspective on space entrepreneurship and space venture finance as presented in this paper requires proper dissemination, discussion and translation into action at the level of the European Commission, ESA, national governments and selected regions, as well as in the space industry.

Finally, it is generally recommended that both space entrepreneurs and venture finance experts and institutions require improved mutual understanding and cross-education in their respective perspectives, goals and constraints. Policymakers can facilitate this process through appropriate measures at European, national and regional levels. The space policy and innovation policy communities should agree on the common goal of developing a strong and sustained portfolio of space enterprises that can operate at higher levels of innovation capacity, risk and capital efficiency than traditional aerospace corporations.

*"The Power of a Singular Idea*
*Is Only as Strong as the Visionaries*
*Who Bring It to Reality"*

[397] CEOS Response to GCOS Requirements: Satellite observation of the climate system. 2006: 54 http://www.ceos.org/pages/pub.html.

[398] The Space Frontier Foundation labels many of these Core + Tourism startup companies the "NewSpace" (or "alt.space") sector, in contrast to risk-averse "OldSpace" aerospace primes. We avoid use of this term to preclude confusion, and for its lack of precision and sound financial criteria for selection of firms which should be included. "NewSpace" firms include both Super Angel companies and other entrepreneurs focusing on the Core+Tourism category.

[399] Most US federal science and technology agencies participate in the Small Business Innovation Research (SBIR) and Small Business Technology Transfer (STTR) Programs. These programs provide an opportunity for small, high technology companies and research institutions to participate in Government-sponsored research and development (R&D) efforts in key technology areas. In 2006, federal agencies awarded a total of 2.3 billion U.S. dollars in SBIR grant funds, with 55% coming from the Department of Defense and 5.4% deriving from NASA. Last year, NASA selected 120 proposals – corresponding to 102 small high technology firms - for negotiation of SBIR Phase 2 contract awards with a total value of approximately 72 million U.S. dollars. SBIR Phase I contract awards in the amount of 25 million U.S. dollars went to 206 companies. NASA SBIR Phase 1 contracts provide a maximum funding level of 100 000 U.S. dollars, and Phase 2 contracts support maximum grant of 600 000 U.S. dollars. The US SBIR/STTR program is viewed internationally as successful in stimulating commercially-valuable innovation by small and medium enterprises.

[400] ©2006 JKIC.

[401] "Space Angels Network"<www.spaceangelsnetwork.com>.

[402] Rep. Dana Rohrbacher (R-CA) in the US House of Representatives Zero Gravity Zero Tax Act H. R. 1024 on amending the Internal Revenue Code of 1986 to provide tax incentives for investing in companies involved in space-related activities. It was later not voted into law. March 2005.

# PART 3

# FACTS AND FIGURES

# 1. Chronology: January 2006–June 2007

Charlotte Mathieu

## 1.1. Access to space

| Europe | Other countries |
|---|---|
| LAUNCH LOG | |
| January 06 | |
| | 19 Atlas V – New Horizons (S)<br>24 H2A – ALOS 1 (R) |
| February 06 | |
| | 15 Zenit 3SL – Echostar X (C)<br>18 H2A – MTSat 2 (M)<br>22 MV – Astro-F Akari (S) Cute 1.7 and APD 1 (D) |
| March 06 | |
| 11 Ariane 5 – Hotbird 7A and Spainsat (C) | 01 Proton M – Arabsat 4A (C)<br>22 Pegasus – Space Technology 5A to C (D)<br>25 Falcon 1 – FalconSat 2 (D)<br>30 Soyuz Soyuz ISS 12S (ISS) |
| April 06 | |
| | 12 Zenit 3SL – JCSat 9 (C)<br>14 Minotaur – Formosat 3 A to F (M)<br>20 Atlas V – Astra 1 KR (C)<br>24 Soyuz – Progress ISS 21P (ISS)<br>25 START 1 – Eros B (R)<br>27 Long March 4B – Yaogan 1 (R)<br>28 Delta 2 – CloudSat and Calipso (S) |
| May 06 | |
| 27 Ariane 5 – SatMex 6 and Thaicom 5 (C) | 03 Soyuz – Kosmos 2420 (I)<br>24 Delta 4 Medium Plus – GOES 13 (M)<br>25 Shtil – Kompass 2 (S) |
| June 06 | |
| | 15 Soyuz – Resurs DK 1(R)<br>17 Proton – KazSat 1 (C) |

| | |
|---|---|
| | 18 Zenit 3SL – Galaxy 16 (C) |
| | 21 Delta 2 – MITEX (D) |
| | 24 Soyuz Progress ISS 22P (ISS) |
| | 25 Cyclone 2 – Kosmos 2421 (I) |
| | 27 Delta 4 Medium-Plus – NRO L-22 (I) |
| **July 06** | |
| | 04 Shuttle Discovery – STS 121 (ISS) |
| | 10 GSLV – Insat 4C (C) |
| | 12 Dnepr 1 – Genesis Pathfinder 1 (D) |
| | 21 Molniya – Kosmos 2422 (I) |
| | 26 Dnepr 1 – BelKA (R) AeroCube 1, Baumanets, ION, PICPot, Polysat 1 and 2, UniSat 4 and Voyager (D) HAUSat 1, ICECube 1 and 2, KuteSat, Merope, Ncube, Rincon, Sacred and SEEDS (S) |
| | 28 Rockot – Kompsat 2 (R) |
| **August 06** | |
| 11 Ariane 5 ECA – JCSAT 10 and Syracuse 3B (C) | 05 Proton M – Hotbird 8 (C) |
| | 21 Zenit 3SL – Koreasat 5 (C) |
| **September 06** | |
| | 09 Shuttle Atlantis – STS 115 (ISS) |
| | 09 Long March 2C – SJ 8 (S) |
| | 11 H 2A – IGS 3A (I) |
| | 13 Long March 3A – Zhongxing 22A (C) |
| | 14 Soyuz – Kosmos 2423 (I) |
| | 18 Soyuz Soyuz ISS 13S (ISS) |
| | 23 M5 – Hinode (S) Hitsat and SSSat (D) |
| | 25 Delta 2 Navstar GPS 2RM-2 (N) |
| **October 06** | |
| 13 Ariane 5 ECA – DirecTV 9S and Optus D1 (C) and LDREX (D) | 19 Soyuz 2 – Metop A (M) |
| | 23 Soyuz – Progress ISS 23P (ISS) |
| | 24 Long March 3B – SJ 6C and 6D (S) |
| | 25 Delta 2 – STEREO A and B (S) |
| | 29 Long March 3B – SINO-Satellite (C) |
| | 30 Zenit 3SL – XM 4 (C) |

| | |
|---|---|
| **November 06** | |
| | 04 Delta 4 Medium – DMSP 5D-3-F17 (M)<br>09 Proton M – BADR-4 (C)<br>17 Delta 2 – Navstar GPS 2RM-3 (N) |
| **December 06** | |
| 08 Ariane 5 ECA – Wildblue 1 and AMC 18 (C) | 08 Long March 3A – Fengyun 2D (M)<br>09 Shuttle Discovery STS 116 (ISS) + ANDE, MARSCom, MEPSI-2 and RAFT 1 (D)<br>12 Proton M – Measat 3 (C)<br>14 Delta 2 – NRO L-21 (Classified)<br>16 Minotaur – TacSat 2 (D) Genesat 1 (D)<br>18 H 2A – ETS 8 (C)<br>19 Kosmos – SAR Lupe 1 (I)<br>24 Soyuz 2 – Meridian (C)<br>25 Proton – Glonass K R4 to R6 (N)<br>27 Soyuz 2 1B – Corot – (S) |
| **January 07** | |
| | 10 PSLV – Cartosat 2 (R) LAPAN-Tubsat, PehuenSat and SRE 1 (D)<br>18 Soyuz Progress ISS 24P (ISS)<br>30 Zenit 3SL – NSS 8 (C) |
| **February 07** | |
| | 03 Long March 3A – Beidou 2A (N)<br>17 Delta 2 – THEMIS 1 to 5 (S)<br>24 H 2A – IGS 3B and IGS Optical 3 Verification (Classified) |
| **March 07** | |
| 11 Ariane 5 ECA – Skynet 5A and Insat 4B (C) | 08 Atlas 5 – Orbital Express 1A, CFESat, FalconSat 3, MIDSTAR 1, Orbital Express 1B, Space Test Program Satellite 1 (D)<br>20 Falcon 1 – Falcon Demosat (D) |
| **April 07** | |
| | 07 Soyuz – Soyuz ISS 14S (ISS)<br>10 Proton M – Anik F3 (C) |

| | |
|---|---|
| | 11 Long March 2C – Haiyang 1B (R)<br>14 Long March 3A – Beidou 2B (N)<br>17 Dnepr 1 – Egypsat (R) SaudiSat 3 (R) SaudiComsat 3 to 7 (C) AeroCube 2, CAPE-1, CSTB 1, Libertad 1, MAST, Polysat 3 and 4 (D)<br>23 PSLV – AGILE (S) AAM (D)<br>24 Minotaur – NFIRE (D)<br>25 Pegasus XL – AIM Explorer (S) |
| May 07 | |
| 04 Ariane 5 ECA – Astra 1L and Galaxy 17 (C) | 04 Ariane 5 ECA – Galaxy 17<br>12 Soyuz – Progress ISS 25P (ISS)<br>13 Long March 3B – Nigcomsat 1 (C)<br>25 Long March 4B – Yaogan II (R)<br>30 Soyuz – Globalstar Replacement 1 to 4 (C) |
| June 07 | |
| | 01 Long March – Sinosat III (C)<br>07 Delta 2 – Cosmo-Skymed 1 (I)<br>08 Shuttle Atlantis – STS 117 (ISS)<br>15 Dnepr 1 – TerraSar-X (R)<br>15 Atlas V – NRO L-30 (I)<br>28 Dnepr 1 – Genesis 2 (D) |
| POLICY AND BUSINESS | |
| *26 February 07* Official opening of the Soyuz launch base construction site | *May 06* Space Launch Acquired By Space Adventures<br>*31 August 06* NASA awarded the Orion prime contract to Lockheed Martin Corp.<br>*August 06* COST contracts awarded by NASA to SpaceX and RpK<br>*1 December 06* Creation of ULA<br>*April 07* NASA Modifies Orion Crew Exploration Vehicle Contract |

C: Communications – D: Development – I: Intelligence – M: Meteorological – N: Navigation – R: Remote Sensing – S: Scientific

## 1.2. Space science and exploration

| Europe | Other countries |
|---|---|
| EARTH SCIENCES | |
| | *28 April 2006* Launch of **Calipso and CloudSat** (NASA)<br>Studies clouds and atmospheric aerosols (airborne particles)<br><br>*25 May 06* Launch of **Kompass 2** (Russia)<br>Monitors seismic disturbances<br><br>*9 September 06* Launch of **SJ 8** (China)<br>Recoverable spacecraft with seeds<br><br>*24 October 06* Launch of **SJ 6C and 6D** (China)<br>Space environment and radiation<br><br>*25 April 2007* Launch of **AIM Explorer** (NASA)<br>Aeronomy of ice in the Mesosphere |
| ASTRONOMY | |
| *27 December 06* Launch of **Corot** (CNES)<br>Telescope to study stellar seismology and search for extra-solar planets<br><br>*23 April 07* Launch of **Agile** (ASI)<br>Gamma ray observatory | *22 February 06* Launch of **Akari/Astro-F** (JAXA)<br>Infrared astronomical telescope<br><br>*25 October 06* Launch of **Stereo** (NASA)<br>Studies the Sun and the Coronal Mass Ejections<br><br>*17 February 07* Launch of **Themis** (NASA)<br>Monitors the under-storms and the reconfiguration of the magnetotail |
| EXPLORATION | |
| *11 April 06* Arrival of **Venus Express** in Venus orbit (ESA)<br>Studies the atmosphere, the plasma environment, and the surface of Venus | *19 January 06* Launch of **New Horizons** (NASA)<br>Mission to Pluto-Kuiper Belt |
| *3 September 06* **Smart-1** impact landing on the Moon (ESA) | *March 06* **Mars Reconnaissance Orbiter** in Mars orbit (NASA) |

| | |
|---|---|
| Used to test solar electric propulsion and performed scientific observations of the Moon<br><br>*31 January 07* Contract for **BepiColombo** (ESA) awarded to Alcatel and Astrium Study of Mercury | Studies water on Mars |
| MANNED SPACEFLIGHT | |
| *4–17 July 06*<br>Tomas Reiter mission (first long mission for an European astronaut onboard the ISS)<br><br>*9–22 December 06*<br>Christer Fuglesang mission to the ISS | *4–17 July 06* Return to flight STS-121 mission<br><br>*9–21 September 06* STS-115 mission<br><br>*9–22 December 06* STS-116 mission<br><br>*8–20 June 07* STS-117 mission |

## 1.3. Applications

| Europe | Other countries |
|---|---|
| EARTH OBSERVATION | |
| | *January 06*<br>Creation of Geoeye<br>*24* Launch of ALOS 1 (Japan)<br><br>*February 06*<br>*18* Launch of MTSat 2 (M) (Japan) |
| *19 October 06*<br>Launch of MetOp-A (M)<br><br>*19 December 06*<br>Launch of SAR-Lupe 1<br><br>*June 07*<br>07 Launch of Cosmo-Skymed 1<br>15 Launch of TerraSAR-X<br>*18* Contract for Sentinel-1 signed between ESA and Thales Alenia Space | *April 06*<br>*14* Launch of Formosat 3 A to F (M) (Taiwan-USA)<br>*25* Launch of Eros B (Israel)<br>*27* Launch of Yaogan 1 (China)<br><br>*May 06*<br>*24* Launch of GOES 13 (M) (USA)<br><br>*June 06*<br>*15* Launch of Resurs DK 1 (Russia)<br><br>*July 06*<br>*26* Failed launch of BelKA (Byelorussia)<br>*28* Launch of Kompsat 2 (Korea)<br><br>*November 06*<br>*04* Launch of DMSP 5D-3-F17 (M) (USA)<br><br>*December 06*<br>*08* Launch of Fengyun 2D (M) (China)<br><br>*January 07*<br>*10* Launch of Cartosat 2 (India)<br><br>*April 07*<br>*11* Launch of Haiyang 1B (China)<br>*17* Launch of Egypsat (Egypt) & SaudiSat 3 (KSA)<br><br>*May 07*<br>*25* Launch of Yaogan II (China) |
| *January 06* End of the Topex/Poseidon mission (NASA/CNES)<br>*April 06* Cooperation agreement between CNES and Eumetsat and NASA and NOAA on Jason-2 | |

| INTELLIGENCE | |
|---|---|
| | *May 06*<br>*03* Launch of Kosmos 2420 (Russia)<br>*June 06*<br>*25* Launch of Kosmos 2421 (Russia)<br>*27* Launch of NRO L-22 (USA)<br>*July 06*<br>*21* Launch of Kosmos 2422 (Russia)<br>*September 06*<br>*11* Launch of IGS 3A (Japan)<br>*14* Launch of Kosmos 2423 (Russia)<br>*December 06*<br>*14* Launch of NRO L-21 (USA)<br>*February 07*<br>*24* Launch of IGS 3B and IGS Optical 3 Verification (Japan)<br>*June 07*<br>*15* Launch of NRO L-30 (USA) |
| NAVIGATION | |
| *January 06*<br>*12* First GIOVE – A signal | *September 06*<br>*25* Launch of Navstar GPS 2RM-2<br>*November 06*<br>*17* Launch of Navstar GPS 2RM-3<br>*December 06*<br>*25* Launch of Glonass K R4 to R6<br>*February 07*<br>*03* Launch of Beidou 2A<br>*April 07*<br>*14* Launch of Beidou 2B |
| TELECOMMUNICATIONS/BROADCASTING | |
| *11 March 06* Launch of Hotbird 7A and Spainsat<br>*20 April 06* Launch of Astra 1 KR | *February 06*<br>*15* Launch of Echostar X<br>*March 06*<br>*01* Launch of Arabsat 4A<br>*April 06*<br>*12* Launch of JCSat 9 |

| | |
|---|---|
| | *May 06*<br>*27* Launch of SatMex 6 and Thaicom 5<br>*June 06*<br>*17* Launch of KazSat 1<br>*18* Launch of Galaxy 16<br>*July 06*<br>*10* Launch of Insat 4C |
| *August 06*<br>*05* Launch of Hotbird 8<br>*11* Launch of Syracuse 3B | *August 06*<br>*11* Launch of JCSAT 10<br>*21* Launch of Koreasat 5 |
| | *September 06*<br>*13* Launch of Zhongxing 22A<br>*October 06*<br>*13* Launch of DirecTV 9S and Optus D1<br>*29* Launch of SINO-Satellite<br>*30* Launch of XM 4 |
| *December 06*<br>04 Globalstar and Alcatel signed a contract for 48 satellites | *November 06*<br>*09* Launch of BADR-4<br>*December 06*<br>*08* Launch of Wildblue 1 and AMC 18<br>*12* Launch of Measat 3<br>*18* Launch of ETS 8<br>*24* Launch of Meridian |
| *March 07*<br>*11* Launch of Skynet 5A<br>28 Contract between ESA and OHB to develop a European Small Geostationary Satellite platform for sitcoms | *January 07*<br>*30* Launch of NSS 8<br>*March 07*<br>*11* Launch of Insat 4B<br>*April 07*<br>*10* Launch of Anik F3<br>*17* Launch of SaudiComsat 3 to 7 |
| *May 07*<br>Astrium and Thales-Alenia space selected to build UAE satcom system<br>*04* Launch of Astra 1L | *May 07*<br>*04* Launch of Galaxy 17<br>*13* Launch of Nigcomsat 1<br>*30* Launch of Globalstar Replacement 1 to 4<br>*June 07*<br>*01* Launch of Sinosat III |

| TECHNOLOGY DEVELOPMENT | |
|---|---|
| | *February 06*<br>*22* Launch of Cute 1.7 and APD 1 (Japan)<br><br>*March 06*<br>*22* Launch of Space Technology 5A to C (USA)<br>*25* Failed launch of FalconSat 2 (USA) |
| *July 06*<br>*26* Failed launch of PICPot (Italy) and NCube (Norway) | *June 06*<br>*21* Launch of MITEX (USA)<br><br>*July 06*<br>*12* Launch of Genesis Pathfinder 1 (USA)<br>*26* Failed launch of AeroCube 1, Baumanets, ION, PICPot, Polysat 1 and 2, UniSat 4 and Voyager (USA)<br><br>*September 06*<br>*23* Launch of Hitsat and SSSat (Japan)<br><br>*October 06*<br>*13* Launch of LDREX<br><br>*December 06*<br>*16* Launch of TacSat 2 and Genesat 1 (USA)<br><br>*January 07*<br>*10* Launch of LAPAN -Tubsat, PehuenSat and SRE 1 (India)<br><br>*March 07*<br>*08* Launch of Orbital Express 1A, CFESat, FalconSat 3, MIDSTAR 1, Orbital Express 1B, Space Test Program Satellite 1 (USA)<br>*20* Failed launch of Falcon Demosat (USA)<br><br>*April 07*<br>*17* Launch of AeroCube 2, CAPE-1, CSTB 1, Libertad 1, MAST, Polysat 3 and 4 (USA)<br>*23* Launch of AAM (India)<br>*24* Launch of NFIRE (USA)<br><br>*June 07*<br>*28* Launch of Genesis 2 (USA) |

# 2. Countries profiles

## AUSTRIA

| | |
|---|---|
| Population[403] | 8.3 millions |
| GDP[404] | 270 billion euros |
| Responsibility | The Austrian Space Programme is funded by the Federal Ministry for Transport, Innovation and Technology (BMVIT) and managed by the Agency for Aeronautics and Space (ALR)[405] of the Austrian Research Promotion Agency (FFG). |
| Activities | In addition to ESA programmes, two main national programmes Austrian Space Applications Programme (ASAP) and Austrian Radionavigation Technology and Integrated Satnav services and products Testbed (ARTIST). |
| Budget | In 2006, 50.02 million euros (ESA 33.46, national space programmes 6.75, Austrian Academy of Sciences 4.02, FFG-ALR 0.97, Eumetsat 4.82) |
| Staff | ALR – 11 |
| Direct employment in the space manufacturing industry[406] | 299 |

## BELGIUM

| | |
|---|---|
| Population[403] | 10.5 milions |
| GDP[404] | 328 billion euros |
| Responsibility | The Belgian Federal Science Policy Office[407] manages Belgian space activities and the Belgian participation in national and international programmes through its Department for space research and applications. |
| Activities | In addition to ESA programmes (mainly telecommunications, Proba, Launchers, Prodex, ISS), there are bilateral cooperation projects with the US on STEREO, France on COROT and Pleiades, Argentina on the SAOCOM programme and Russia on MIRAS and SPICAM. |
| Budget | The Department for space research and applications has a budget in 2006 of 181.6 million euros (166.7 for ESA and bilateral cooperation, 13.3 for EU, Eumetsat and ESO and 1.6 for the Department)[408] |
| Staff | Department for space research and applications – About 20 |
| Direct employment in the space manufacturing industry[406] | 1,189 people |

## CZECH REPUBLIC

| | |
|---|---|
| Population[403] | 10.2 millions |
| GDP[404] | 123 billion euros |
| Responsibility | The Ministry of Education, Youth and Sports supervises space activities and the cooperation with ESA. The Czech Space Office[409] (CSO), a private organisation, coordinates space activities. |
| Activities | In addition to the ESA PECS programme, the Czech space activities focus on astronomy, magnetospheric, ionospheric and atmospheric research, microgravity research experiments, scientific instruments and micro-satellites (Magion and MIMOSA). |
| Budget | In 2005, 2.445 million euros (ESA PECS 0.905, Eumetsat 0.240, national activities 0.700, Galileo-related national activities 0.600) |
| Staff | CSO-11 |

## DENMARK

| | |
|---|---|
| Population[403] | 5.4 millions |
| GDP[404] | 232 billion euros |
| Responsibility | The Ministry of Science, Technology and Innovation is responsible for the national space policy and space activities. The Danish National Space Center (DNSC)[410], which is a research centre belonging to the Technical University of Denmark, is heading the Danish Space Consortium (all major Danish space-related institutions and companies). |
| Activities | In addition to ESA programmes, bilateral cooperation is undertaken with the US (HEAO-3 NuStar), Sweden (Viking), Russia (Spectrum) and France (Granat). |
| Budget | 29.92 million euros (ESA 24.92 and national projects 5) |
| Staff | DNSC-105 |
| Direct employment in the space manufacturing industry[406] | 155 |

## FINLAND[411]

| | |
|---|---|
| Population[403] | 5.2 millions |
| GDP[404] | 175 billion euros |
| Responsibility | The Ministry of Trade and Industry, the Funding Agency for Technology and Innovation (Tekes) and the Academy of Finland are funding space activities in Finland. The Finnish Space Committee consists of representatives of all stakeholders and coordinates all the activities. Tekes is the executive body for space activities and, together with the Academy of Finland for basic research, manages the Finnish participation within ESA programmes and other international projects. |
| Activities | In addition to ESA programmes, Finland has bilateral activities with the USA (TWINS, Mars Science Laboratory, Phoenix, ISS), Russia (MetNet), France (Pleaiades) Germany (TanDEM-X and TerraSAR-X) and Japan (ISS), as well as a national space technology programme. |
| Budget | In 2006, 39.5 million euros (including 15.8 for ESA and about 5 for Eumetsat, ESO, ESICAT and NOT). |
| Staff | Within Tekes, space technology – 7 |
| Direct employment in the space manufacturing industry[406] | 131 |

## FRANCE

| | |
|---|---|
| Population[403] | 63.0 millions |
| GDP[404] | 1,871 billion euros |
| Responsibility | The Centre National d'Etudes Spatiales[412] (CNES) is responsible for the French space activities. It is under the shared responsibility of the Ministry of Education and Research and of the Ministry of Defense. The Office National d'Etudes et de Recherches Aérospatiales (ONERA)[413] is also responsible for space-related research. |
| Activities | In addition to ESA programmes, national civil and military programmes are undertaken (Pleiades, Syracuse, Helios, Essaim) as well as bilateral cooperation with the USA (Calipso, Jason), India (Megha Tropiques), and Japan (Polder). |
| Budget | In 2006, CNES had a budget of 1,760 million euros (including ESA 742). In 2006, ONERA had a budget of 188 million euros and 18% of the ONERA's revenues came from space-related research activities. |
| Staff | CNES – 2,423 |
| Direct employment in the space manufacturing industry[406] | 11,099 |

# GERMANY

| | |
|---|---|
| Population[403] | 82.4 millions |
| GDP[404] | 2,422 billion euros |
| Responsibility | The German Space Agency within the German Aerospace Center (DLR)[414] is responsible for German space activities. It is under the responsibility of the Ministry of Economics and Technology. |
| Activities | In addition to ESA programmes, Germany has national civil and commercial programmes that include:<br>• Earth Observation (RapidEye, TerraSar-X, Lapan-Tubsat, EnMAP)<br>• Smallsat platforms (Astrobus)<br>• Telecommunications (Tubsat)<br>• Manned space flight (ISS, microgravity experiments)<br>• Launch services (Eurockot, OHB-Cosmos)<br>• Associated ground systems<br>• Space technologies (such as intersatellite links)<br>• Bilateral cooperation with the US (GRACE)<br>Military programmes include remote sensing satellites (Sar-Lupe radar satellites) and satcoms (Satcom BW). |
| Budget | In 2006, 982 million euros including ESA 555, national space programme about 170, for DLR other national organisations 257. To this amount can be added about 60 million euros for EUMETSAT. |
| Staff | DLR for space activities – about 1500 |
| Direct employment in the space manufacturing industry[406] | 4,495 |

# GREECE[415]

| | |
|---|---|
| Population[403] | 11.1 millions |
| GDP[404] | 209 billion euros |
| Responsibility | The General Secretariat for Research and Technology (GSRT) of the Ministry of Development is responsible for Greek space activities. |
| Activities | The Greek space activities cover mainly the fields of space physics, remote sensing and telecommunications. |
| Budget | In 2006, about 10 million euros for ESA, to which should be added the contributions to Eumetsat, about 3 million euros. |
| Staff | Space < 5 |

## HUNGARY[416]

| | |
|---|---|
| Population[403] | 10.1 million |
| GDP[404] | 104 billion euros |
| Responsibility | The Hungarian Space Office, under the responsibility of the Ministry of Environment and Water, manages Hungarian space activities. |
| Activities | In ESA PECS projects and national activities, the main fields are Earth observation applications, space physics, life sciences and space technology. |
| Budget | The budget of the Hungarian Space Office is about 2 million euros (ESA PECS about 1, national projects about 0.6), to which should be added a contribution to Eumetsat of about 0.2 million euros. |
| Staff | HSO – 3 |

## IRELAND[417]

| | |
|---|---|
| Population[403] | 4.2 millions |
| GDP[404] | 189 billion euros |
| Responsibility | Enterprise Ireland, in association with the Office of Science and Technology of the Department of Enterprise, Trade and Employment, manages and coordinates space activities in Ireland. |
| Activities | Irish space activities are in the fields of software systems and services, precision mechanical components, advanced materials, electronics/ microelectronics and telecommunications systems and service engineering. |
| Budget | In 2006, about 11 million euros to ESA |
| Staff | For space<5 |
| Direct employment in the space manufacturing industry[406] | 41 |

## ITALY[418]

| | |
|---|---|
| Population[403] | 58.7 millions |
| GDP[404] | 1534 billion euros |
| Responsibility | The Italian Space Agency (ASI), under the Ministry of University and Research, manages the Italian space activities. |
| Activities | Italian civil space activities include both ESA programmes and a national civil programme according to the aerospace plan 2006–2008. Italy's national activities include small scientific missions (AGILE, PRIMA platforms), dual-use Earth observation satellites (Cosmo-Skymed) and military satcoms (Sicral), as well as commercial telecommunications and radar satellites. Italy has developed bilateral cooperation with France (Vega, Pleiades) and Argentina (Siage). |
| Budget | In 2006, 794 million euros including ESA 365 and 429 for ASI |
| Staff | ASI – About 200 |
| Direct employment in the space manufacturing industry[406] | 3,762 |

## LUXEMBOURG[419]

| | |
|---|---|
| Population[403] | 0.5 millions |
| GDP[404] | 37 billion euros |
| Responsibility | Luxinnovation, the National Agency for Innovation and Research, under the responsibility of the Ministry of Culture, Higher Education and Research, coordinates space activities in Luxembourg. |
| Activities | Luxembourg focuses mainly on telecommunications with a major player in the field, SES Astra. |
| Budget | In 2006, 4 million euros to ESA |
| Staff | Luxinnovation Space<5 |

## THE NETHERLANDS

| | |
|---|---|
| Population[403] | 16.3 millions |
| GDP[404] | 559 billion euros |
| Responsibility | The Netherlands Agency for Aerospace Programmes (NIVR)[420], under the Ministry of Economic Affairs, is responsible for the industrial space activities, while the Institute for Space Research (SRON)[421] manages research activities. |
| Activities | The space research fields include astrophysics, astronomy, microgravity and Earth Observation. |
| Budget | In 2006, about 102 million euros (ESA 64, national 24, Eumetsat 14) |
| Staff | NVIR Space Division – 12 |
| Direct employment in the space manufacturing industry[406] | 540 |

## NORWAY[422]

| | |
|---|---|
| Population[403] | 4.6 millions |
| GDP[406] | 285 billion euros |
| Responsibility | The Norwegian Space Centre (NSC), under the Ministry of Trade and Industry, manages Norwegian space activities. |
| Activities | In addition to ESA programmes (in particular Earth observation, telecommunications, and launchers), Norway has national support programmes and commercial activities (Telenor). Moreover, Norway operates the Andoya rocket range and the Svalbard ground station. Norway has also a bilateral agreement with Canada on the use of Radarsat data. |
| Budget | In 2006, 35 million euros, including ESA 28 |
| Staff | NSC – 21 |
| Direct employment in the space manufacturing industry[406] | 303 |

## POLAND

Polskie Biuro do spraw Przestrzeni Kosmicznej

| | |
|---|---|
| Population[403] | 38.2 millions |
| GDP[404] | 301 billion euros |
| Responsibility | The Space Research Centre[423] coordinates space activities and hosts the Polish space office (PSO).[424] |
| Activities | The Polish space activities are mainly in the fields of space science, navigation and remote sensing applications |
| Budget | In 2006, 4.6 million euros (including Eumetsat 0.6) |
| Staff | PSO – 5 |

## PORTUGAL

gpe

| | |
|---|---|
| Population[403] | 10.6 millions |
| GDP[404] | 162 billion euros |
| Responsibility | Portuguese space activities are coordinated by a Portuguese Space Office within the International Relation Department of the Ministry of Science and Higher Education (GRICES).[425] |
| Activities | Mainly participation in ESA programmes (telecommunications systems, technology developments, Earth observation, exploration). |
| Budget | In 2006, 16 million euros including ESA 15 and about 1 for national activities. |
| Staff | GPE – 7 |
| Direct employment in the space manufacturing industry[406] | 64 |

## ROMANIA

| | |
|---|---|
| Population[403] | 21.6 millions |
| GDP[404] | 121 billion euros |
| Responsibility | The Romanian Space Agency (ROSA),[426] under the responsibility of the Ministry of Education and Research, is managing the Romanian space activities. |
| Activities | ESA PECS and national activities defined by the R&D programme are covering the fields of space science (space physics and astronomy), space systems (construction of nanosatellites and microgravity experiments) and space applications (telemedicine, Earth observation and navigation space-based services). |
| Budget | In 2006, about 1.6 million euros |
| Staff | ROSA – about 35 |

## SPAIN[427]

| | |
|---|---|
| Population[403] | 43.8 millions |
| GDP[404] | 1,049 billion euros |
| Responsibility | The Centre for the Development of Industrial Technology (CDTI), under the Ministry of Industry, Commerce and Tourism, is funding and coordinating Spanish space activities. |
| Activities | In addition to ESA programmes, Spain has a national space programme including governmental and commercial programmes, especially in civil and military telecommunications (Hispasat, Spainsat), and is involved in bilateral cooperation projects, like with France in military observation systems (Helios). Moreover, Spain manages national, ESA and NASA ground facilities. |
| Budget | The total budget of the 2007–2011 Strategic Plan for the Space Sector amounts to 1,071 million euros.[428] In addition, the ESA contribution represented 132 million euros in 2006. |
| Staff | n.a. |
| Direct employment in the space manufacturing industry[406] | 1,956 |

## SWEDEN

RYMDSTYRELSEN
Swedish National Space Board

| | |
|---|---|
| Population[403] | 9.0 millions |
| GDP[404] | 327 billion euros |
| Responsibility | The Swedish National Space Board (SNSB)[429], under the Ministry of Industry, Employment and Communication, is responsible for space activities in Sweden. Basic research is funded via the Ministry of Education and Research. |
| Activities | In addition to ESA programmes, Sweden has national programmes (subsystems, satellites and sounding rockets), and bilateral cooperation mainly with France (Odin, Pleiades) and Germany. Both countries are partners of the technology demonstration project Prisma. |
| Budget | In 2006, about 76 million euros including about 50 million euros for ESA |
| Staff | SNSB – 18 |
| Direct employment in the space manufacturing industry[406] | 694 |

## SWITZERLAND[430]

Schweizerische Eidgenossenschaft
Confédération suisse
Confederazione Svizzera
Confederaziun svizra

| | |
|---|---|
| Population[403] | 7.5 millions |
| GDP[404] | 305 billion euros |
| Responsibility | The Space affairs division (or Swiss Space Office) of the State Secretariat for Education and Research of the Federal Department of Home Affairs is responsible for Swiss space activities and cooperates closely with the Swiss Department of Foreign Affairs on that topic. |
| | The Federal Commission for space affairs (CFAS) is preparing a new Swiss space policy. The Interdepartmental coordination committee for space (IKAR) is responsible for the coordination of the activities. |
| Activities | Most of the Swiss activities are undertaken within ESA programmes (space science, human spaceflight, launchers, Earth observation, Prodex). |
| Budget | In 2006, about 84 million euros, including 82 for ESA |
| Staff | n.a. |
| Direct employment in the space manufacturing industry[406] | 634 |

## United Kingdom

BNSC

| | |
|---|---|
| Population[403] | 60.4 millions |
| GDP[404] | 2,024 billion euros |
| Responsibility | The British National Space Centre (BNSC)[431] coordinates UK civil space policy and programmes. It is formed of 7 UK government departments and 2 research councils and is hosted by the Department for Innovation, Universities and Skills (DIUS). |
| Activities | In addition to ESA programmes, the main fields of activities are space science, Earth observation systems (Topsat), military and commercial communications systems and microsatellites. |
| Budget | In 2005/2006, BNSC's partners spent 207 million pounds on space programmes (about 300 million euros, including 200 million euros for ESA) |
| Staff | BNSC headquarters – about 40 |
| Direct employment in the space manufacturing industry[406] | 3,501 |

## ESA[432]

| Responsibility | The European Space Agency is an inter-governmental organisation with the mission to provide and promote, for exclusively peaceful purposes, the exploitation of space science, research and technology, as well as space applications. |
|---|---|
| | ESA achieves this through:<br>• Space activities and programmes<br>• Long term space policy<br>• A specific industrial policy<br>• Coordinating European with national space programmes |
| Activities | ESA activities are divided into mandatory programmes (mainly scientific programmes) and optional programmes including:<br>• Telecommunications<br>• Earth Observation<br>• Launchers<br>• Human Spaceflight, Microgravity and Exploration<br>• Navigation |
| Budget | In 2006, 2,903 million euros (including 2,479 from Member States' contributions) |

*2006 Contributions from ESA's Member States and Canada*

| | Contribution in thousand euros | Percentage |
|---|---|---|
| France | 742,531 | 30.0% |
| Germany | 555,000 | 22.4% |
| Italy | 344,029 | 13.9% |
| Great Britain | 199,363 | 8.0% |
| Belgium | 148,185 | 6.0% |
| Spain | 132,125 | 5.3% |
| Switzerland | 82,364 | 3.3% |
| The Netherlands | 63,419 | 2.6% |
| Sweden | 50,575 | 2.0% |
| Austria | 33,461 | 1.3% |
| Norway | 28,473 | 1.1% |
| Denmark | 24,920 | 1.0% |

| | | |
|---|---|---|
| Canada | 20,172 | 0.8% |
| Finland | 16,482 | 0.7% |
| Portugal | 12,167 | 0.5% |
| Ireland | 11,446 | 0.5% |
| Greece | 10,131 | 0.4% |
| Luxemburg | 4,274 | 0.2% |
| Total | 2,479,117 | 100.0% |

| | |
|---|---|
| Staff | In 2006, about 1900 |

## European Commission[433]

| | |
|---|---|
| Responsibilities | The European Commission responsibilities include:<br>– Defining the priorities and requirements for space based systems at the service of the EU's main objectives and policies and citizens' needs<br>– Aggregating the political will and user demand in support of these<br>– Ensuring the availability and continuity of services supporting EU policies<br>– Ensuring integration of space-based systems with related ground and in-situ systems in promoting the development of user-driven application services supporting EU policies<br>– Creating an optimum regulatory environment to facilitate innovation<br>– Promoting coordination of the European position in international cooperation |
| Activities | Space-related activities in the Commission taken lace in different Directorates. The Directorate Aerospace, Security, Defence and Equipment of the Enterprise and Industry Directorate-General is responsible for the coordination of the Europena Commission's space policy. The Directorate for Research is responsible for space-related research activities. The Directorate for Energy and Transport is responsible for most Galileo-related activities. |
| Budget | The Framework Programme 7 has a budget dedicated to the Theme "Space" of 1,430 million euros[434] over the period 2007–2013. This budget covers mainly GMES-related activities.<br>For Galileo activities, until 2007, the total EC budget for Galileo was 1,425 million euros (535 million euros from the FP5, 6 and 7 and 890 from the DG Transport and Energy). |

## EUMETSAT

EUMETSAT

| | |
|---|---|
| Responsibility | Formed in 1986, the European Organisation for the Exploitation of Meteorological Satellites (EUMETSAT) is an inter-governmental organisation. EUMETSAT operates a fleet of meteorological satellites and their related ground systems to deliver reliable and cost-efficient data, images and products to national meteorological services. |
| Activities | EUMETSAT's main purpose is to continuously deliver weather and climate-related satellite data, images and products. This information is supplied to the National Meteorological Services of the organisation's 20 Member and 10 Cooperating States in Europe , as well as other users world-wide. |
| | Through EUMETCast (EUMETSAT's Broadcast System for Environmental Data), EUMETSAT disseminates data and products to a wide user community. In addition, EUMETSAT provides training to help users exploit satellite data. |
| Budget | In 2006, 251.9 million euros |
| | The Member States' contributions are based on a scale which is proportional to the gross national income of the individual Member States. 10 States have signed cooperation agreements with EUMETSAT, and each of them contributes 50% of the full membership fee. |

*Contributions of EUMETSAT's Member States (in %)*

| | Percentage |
|---|---|
| Germany | 21.5% |
| UK | 16.7% |
| France | 15.7% |
| Italy | 12.7% |
| Spain | 7.3% |
| The Netherlands | 4.4% |
| Switzerland | 3.1% |
| Belgium | 2.7% |
| Sweden | 2.6% |
| Austria | 2.2% |
| Norway | 2.0% |

| | |
|---|---|
| Turkey | 1.9% |
| Denmark | 1.8% |
| Greece | 1.5% |
| Finland | 1.4% |
| Portugal | 1.3% |
| Ireland | 1.1% |
| Slovakia | 0.3% |
| Luxemburg | 0.2% |
| Total | 100.0% |

| | |
|---|---|
| Staff | In 2006, about 500 |

## CANADA

| | |
|---|---|
| Population[435] | 32.6 millions |
| GDP[436] | 969 billion euros |
| Responsibility | The Canadian Space Agency (CSA)[437] manages Canadian space activities. It reports to the Parliament through the Ministry of Industry. |
| Activities | The CSA activities include:<br>• Four key programmes: Earth Observation (Radarsat), Space Science and Exploration (MOST, ePOP on Cassiope), Satellite Communications (Anik), and Space Awareness and Learning.<br>• Manned spaceflight (Canada has an astronaut corps)<br>• Cooperation on a variety of projects with NASA (ISS, CloudSat)<br>• Participation in ESA programmes as a Cooperating State Canada also develops and cooperates with the US on military space capabilities. |
| Budget | In 2006/2007, CSA had a budget of 308.2 million Canadian dollars, i.e. about 200 million euros (including about 20 million euros for ESA). |
| Staff | In 2007, CSA – 635 |
| Employment in the space sector[438] | 6,100 |

## CHINA

| | |
|---|---|
| Population[435] | 1,314 millions |
| GDP[436] | 2,000 billion euros |
| Responsibility | The CNSA (Chinese National Space Administration) under the responsibility of the COSTIND (Commission on Science, Technology and Industry for National Defence) coordinates the civilian space programmes and the cooperation with foreign space agencies. The GAD (General Armament Department), under the auspices of the Central Military Commission (CMC), runs the military space programmes, the launch infrastructure and manned spaceflight activities. The Chinese Academy of Science (CAS) is responsible for space research and the elaboration of the National Programmes. |
| Activities | The Chinese activities include:[439]<br>• Satellites for science and technology demonstration (Shi Jian), Earth observation (Zi Yuan), navigation (Beidou), meteorological (Feng Yun), telecommunications (DFH) as well as recoverable capsules.<br>• Lunar orbiter (Chang'e 1)<br>• Commercial telecommunications satellites (sold for instance to Nigeria and Venezuela)<br>• Ground facilities for satellites<br>• Launch services, launchers (Long March) and launch facilities<br>• Manned spaceflight<br>• Bilateral cooperation with Brazil on CBERS satellites, and with ESA on the scientific mission Double Star and on the Dragon programme in applications.<br>China is hosting the Asia-Pacific Space Cooperation organisation (APSCO) Headquarters. |
| Budget | Estimated between 1.5 and 2.5 billion US dollars, i.e. between 1 and 2 billion euros |

## INDIA

| | |
|---|---|
| Population[435] | 1,113 millions |
| GDP[436] | 663 billion euros |
| Responsibility | The Space Commission defines Indian space policy and the Department of Space is responsible for India's space activities. The Indian Space Research Organisation (ISRO)[440] implements the space programmes. |
| Activities | The Indian activities include:<br>• Remote sensing satellites (IRS) and multi-purpose satellites (INSAT) with telecommunications and meteorological functions<br>• Dual-use satellites (CartoSat)<br>• Launch vehicles (PSLV and GSLV) and services<br>• Sounding Rockets<br>• Associated ground systems<br>• Cooperation mainly with Bulgaria, ESA and NASA on Chandrayaan-1, with CNES on Megha-Tropiques and Oceansat-3, with ASI on Oceasat-2, with Russia on Coronas-Foton, with Israel on TAUVEX and with Canada on UVIT |
| Staff | n.a. |
| Budget | In 2006/2007, about 35 billion rupees (about 600 million euros) |

## JAPAN[441]

| | |
|---|---|
| Population[435] | 127.7 millions |
| GDP[436] | 3,318 billion euros |
| Responsibility | The Japanese Aerospace Exploration Agency (JAXA), under the responsibility of the Ministry of Education, Culture, Sports, Science and Technology (MEXT), manages Japanese space activities. |
| Activities | The Japanese activities include:<br>• Scientific missions (Selene, Solar-B, Planet-C, Astro)<br>• ISS and manned spaceflight (Kibo module)<br>• Earth observation (GOSAT, GPM, GCOM)<br>• Telecommunications (ETS-VII, Winds, QZSS)<br>• Launch services (HII-A and M-V)<br>• New launchers (Galaxy Express)<br>• Cooperation with ESA on Bepi-Colombo and ALOS |
| Budget | In 2006, 1.8 billion euros including 1.1 billion euros for JAXA |
| Staff | JAXA – 1,700 |

# RUSSIA

ROSCOSMOS

| | |
|---|---|
| Population[435] | 142.8 millions |
| GDP[436] | 747 billion euros |
| Responsibility | Russia's Federal Space Agency, Roskosmos[442], under the direct responsibility of the Government, manages the Russian civil space activities, while the Military Space Forces (VKS), under the Ministry of Defence, manages the military space programmes. |
| Activities | Russian civil activities defined in the Federal Space Programme include:<br>• Satellites for science (Kompass), remote sensing (Resurs)<br>• Launchers (Soyuz, Rockot, Cosmos, Start, Shtil, and Cyclon, Zenit and Dnepr with Ukraine, and in development Angara)<br>• Cooperation with international partners in commercial ventures to commercialise launch services (ILS, Starsem, Rockot, Cosmos, Sea Launch)<br>• Cooperation with CNES on future launchers (Oural)<br>• Manned spaceflight (ISS) in cooperation with NASA and ESA including commercial activities (space tourists)<br>• Associated ground facilities<br>Military activities include dual-use navigation systems (Glonass), surveillance, early-warning, ELINT, and communications missions. |
| Staff | n.a. |
| Budget | In 2006, 25 billion rubles (about 700 million euros) and 305 billion rubles (about 9 billion euros) for the 10-year plan 2006–2015. |

## UKRAINE

| | |
|---|---|
| Population[435] | 46.6 millions |
| GDP[436] | 80 billion euros |
| Responsibility | The National Space Agency of Ukraine (NSAU)[443] is responsible for space activities in Ukraine, as well as about 30 design offices, research institutes and enterprises that represent most of the sector in Ukraine. The National Academy of Sciences of Ukraine supervises space research in its institutes and has joint institutions with NSAU. |
| Activities | Ukraine has a National Space Programme that includes:<br>• Scientific research (astrophysics and astronomy, ionosphere and magnetosphere research, microgravity and life sciences)<br>• Launch vehicles and launch services<br>• Operational launch vehicles (Zenit, Cyclone, Dnepr)<br>○ Launch vehicles under development (Cyclone 4, Dnepr-M, Mayak)<br>○ Remote sensing satellites (Okean and Sich series), telecommunications satellites (Lybid), automatic multi-purpose space platforms,<br>• Rocket and spacecraft engines, as well as advanced materials and technologies<br>• Ground facilities (National Space Facilities Control and Test Centre in Yevpatoriya) |
| Budget | In 2007, 52 million euros |
| Staff | NSAU HQ – 115 |

# USA

| | |
|---|---|
| Population[435] | 299 millions |
| GDP[436] | 10,030 billion euros |
| Responsibility | The National Aeronautics and Space Administration (NASA)[444] is responsible for most of the civil space programmes. The National Oceanic and Atmospheric Agency (NOAA), under the Department of Commerce, manages meteorological and oceanographic programmes.<br>The Department of Defense (DoD), and in particular the Air Force, manages most of the military space programmes. The National Reconnaissance Office (NRO) is responsible for the intelligence programmes. |
| Activities | American civil space activities include:<br>• Manned space flight (ISS)<br>• Science (Stereo, Dawn, GLAST, Kepler, SDO)<br>• Exploration (New Horizons, Phoenix, LCROSS)<br>• Earth observation (CloudSat, OCO) and meteorological satellites (GOES,)<br>• Commercial telecommunications satellites<br>• Launchers and launch services (Shuttle, Pegasus, Minotaur, Falcon)<br>• Associated ground systems<br>Military systems includes launchers (Atlas, Delta), as well as dual-use navigation (GPS), meteorological (DMSP), surveillance, intelligence, communications, early-warning, and technology demonstration systems. |
| Staff | n.a. |
| Budget | Most of the US space budget comes from NASA and the DoD.<br>In 2006, NASA's budget amounted to 16.6 billion US dollars (about 12.6 billions euros), which includes about 5% for aeronautic activities. |

[403] Eurostat. Average population 2006.
[404] Eurostat. Gross Domestic Product at market prices, Forecast 2007.
[405] FFG website. www.ffg.at/content.php?cid=232.
[406] Eurospace. Direct employment in the space sector 2006.
[407] Belgian Federal Science Policy Office website. www.belspo.be/belspo/res/rech/spatres_en.stm.
[408] ESA. December 2005.
[409] Czech Space Office website. www.czechspace.cz.
[410] Danish National Space Center website. www.spacecenter.dk.
[411] Tekes website. www.tekes.fi/eng/interests/interests.asp?aihe=Avaruus&eng=Space.
[412] CNES website. www.cnes.fr.
[413] ONERA website. www.onera.fr
[414] DLR website. www.dlr.de.
[415] GSRT website. www.gsrt.gr.
[416] Hungarian Space Office website. www.hso.hu.
[417] Entreprise Ireland. www.enterprise-ireland.com.
[418] ASI website. www.asi.it.
[419] Luxinnovation website. www.innovation.public.lu.
[420] NVIR website. www.nivr.nl.
[421] SRON website. www.sron.nl.
[422] Norwegian Space Center website. www.spacecentre.no.
[423] Space Research Centre website. e.cbk.waw.pl/cbk.
[424] Polish Space Office website. www.kosmos.gov.pl.
[425] GRICES website. www.grices.mctes.pt/gpe.
[426] ROSA website. www.rosa.ro.
[427] CDTI website. www.cdti.es/index.asp?MP=15&MS=192&MN=3.
[428] Prospacio website. www.proespacio.org.
[429] Swedish National Space Board. www.snsb.se.
[430] Space Affairs Division website. www.sbf.admin.ch/htm/themen/weltraum_fr.html.
[431] UK Space Activities 2007. http://www.bnsc.gov.uk/assets//channels/resources/publications/pdfs/BNSC%20SpaceActivities_2007.pdf.
[432] ESA website. www.esa.in t.
[433] DG Enterprise Space website. ec.europa.eu/enterprise/space.
[434] FP7 website. Space" theme budget as of December 2006, cordis.europa.eu/fp7.
[435] IMF. Population 2006.
[436] IMF. Gross Domestic Product at market price 2006.
[437] CSA website. www.space.gc.ca.
[438] CSA. 2003.
[439] CNSA website. www.cnsa.gov.cn.
[440] ISRO website. www.isro.org.
[441] JAXA website. www.jaxa.jp/index_e.html.
[442] Roscosmos website. www.roscosmos.ru.
[443] NSAU website. www.nkau.gov.ua.
[444] NASA website. www.nasa.gov.

# 3. Bibliography of space policy publications. January 2006–June 2007

Blandina Baranes

## 3.1. Monographs

Baker, Philip. The Story of Manned Space Stations: An Introduction. Berlin, New York: Springer; Chichester, UK: Praxis Publishing, 2007.
Barbree, Jay. "Live from Cape Canaveral": Covering the Space Race, from Sputnik to Today. New York: Smithsonian Books/Collins, 2007.
Basalla, George. Civilized Life in the Universe: Scientists on Intelligent Extraterrestrials. Oxford, New York: Oxford University Press, 2006.
Battrick, Bruce, ed. The Changing Earth – New Scientific Challenges for ESA's Living Planet Programme. Noordwijk: ESA Publications, 2006.
Beattie, Donald A. ISScapades: The Crippling of America's Space Program. Burlington, Ontario: Apogee Books, 2006.
Belbruno, Edward. Fly Me to the Moon: An Insider's Guide to the New Science of Space Travel. Princeton, New Jersey: Princeton University Press, 2007.
Belfiore, Michael. Rocketeers: How a Visionary Band of Business Leaders, Engineers, and Pilots is Boldly Privatizing Space. New York: Smithsonian Books/Collins, 2007.
Bizony, Piers. Space 50. New York: Smithsonian Books/Collins, 2006.
—. The Man Who Ran the Moon: James E. Webb, NASA, and the Secret History of Project Apollo. New York: Thunder's Mouth Press, 2006.
Bond, Peter. Distant Worlds: Milestones in Planetary Exploration. New York: Copernicus Books in association with Praxis Publishing, 2007.
Bonnal, Christophe. Position Paper on Space Debris Mitigation. Paris: International Academy of Astronautics, 2006.
Brisibe, Tare. Aeronautical Public Correspondence by Satellite. Utrecht: Eleven International Publishing, 2006.
Brzezinski, Matthew. Red Moon Rising: Sputnik and the Hidden Rivalries That Ignited the Space Age. Waterville, Maine: Thorndike Press, 2007.
Cadbury, Deborah. Space Race: The Epic Battle Between America and the Soviet Union for Dominion of Space. New York: Harper/Collins, 2006.
—. Space Race: The Battle to Rule the Heavens. New York: Harper/Collins, 2006.
Caldicott, Helen and Eisendrath, Craig. War in Heaven: The Arms Race in Outer Space. New York: The New Press, 2007.
Chien, Philip. Columbia – Final Voyage: The Last Flight of NASA's First Space Shuttle. New York: Copernicus Books, 2006.
Chouinard, Vicky. The Legal Framework Related to the Privatization and Commercialization of Remote Sensing Satellites in the United States and in Canada. Montreal: Mc Gill University, 2006.
Chun, Clayton K.S. Defending Space. US Anti-Satellite Warfare and Space Weaponry. Oxford: Osprey Publishing, 2006.
Clarke, Jonathan D.A., ed. Mars Analog Research. San Diego, California: Univelt, 2006.
Cockell, Charles S. Space on Earth: Saving our World by Seeking Others. New York: Macmillan, 2007.

Collins, Martin, ed. After Sputnik: 50 Years of the Space Age. New York: Smithsonian Books, 2007.
Comins, Neil F. The Hazards of Space Travel: A Tourist's Guide. Villard, 2007.
Contant-Jorgenson, Corinne, Lála Petr, and Schrogl, Kai-Uwe, eds. Space Traffic Management: Cosmic Study by the International Academy of Astronautics. Paris: International Academy of Astronautics, 2006.
Cook, Richard C. Challenger Revealed: An Insider's Account of How the Reagan Administration Caused the Greatest Tragedy of the Space Age. New York: Thunder's Mouth Press, 2007.
D'Antonio, Michael. A Ball, A Dog, and A Monkey: 1957 – The Space Race Begins. New York: Simon and Schuster, 2007.
De Groot, Gerard J. Dark Side of the Moon: The Magnificent Madness of American Lunar Quest. New York: New York University Press, 2006.
Desingly, Aurélien. Galileo, la navigation par satellite européenne: Questions juridiques et politiques au temps de la concession. Paris: Institut français des relations internationales, 2006.
Dick, Steven J. and Launius, Roger D., eds. Critical Issues in the History of Spaceflight. Washington DC: National Aeronautics and Space Administration, Office of External Relations, History Division, 2006.
—. Societal Impact of Spaceflight. Washington DC: National Aeronautics and Space Administration, Office of External Relations, 2007.
Feuerbacher, Berndt and Stoewer, Heinz, eds. Utilization of Space – Today and Tomorrow. Berlin, New York: Springer, 2006.
French, Francis and Burgess, Colin. In the Shadow of the Moon: A Challenging Journey to Tranquility, 1965–1969. Lincoln, Nebraska: University of Nebraska Press, 2007.
—. Into that Silent Sea: Trailblazers of the Space Era, 1961–1965. Lincoln, Nebraska: University of Nebraska Press, 2007.
Furniss, Tim, Shayler, David J., and Shayler, Michael D. Praxis Manned Spaceflight Log 1961–2006. New York: Springer, 2007.
—. A History of Space Exploration. London: Mercury Books, 2006.
Gethmann, Carl Friedrich, Rohner, Nicola, and Schrogl, Kai-Uwe, eds. Die Zukunft der Raumfahrt. Ihr Nutzen und ihr Wert. Bad Neuenahr-Ahrweiler: Europäische Akademie zur Erforschung von Folgen wissenschaftlich-technischer Entwicklungen, 2007.
Godwin, Robert, ed. Surveyor Lunar Exploration Program: The NASA Mission Reports. Burlington Ontario: Apogee Books, 2006.
—. Project Apollo: Exploring the Moon. Burlington Ontario: Apogee Books, 2006.
—. First Men at the Moon: Apollo 11. Burlington Ontario: Apogee Books, 2006.
—. Russian Spacecraft. Burlington Ontario: Apogee Books, 2006.
—. Space Shuttle Fact Archive. Burlington Ontario: Apogee Books, 2007.
Goh, Gérardine Meishan. Dispute Settlement in International Space Law. Leiden, Boston: Martinus Nijhoff Publishers, 2007.
Handberg, Roger and Li, Zhen. Chinese Space Policy. A Study in Domestic and International Politics. New York: Routledge, 2007.
—. International Space Commerce: Building from Scratch. Gainesville, Florida: University Press of Florida, 2006.
Harland, David M. The First Men on the Moon: The Story of Apollo 11. Berlin, New York: Springer; Chichester, UK: Praxis Pub., 2006.
Harland, David M. Space Exploration 2008, New York: Springer 2007.
Harvey, Brian. Russian Planetary Exploration: History, Development, Legacy, Prospects. Berlin, New York: Springer; Chichester, UK: Praxis Publishing, 2007.
Heiken, Grant and Jones, Eric. On the Moon: The Apollo Journals. Berlin, New York: Springer; Chichester, UK: Praxis Publishing, 2007.
Hey, Nigel. The Star Wars Enigma: Behind the Scenes of the Cold War Race for Missile Defense. Washington DC: Potomac Books, 2006.
Hitchens, Theresa. European Military Space Capabilities: A Primer. Washington DC: Center for Defense Information, 2006.

Hobe, Stephan, et al. Rechtliche Rahmenbedingungen einer zukünftigen kohärenten Struktur der europäischen Raumfahrt. Münster: LIT, 2006.

Hobe, Stephan, Schmidt-Tedd, Bernhard, and Schrogl, Kai-Uwe, eds. "Project 2001 Plus" – Global and European Challenges for Air and Space Law at the Edge of the 21st Century. Köln: Carl Heymanns Verlag, 2006.

Holtzmann-Kevles, Bettyann. Almost Heaven: The Story of Women in Space. 1st MIT Press ed. Cambridge Massachusetts: MIT Press Cambridge, 2006.

Jakosky, Bruce. Science, Society, and the Search for Life in the Universe. Tucson: University of Arizona Press, 2006.

Jensen, John R. Remote Sensing of the Environment: an Earth Resource Perspective. 2nd ed. Upper Saddle River, New Jersey: Pearson Prentice Hall, 2007.

Johnson-Freese, Joan. Space as a Strategic Asset. New York: Columbia University Press, 2007.

Jones, Chris. Too Far from Home: A Story of Life and Death in Space. New York: Doubleday, 2007.

Kahn, Nicholas and Selesnick, Richard. The Apollo Prophecies. New York: Aperture Foundation, 2006.

Kanipe, Jeff. Chasing Hubble's Shadows: The Search for Galaxies at the Edge of Time. 1st ed. New York: Hill and Wang, 2006.

Kemp, Kenny. Destination Space: How Space Tourism is Making Science Fiction a Reality. London: Virgin Books, 2007.

Kitmacher, Gary. Reference Guide to the International Space Station. Burlington, Ontario: Apogee Books, 2006.

Klein, John J. Space Warfare: Strategy, Principles and Policy. London, New York: Routledge, 2006.

Klinkrad, Heiner. Space Debris: Models and Risk Analysis. Berlin: Springer Praxis Books, 2006.

Krone, Bob, ed. Beyond Earth: the Future of Humans in Space. Burlington, Ontario: CG Publishing, 2006.

Lamb, Lawrence E. Inside the Space Race: A Space Surgeon's Diary. Austin, Texas: Synergy Books, 2006.

Lewis, James A. Waiting for Sputnik. Basic Research and Strategic Competition. Washington DC: CSIS Press, 2006.

Lilensten, Jean and Bornarel, Jean. Space Weather, Environment and Societies. Dordrecht: Springer, 2006.

Lipartito, Kenneth and Butler, Orville R. A History of the Kennedy Space Center. Gainesville, Florida: University Press of Florida, 2007.

Maini, Anil Kumar and Agrawal, Varsha. Satellite Technology: Principles and Applications. Chichester, UK, Hoboken, New Jersey: John Wiley, 2006.

Matloff, Gregory L., Johnson, Les, and Bangs, Constance. Living Off the Land in Space: Green Roads to the Cosmos. New York: Springer/Praxis, 2007.

Meadows, Arthur J. The Future of the Universe. London: Springer, 2007.

Michaud, Michael A.G. Contact With Alien Civilizations: Our Hopes and Fears about Encountering Extraterrestrials. New York: Springer, 2007.

Moulin, Hervé. La France dans l'Espace 1959–1979: Contribution à l'effort spatial européen. Noordwijk: ESA Publications, 2006.

Netzband, Maik, Stefanov, William L., and Redman, Charles, eds. Applied Remote Sensing for Urban Planning, Governance and Sustainability. Berlin, New York: Springer, 2007.

Orloff, Richard W. and Harland, David M. Apollo: The Definitive Sourcebook. Berlin: Springer, 2006.

Pop, Virgiliu. Unreal Estate: The Men who Sold the Moon. Liskeard: Exposure, 2006.

Powell-Willhite, Irene E., ed. The Voice of Dr. Wernher von Braun. Burlington, Ontario: Collectors Guide Pub., 2007.

Reidy Michael S., Kroll, Gary, and Conway, Erik M. Exploration and Science: Social Impact and Interaction. Santa Barabra, California: Abc-Clio, 2007.

Rieke, George H. The Last of the Great Observatories: Spitzer and the Era of Faster, Better, Cheaper at NASA. Tucson, Arizona: University of Arizona Press, 2006.

Sankar, U. The Economics of India's Space Programme: An Exploratory Analysis. New Delhi: Oxford University Press, 2007.
Sawyer, Kathy. The Rock from Mars: A Detective Story on Two Planets. New York: Random House, 2006.
Schmitt, Harrison H. Return to the Moon: Exploration, Enterprise, and Energy in the Human Settlement of Space. New York: Copernicus Books in association with Praxis Publishing, 2006.
Schrogl, Kai-Uwe and Trischler, Helmuth, eds. Ein Jahrhundert im Flug: Luft- und Raumfahrtfoschung in Deutschland 1907–2007. Frankfurt: Campus, 2007.
Schrunk, David G. The Moon: Resources, Future Development and Settlement. New York: Springer, 2007.
Space Security, 2006. Waterloo, Ontario: Space Security Org, 2006.
Sparrow, Giles. Spaceflight: The Complete Story from Sputnik to Shuttle and Beyond. New York: DK Publishing, 2007.
Taylor, George and Blewitt, Geoff. Intelligent Positioning: GIS-GPS Unification. Chichester, UK: Wiley, 2006.
Tumminia, Diana G. ed. Alien Worlds: Social and Religious Dimensions of Extraterrestrial Contact. Syracuse, NY: Syracuse University Press, 2007.
Tyson – de Grasse, Neil. Death by Black Hole: And Other Cosmic Quandaries. New York: W.W. Norton, 2007.
van Pelt, Michel. Space Invaders: How Robotic Spacecraft Explore the Solar System. New York: Springer, 2007.
von der Dunk, Frans and Brus, Marcel M.T.A., eds. The International Space Station: Commercial Utilisation from a European Legal Perspective. Leiden, Boston: Martinus Nijhoff, 2006.
Weintraub, David A. Is Pluto a Planet? A Historical Journey Through the Solar System. Princeton New Jersey: Princeton University Press, 2007.
Werth, Karsten. Ersatzkrieg im Weltraum: Das US-Raumfahrtprogramm in der Öffentlichkeit der 1960er Jahre. Frankfurt/Main: Campus, 2006.
Westwick, Peter J. Into the Black: JPL and the American Space Program, 1976–2004. New Haven, Connecticut: Yale University Press, 2006.
Williamson, Mark. Space: The Fragile Frontier. Reston, Virginia: American Institute of Aeronautics and Astronautics Press, 2006.
Wolter, Detlev. Common Security in Outer Space and International Law. Geneva: UNIDIR, 2006.
Wong, Wilson. Weapons in Space: Strategic and Policy Implications. Winnipeg: University of Manitoba, 2006.

## 3.2. Articles

Ailor, William H. "Space Traffic Management: Implementations and Implications." Acta Astronautica 58 (2006): 279–286.
Autret, Florence. "Quelle organisation pour l'Europe spatiale? Politique étrangère 2 (2007): 281–292.
Avnet, Mark S. "The Space Elevator in the Context of Current Space Exploration Policy." Space Policy 22 (2006): 133–139.
Bashor, Harold. "Interpretation of the Moon Treaty: Recourse to Working Papers and Related International Documents." Annals of Air and Space Law XXXII (2007).
Billings, Linda. "Exploration for the Masses? Or Joyrides for the Ultra-rich? Prospects for Space Tourism." Space Policy 22 (2006): 162–164.

——. "How Shall We Live in Space? Culture, Law and Ethics in Spacefaring Society." Space Policy 22 (2006): 249–255.

Bini, Antonella. "Export Control of Space Items: Preserving Europe's Advantage." Space Policy 23 (2007): 70–72.

Brachet, Gérard and Deloffre, Bernard. "Space for Defence: A European Vision." Space Policy 22 (2006): 92–99.

Brearley, Andrew. "Mining the Moon: Owning the Night Sky?" Astropolitics 4 (2006): 43–67.

Brisibe, Tare. "Law and Regulation of Activities Related to Outer Space in Nigeria." ZLW – Zeitschrift für Luft- und Weltraumrecht (German Journal of Air and Space Law) 55 (2006): 554–566.

Brünner, Christian and Soucek, Alexander. "Regulating ISS – An Interdisciplinary Essay." Acta Astronautica 60 (2007): 594–598.

Bujon de l'Estang, Francois and de Montluc, Bertrand. "Making Space the Key to Security and Defence Capabilities in Europe: What Needs to be Done." Space Policy 22 (2006): 75–78.

Burzykowska, Anna. "ESDP and the Space Sector – Defining the Architecture and Mechanisms for Effective Cooperation." Space Policy 22 (2006): 35–41.

Casini, Silvano. "Dealing with the International Implications of Space Exploration." Space Policy 22 (2006): 155–157.

Chatzipanagiotis, Michael. "Registration of Space Objects and Transfer of Ownership in Orbit." ZLW – Zeitschrift für Luft- und Weltraumrecht (German Journal of Air and Space Law) 56 (2007): 229–238.

Choi, Eric and Niculescu, Sorin. "The Impact of US Export Controls on the Canadian Space Industry." Space Policy 22 (2006): 29–34.

Cockell, Charles S. and Horneck, Gerda. "Planetary Parks – Formulating a Wilderness Policy for Planetary Bodies." Space Policy 22 (2006): 256–261.

Correll, Randall R. "U.S.- India Space Partnership: The Jewel in the Crown." Astropolitics 4 (2006): 159–177.

Couston, Mireille. "Eléments de réflections sur le principe de l'utilisation pacifique de l'espace." Revue Française de Droit Aérien et Spatial, 238 (2006): 130.

——. "Les leçons de Galilée. "Revue Française de Droit Aérien et Spatial, 238 (2006): 128.

Cremins, Thomas and Spudis, Paul D. "The Stratgic Context of the Moon Echoes of the Past, Symphony of the Future." Astropolitics 5 (2007): 87–104.

David, James. "Astronaut Photography and the Intelligence Community: Who Saw What?" Space Policy 22 (2006): 185–193.

Dick, Steven J. "Assessing the Impact of Space on Society." Space Policy 23 (2007): 29–32.

Dorado, José M. "The First Spanish Space Programme 1968–1974." Acta Astronautica 61 (2007): 534–547.

Dupas, Alain and Logsdon, John M. "Creating a Productive International Partnership in the Vision for Space Exploration." Space Policy 23 (2007): 24–28.

Eichler, Peter, et al. "Astronaut Training for the European ISS Contributions Columbus Module and ATV." Acta Astronautica 59 (2006): 1111–1162

Finarelli, Peggy and Pryke, Ian. "Building and Maintaining the Constituency for Long-term Space Exploration." Space Policy 23 (2007): 13–19.

——. Implementing International Co-operation in Space Exploration. Space Policy 22 (2006): 23–28.

Finch, Edward R. Jr. "World Economics for Mankind's Frontier." Acta Astronautica 60 (2007): 780–782.

Froehlich, Annette. "Verletzung von Menschenrechten durch TV-Satellitenprogramme aufgrund unterschiedlicher kultureller Wertvostellungen. Zugleich Anmerkungen zum französischen Fall *Al-Manar*." ZLW – Zeitschrift für Luft- und Weltraumrecht (German Journal of Air and Space Law) 55 (2006): 541–553.

Gainor, Christopher. "Canada's Space Program, 1958–1989: A Program Without an Agency." Acta Astronautica 60 (2007): 132–139.

Gangale, Thomas. "Who Owns the Geostationary Orbit?" Annals of Air and Space Law XXXI (2006).
Gaubert, Alain. "Is There Really Any Duplication in Europe's Space Activities?" Space Policy 22 (2006): 1–2.
Genta, Giancarlo and Rycroft, Michael. "Will Space Actually Be the Final Frontier of Humankind?" Acta Astronautica 58 (2006): 287–295.
Gerhard, Michael and Marenkov, Dmitry. "Zur Lizenzierung von Weltraumaktivitäten in Russland." ZLW – Zeitschrift für Luft- und Weltraumrecht (German Journal of Air and Space Law) 56 (2007): 211–228.
Gilruth, Peter T., et al. "Measuring Performance: Moving NASA Earth Science Products into the Mainstream User Community." Space Policy 22 (2006): 165–175.
Gleason, Michael P. "European Union Space Initiatives: The Political Will for Increasing European Space Power." Astropolitics 4 (2006): 7–41.
Gomes, Vera. "Portugal in Space." Acta Astronautica 61 (2007): 526–533.
Harris, Alexandra and Harris, Ray. "The Need for Air Space and Outer Space Demarcation." Space Policy 22 (2006): 3–7.
Harsch, Viktor. "Historical Aspects of Human Presence in Space." Acta Astronautica, 60 (2007): 607–609.
Hempsell, Mark. "Space Power as a Response to Global Catastrophes." Acta Astronautica 59 (2006): 524–530.
Hermida, Julian. "A Legal and Criminological Approach to Criminal Acts in Outer Space." Annals of Air and Space Law XXXI (2006).
Hertzfeld, Henry R., Williamson, Ray A., and Peter, Nicolas. "The Relevance of Economic Data in the Decision-making Process for Orbital Launch Vehicle Programs, a U.S. Perspective." Acta Astronautica 61 (2007): 1076–1084.
Hertzfeld, Henry R. and Peter, Nicolas. "Developing New Launch Vehicle Technology: The Case for Multi-national Private Sector Cooperation." Space Policy 23 (2007): 81–89.
Hobe, Stephan. "Adequacy of the Current Legal and Regulatory Framework Relating to the Extraction and Appropriation of Natural Resources in Outer Space." Annals of Air and Space Law XXXII (2007).
Hwang, Young, Chin. "Space Activities in Korea – History, Current Programs and Future Plans." Space Policy 22 (2006): 194–199.
Ingold, Olivier. "Soyuz in French Guiana: A Strategic Perspective." Space Policy 22 (2006): 140–148.
Ionine, Andreï. "Une heure stratégique pour le spatial russe." Politique étrangère 2 (2007): 267–279.
Johnson-Freese, Joan and Erickson, Andrew S. "The Emerging China-EU Space Partnership: A Geotechnological Balancer." Space Policy 22 (2006):12–22.
Johnson-Freese, Joan. "A New US-Sino Space Relationship: Moving Toward Cooperation." Astropolitics 4 (2006): 131–158.
Jones, Harriet, Yeoman, Kay, and Cockell, Charles. "A Pilot Survey of Attitudes to Space Sciences and Exploration among British School Children." Space Policy 23 (2007): 20–23.
Jurist, John, Dinkin, Sam, and Livingston, David. "Low Cost Earth Orbit Access: A Look at Physics, Economics, and Reality." Astropolitics 4 (2006): 295–331.
Kanas, Nick A., et al. "Human Interactions in Space: ISS vs. Shuttle/Mir." Acta Astronautica 59 (2006): 413–419.
Kaul, Ranjana. "Control of Space Assets – Ethics in International Diplomacy." Annals of Air and Space Law XXXI (2006).
Kerstein, Aleksander and Matko, Drago. "Eugen Sänger: Eminent Space Pioneer." Acta Astronautica 61 (2007): 1085–1092.
Kim, Doo Hwan. "Korea's Space Development Programme: Policy and Law." Space Policy 22 (2006): 110–117.
Kleinberg, Howard. "On War in Space." Astropolitics 5 (2007): 1–27.
Komerath, Narayanan, Nally, James, and Zilin Tang, Elizabeth. "Policy Model for Space Economy Infrastructure." Acta Astronautica 61 (2007): 1066–1075.

Koudelka, Otto and Schrotter, Peter. "Satellite Services for Disaster Management and Security Applications." Acta Astronautica 60 (2007): 986–991.

Lafferranderie, Gabriel. "Faut-il toujours enseigner le droit de l'espace? "ZLW – Zeitschrift für Luft- und Weltraumrecht (German Journal of Air and Space Law) 55 (2006): 517–540.

Lambright, William Henry. "Leading Change at NASA: The Case of Dan Goldin." Space Policy 23 (2007): 33–43.

Launius, Roger D. and Jenkins, Dennis R. "Is It Finally Time for Space Tourism?" Astropolitics 4 (2006): 253–280.

Launius, Roger D. "Assessing the Legacy of the Space Shuttle." Space Policy 22 (2006): 226–234.

Lautenbacher, Conrad C. "The Global Earth Observation System of Systems: Science Serving Society." Space Policy 22 (2006): 8–11.

Lee, Ricky J. "The Liability Convention and Private Space Launch Services – Domestic Regulatory Responses." Annals of Air and Space Law XXXI (2006).

Lee, Yoon. "Registration of Space Objects: ESA Member States' Practice." Space Policy 22 (2006): 42–51.

Lenard, Roger X. "Nuclear Safety, Legal Aspects and Policy Recommendations for Space Nuclear Power and Propulsion Systems." Acta Astronautica 59 (2006): 398–412.

Lewis, James A. "La dynamique de l'arsenalisation de l'espace." Politique étrangère 2 (2007): 253–265.

Lin, Patrick. "Look Before Taking Another Leap for Mankind-Ethical and Social Considerations in Rebuilding Society in Space." Astropolitics 4 (2006): 281–294.

Macauley, Molly K. and Shih, Jhih-Shyang. "Satellite Solar Power: Renewed Interest in an Age of Climate Change?" Space Policy 23 (2007): 108–120.

Macauley, Molly K. "The Value of Information: Measuring the Contribution of Space-derived Earth Science Data to Resource Management." Space Policy 22 (2006): 274–282.

Maccone, Claudio. "Planetary Defense From Space: Part 2 (Simple) Asteroid Deflection Law." Acta Astronautica 58 (2006): 662–670.

Madders Kevin and Thiebaut, Walter. "Carpe Diem: Europe Must Make a Genuine Space Policy Now." Space Policy 23 (2007): 7–12.

Marenkov, Dmitry. "Zum Russischen Weltraumgesetz in seiner aktualisierten Fassung vom 2. Februar 2006." ZLW – Zeitschrift für Luft- und Weltraumrecht (German Journal of Air and Space Law) 56 (2007): 58–85.

Mathurin, Jeph and Peter, Nicolas. "Private Equity Investments Beyond Earth Orbits: Can Space Exploration Be the New Frontier for Private Investments?" Acta Astronautica 59 (2006): 438–444.

Matsumoto Kohtaro, et al. "Japanese Lunar Exploration Long-term Plan." Acta Astronautica 59 (2006): 68–76.

Moltz, James Clay. "Preventing Conflict in Space: Cooperative Engagement as a Possible U.S. Strategy." Astropolitics 4 (2006): 121–129.

Müller, Hartmut, Heidmann, Hans-Jörg and Apel, Uwe. "Autonomous European Lunar Exploration-Entry Point for a Global Cooperation." Acta Astronautica 61 (2007): 88–94.

Nardon, Laurence. "Cold War Space Policy and Observation Satellites." Astropolitics 5 (2007): 29–62.

—. "Où va le programme spatial français?" Politique étrangère 2 (2007): 293–305.

Nase, Vernon. "The Questionable Legality of the US Space Elevator Concept". ZLW – Zeitschrift für Luft- und Weltraumrecht (German Journal of Air and Space Law) 55 (2006): 118–136

Neufeld, Michael J. "Space Superiority: Wernher von Braun's Campaign for a Nuclear-armed Space Station, 1946–1956." Space Policy 22 (2006): 52–62.

Neuneck, Götz and Rothkirch, André. "The Possible Weaponization of Space and Options for Preventive Arms Control." ZLW – Zeitschrift für Luft- und Weltraumrecht (German Journal of Air and Space Law) 55 (2006): 501–516.

Nylund, Amund and Rønningen, Jan-Erik. "Technical and Educational Improvements of the Student Rocket Program at NAROM and Andøya Rocket Range." Acta Astronautica 61 (2007): 506–513.

Oliver, Carol A. and James, Fergusson. "Astrobiology: A Pathway to Adult Science Literacy?" Acta Astronautica 61 (2007): 716–723.
Oliver, Carol A. "The Virtual Space Exploration Education Portal." Acta Astronautica 61 (2007): 548–552.
Omarova, Gulnara and Omarova, Zhuldis. "Kazakhstan's Space Policy in a Changing National and Global Context." Space Policy 22 (2006): 200–204.
Paikowsky, Deganit. "Israel's Space Program as a National Asset." Space Policy 23 (2007): 90–96.
Pass, Jim. "Astrosociology as the Missing Perspective." Astropolitics 4 (2006): 85–99.
Paxton, Larry J. "Faster, Better, and Cheaper at NASA: Lessons Learned in Managing and Accepting Risk." Acta Astronautica 61 (2007): 954–963.
Pelton, Joseph N. "Revitalizing NASA? A Five-point Plan." Space Policy 22 (2006): 221–225.
Perek, Luboš. "Between a Celestial Body and a Spacecraft: Making the Space Elevator a Success." Space Policy 23 (2007): 3–6.
Peter, Nicolas, et al. "Space Technology, Sustainable Development and Community Applications: Internet as a Facilitator." Acta Astronautica 59 (2006): 445–451.
Peter, Nicolas. "The Changing Geopolitics of Space Activities." Space Policy 22 (2006): 100–109.
—. "The EU's Emergent Space Diplomacy." Space Policy 23 (2007): 97–107.
Pierson, Thomas. "SETI Institute as a Model for Managing Interdisciplinary Science." Acta Astronautica 58 (2006): 478–484.
Pinkston, Daniel A. "North and South Korean Space Development: Prospects for Cooperation and Conflict." Astropolitics 4 (2006): 207–227.
Pisano, Marco. "Moving Europe Towards a More Effective Procurement of Space-based Assets." Space Policy 22 (2006): 176–184.
Plattard, Serge. "View from the Top: Space Policy: Well It's a Start". Research Europe 231 (2007): 8.
—. "La France et l'Europe face aux défis de l'espace" Géopolitique 98 (2007): 58–70.
Pop, Virgiliu. "The Nation of Celestial Space." Space Policy 22 (2006): 205–213.
Raftery, Michael and Fox, Todd. "The Crew Exploration Vehicle (CEV) and the Next Generation of Human Spaceflight." Acta Astronautica 61 (2007): 185–192
Rao, Mukund and Murthi, K.R. Sridhara. "Keeping up with Remote Sensing and GI Advances – Policy and Legal Perspectives." Space Policy 22 (2006): 262–273.
Rapp, Donald. "Solar Power Beamed from Space." Astropolitics 5 (2007): 63–86.
Ravillon, Laurence. Jurisprudence spatiale – actualités 2005/2006. Revue Française de Droit Aérien et Spatial 239 (2006): 262.
Robinson, Julie A., Thumm, Tracy L. and Thomas, Donald A. "NASA Utilization of the International Space Station and the *Vision for Space Exploration*." Acta Astronautica 61 (2007): 176–184.
Robinson, George S. "Forward Contamination of Interstitial Space and Celestial Bodies: Risk Reduction, Cultural Objectives, and the Law." ZLW – Zeitschrift für Luft- und Weltraumrecht (German Journal of Air and Space Law) 55 (2006): 380–399.
—. "The U.S. National Space Policy: Pushing the Limits of Space Treaties?" ZLW – Zeitschrift für Luft- und Weltraumrecht (German Journal of Air and Space Law) 56 (2007): 45–57.
Rogers, Thomas F. "Magnifying Our World: Why we Must Extend Civilization to the Moon." Space Policy 22 (2006): 128–132.
Ryan, Mike H. and Kutschera, Ida. "Lunar-based Enterprise Infrastructure – Hidden Keys for Long-term Business Success." Space Policy 23 (2007): 44–52.
Rycroft, Michael J. "Space Exploration Goals for the 21st Century." Space Policy 22 (2006): 158–161.
Sadeh, Eligar. "Management Dynamics of NASA's Human Spaceflight Programs." Space Policy 22 (2006): 235–248.
—. "Public Management Dynamics of NASA: Interview with NASA Associate Administrator Rex Geveden." Astropolitics 4 (2006): 101–119.
Sarkissian, John M. "Return to the Moon: A Sustainable Strategy." Space Policy 22 (2006): 118–127.
Scheffran, Jürgen. "Dual – Use in a New Security Environment. The Case of Missiles and Space." INESAP Information Bulletin 26 (2006): 48

Sevastiyanov, N. Nikolay and Bryukhanov, Nikolay A. "Space Stations: Evolution and New Programs." Acta Astronautica 61 (2007): 193–197.
Sheldon, John B. "A Really Hard Case: Iranian Space Ambitions and the Prospects for U.S. Engagement." Astropolitics 4 (2006): 229–251.
Skoog, Ingemar A. and Abramov, Isaak P. "The Soviet/Russian Spacesuit History: Part III – The European Connection." Acta Astronautica 60 (2007): 1002–1014.
Smith, Lesley Jane and Hörl, Kay-Uwe. "Institutional Challenges for Space Activities in Europe ." Acta Astonautica 60 (2007): 210–220.
Sterken, Veerle J. "Sir Hermann Bondi: A Journey Through his Life and the Early Endeavours of Europe into Space." Acta Astronautica 61 (2007): 514–525.
Suedfeld, Peter. "Space memoirs: Value Hierarchies Before and After Missions – A Pilot Study." Acta Astronautica 58 (2006): 583–586.
Suzuki, Kazuto. "Transforming Japan's Space Policy-making." Space Policy 23 (2007): 73–80.
Suzuki, Minoru. "Alternative International Cooperation in Space Development for Japan-Need for More Cost-effective Space Application Projects." Acta Astronautica 59 (2006): 430–437.
Swan, Cathy W. and Swan, Peter A. "Why We Need a Space Elevator." Space Policy 22 (2006): 86–91.
Trepczynski, Susan. "The Benefits of Granting Immunity to Private Companies Involved in Commercial Space Ventures." Annals of Air and Space Law XXXI (2006).
von der Dunk, Frans G. "The Moon Agreement and the Prospect of Commercial Exploitation of Lunar Resources." Annals of Air and Space Law XXXII (2007).
—. "Towards Monitoring Galileo: the European GNSS Supervisory Authority in *statu nascendi*" ZLW – Zeitschrift für Luft- und Weltraumrecht (German Journal of Air and Space Law) 55 (2006): 100–117.
Wertz, Julie and Miller, David. "Expected Productivity-based Risk Analysis in Conceptual Design." Acta Astronautica 59 (2006): 420–429.
Worden, Simon P. and Sponable, Jess. "Access to Space: A Strategy for the Twenty-First Century." Astropolitics 4 (2006): 69–83.
Worms, Jean-Claude and Walter, Nicolas. "Enhancing Europe 's Capabilities in Space Research and Technology: Towards a European Space Board." Space Policy 22 (2006): 79–85.
Zaborskiy, Victor. "Space Engagement with Russia and Ukraine: Preventing Conflicts and Proliferation." Astropolitics 4 (2006): 179–206.
Zhao, Yun. "National Space Legislation, with Reference to China's Practice." Annals of Air and Space Law XXXII (2007).

# About the authors

**Blandina Baranes** joined the European Space Policy Institute (ESPI) in Vienna in February 2005. She is the librarian and documentalist of ESPI. Before she was the chief librarian of the Jewish Studies Department at the University of Vienna. During the past years she has been working as documentalist and librarian for different institutions such as, The Austrian Broadcasting Corporation, DER SPIEGEL and others. She conducted her studies and research in Austria and Israel and graduated with a Master from Vienna University, Faculty of Philosophy, Department of Social and Cultural Anthropology.

**Klaus Becher** is Managing Partner of the security, defence and space consultancy Knowledge & Analysis LLP in London and Research Director at the International Institute for Liberal Policy (IILP) in Vienna, Austria. His past positions include Associate Director at the Foreign and Commonwealth Office's conference centre Wilton Park in Steyning, West Sussex; Head of the European security programme at the International Institute for Strategic Studies (IISS) in London; Senior research fellow for strategic developments, arms control and technological trends at the Stiftung Wissenschaft und Politik (SWP) in Ebenhausen near Munich, and executive editor of the German Council on Foreign Relation's international affairs yearbook.

**Jean-Louis Fellous** graduated from Paris University with a doctorate degree in Meteorology (1972). He worked as an atmospheric scientist for ten years before joining the French space agency, Cnes, in 1982 as program manager of the U.S.-French Topex/Poseidon oceanography satellite. He has headed the Earth Observation Programmes at Cnes since 1998 and was director for ocean research at Ifremer, the French ocean research institute, from 2001 until 2005. Jean-Louis Fellous presently works with the European Space Agency as coordinator of Earth observation satellite programs related to climate, environment and security. He is the Executive Officer of the Committee on Earth Observation Satellites (Ceos) and a co-president of the Joint Technical Commission on Oceanography and Marine Meteorology (Jcomm) of the World Meteorological Organization (WMO) and the Intergovernmental Oceanographic Commission (IOC). Jean-Louis Fellous has published numerous scientific papers, and several books in the field of climate change and space observations. He was awarded the NASA Public Service Medal in 1994, the William T. Pecora Award (1998) as part of the International Topex/Poseidon team, and has received the medals of chevalier of the French Ordre national du mérite (2003) and Légion d'honneur (2007). Starting January 2008 he will become the Executive Director of COSPAR, the International Committee on Space Research.

**Joerg Kreisel** is CEO of JKIC (Joerg Kreisel International Consultant – Space Business & Finance Advisors) in Germany. Mr. Kreisel specializes in technology commercialization – with a strong focus on commercial space – since 1987. After pursuing a career in space business (research and technology management), he became a venture capitalist in the early 1990s, with a focus on international activities in both space and early-stage equity finance. Today Mr. Kreisel (JKIC) advises space institutions, industry, SMEs and selected investors worldwide in strategy development and business matters. JKIC's major business activities are aimed at creating and supporting space ventures, business partnerships, equity finance and global links. In addition JKIC provides leading solutions in data security for the business continuity via a Swiss partnership. Mr. Kreisel holds a degree in aerospace engineering from Aachen University of Technology (RWTH Aachen) and is an alumnus of the International Space University (ISU). He is member of the European Private Equity & Venture Capital Association (EVCA), the American Institute for Aeronautics & Astronautics (AIAA), the German Aerospace Association (DGLR), as well as several other regional and international networks. Mr. Kreisel is also a lecturer at two international post-graduate masters degree programs. He is married with two children.

**Burton H. Lee** is Managing Partner of Innovarium Ventures, Managing Director of the Space Angels Network, and an accredited angel investor. Innovarium Ventures, based in Silicon Valley and Washington DC, provides senior strategic, financial and technical advisory services to technology startup companies, venture capital and private equity firms, angel networks, major corporations and governments in the commercial space, advanced computing and IT, robotics, alternative energy and nanotechnology industries. Burton's professional experience spans 15 years in strategy consulting, high tech industry, government and venture-backed startups working in corporate development and strategy, business development, technology commercialization, and advanced computing and space systems research. His management and technical experience includes senior positions with leading organizations such as GE Global Research, Hewlett Packard, DaimlerChrysler AG and NASA in the United States, Europe and Japan. Burton H. Lee served as Chairman of the highly successful May 2007 Space Venture Finance Symposium, which was held in Dallas. He also recently acted as Technical Consultant to the New Mexico Spaceport Authority. In 2006, Lee was appointed a Senior Science and Technology Policy Fellow at the National

Academy of Sciences in Washington D.C. He holds a doctorate degree in mechanical and electrical engineering, an MBA in finance and entrepreneurship, and has been a guest lecturer and graduate of the founding class of the International Space University (ISU SSP 1988).

**John M. Logsdon** is Director of the Space Policy Institute at George Washington University's Elliott School of International Affairs, where he is also Research Professor and Professor Emeritus of Political Science and International Affairs. He holds a B.S. in Physics from Xavier University (1960) and a Ph.D. in Political Science from New York University (1970). John M. Logsdon's research interests focus on the policy and historical aspects of U.S. and international space activities. John M. Logsdon is the author of The Decision to Go to the Moon: Project Apollo and the National Interest and is general editor of the eight-volume series Exploring the Unknown: Selected Documents in the History of the U.S. Civil Space Program. He has written numerous articles and reports on space policy and history. He is frequently consulted by the electronic and print media for his views on space issues. John M. Logsdon is a member of the NASA Advisory Council and of the Commercial Space Transportation Advisory Committee of the Department of Transportation. In 2003, he served as a member of the Columbia Accident Investigation Board. He is a recipient of the NASA Distinguished Public Service and Public Service Medals, the 2005 John F. Kennedy Award from the American Astronautical Society, and the 2006 Barry Goldwater Space Educator Award of the American Institute of Aeronautics and Astronautics. He is a Fellow of the American Institute of Aeronautics and Astronautics and the American Association for the Advancement of Science. He is a member of the International Academy of Astronautics.

**Kevin Madders** is Director of Policy Studies at the Interdisciplinary Centre for Space Studies (ICSS) at the Catholic University of Leuven, Belgium, and the co-organizer of the European Space Policy Workshop series at Leuven since 2002. He is a specialist in space and information and communications technology policy and law, and heads the independent space and telecommunications consultancy Systemics Network International. Over his career he has combined academic with professional activities in various contexts – as an editor, barrister, European Space Agency official, company director, and government and corporate adviser and negotiator. His research has included general and applied systems theory and, in the space field, subjects ranging from the Strategic Defense Initiative to the regulation of satellite communications. He published a comprehensive analysis and critique of Europe's development in the space field in his book A New Force at a New Frontier (1997, in paperback 2006, Cambridge Univ. Press). Recent research projects have included managing a major study on potential space cooperation with China, referenced below. His recent university teaching has included space policy and law, EU law and ICT law. He has chaired various groups and fora, including the final drafting session for the International Space Station IGA in 1989, the legal group reviewing Eutelsat's privatization arrangements in 2001, and the governance working group at the first ESPI space conference in 2005 which defined institutional and management scenarios for reform of European space policy structures and their functioning. Among his professional affiliations, he founded and chaired for eight years a commercial ICT forum. He holds a bachelors from London, a masters from Yale and a doctorate from Cambridge.

**Charlotte Mathieu** has been a Research Fellow at the European Space Policy Institute (ESPI) since October 2006. Before joining ESPI, she was a research assistant at the Massachusetts Institute of Technology (MIT) where she conducted research on "fractionated" satellite architectures for DARPA for two years. In 2003–04, she was a Young Graduate Trainee in the Cost Engineering Division at ESTEC. She interned at the CNES Washington Office, EADS Launchers and SNECMA. She holds a double Masters degree in aerospace engineering from the Ecole Centrale Paris, France, and the Royal Institute of Technology (KTH), Sweden, as well as a second Masters degree in Technology and Policy from the Massachusetts Institute of Technology (MIT), USA. She also holds a degree in Economics from Université Pierre Mendès France, France.

**Götz Neuneck** is since 2003 the Head of the Interdisciplinary Research Group on Disarmament, Arms Control and Risk Technologies (IFAR) at the Institute for Peace Research and Security Policy at the University of Hamburg (IFSH). From 1976 to 1983, he studied Physics and Philosophy at the University of Düsseldorf (Grade: Dipl.-Phys.). From 1984–1987, he was Scientific Associate with Horst Afheldt and C.-F. von Weizsäcker in the Max-Planck-Society in Starnberg near Munich (Non-Offensive Defense). From 1988 to 1989 he was awarded a scholarship from the Volkswagen Foundation at IFSH and carried out research at the School of Public Affairs/University of Maryland (Prof. Catherine Kelleher, Prof. Steve Fetter). Since 1989, he is a Science Fellow at IFSH. His main research subjects are New Military Technologies, Proliferation of Weapons of Mass Destruction, Preventive Arms Control, Missile Defense, Missile Proliferation, and Space Weaponization. In 1995, he defended his dissertation in Mathematics at the University of Hamburg; Dr. rer. nat From 1988 to 2003 he was a member in the CENSIS-Working Group "Modernisation and Stability: The Impact of New

Weapon Technologies" and since 1996, he is an Elected Member of the German Physical Society (DPG) Council. Since 2001, he has lectured in the Master Programme "Peace and Security Studies" at the University of Hamburg. Since 2003, he has chaired the Working group "Physics and Disarmament" (AKA) of the DPG. In May 2005, he was a member of the German Delegation at the 7th NPT Review Conference in New York. He is a member of the Council of the "Pugwash Conferences on Science and World Affairs", of the Scientific Advisory Board of the German Foundation of Peace Research (DSF), and of the Advisory Board of International Physicians for the prevention of Nuclear War (IPPNW). He is the Vice-Chairman of the "Research Association Science, Disarmament, and International Security" (FONAS) and a Pugwash-Representative of the Federation of German Scientists (VDW).

**Xavier Pasco** (Political Science-University of Paris-Sorbonne) is a Senior Research Fellow at the Foundation pour la Recherche Stratégique (FRS) based in Paris where he is in charge of the Department "Technology, Space and Security". Previous to 1997, he was researcher at CREST (Center for Research and Evaluation of the relationships between Strategies and Technology) associated to Ecole Polytechnique. His research is currently focused on space and high technology policies and decision-making processes associated with national security strategies. He is working more specifically on the U.S. policies and on their impact on the transatlantic relationship in the space activity, both in the civilian and military domains. He has also conducted work on the NATO-European defense structure relationship in the domain of interoperability and coalition warfare. He is also involved in a number of projects studying the use of space for the security of Europe, notably in relation with the "Preparatory Action for Security Research" started in 2004 by the European Commission to structure the future Framework Programme (2007–2013). Xavier Pasco is also associate Professor at the University of Marne-la-Vallée and is associate research fellow at the Space Policy Institute in the George Washington University (Washington D.C., USA). He is also giving lectures in the French Military School in Paris. He is also the European Editor of the international academic review Space Policy. He is permanent Invited Member of the French National Marine Academy. In 2006, Xavier Pasco has also been elected corresponding member of the International Academy of Astronautics. He has published numerous works (books, articles, papers) on these topics among which: "La politique spatiale des Etats-Unis", 1958–1997, Technologie, Intérêt national et débat public, Paris, L'Harmattan, 300 p., 1997; "Espace et puissance", FRS, Ellipse, Paris 1999 (in collaboration); and as a co-author, "L'espace nouveau territoire, Atlas des satellites et des politiques spatiales", Paris, Belin, 2002, 384 p. and "The Cambridge Encyclopaedia of Space", Missions, Applications and Exploration, Cambridge University Press, 2003, 418 p.

**Nicolas Peter** is currently Research Fellow at the European Space Policy Institute (ESPI). He has been a Lockheed Martin Fellow for two years at the Space Policy Institute at the George Washington University (GWU) and has worked for the X Prize Foundation in Washington DC on future space prizes. Mr. Peter has also been a Trainee in the Science, Technology and Education Section of the European Union Delegation of the European Commission to the USA, as well as Teaching Associate for the International Space University's Master programme and a Faculty and Team Project Co-chair for the Summer Session Programme. Nicolas Peter has completed various research activities in Europe (France and Austria) North America (Canada and USA) and Asia-Pacific (Australia and Japan). His primary research interests are in space policy and international relations. Mr. Peter has published and presented over 50 articles in peer-reviewed journals and international conferences related to space activities, particularly on space policy issues. He has also been invited to be rapporteur for sessions dealing with space policy affairs held in the framework of international space conferences in Canada, Spain and India. Mr. Peter holds a Bachelor of Geography from the Louis Pasteur University in Strasbourg, France. He holds also his first Masters Degree in Space Systems and Environment and second Masters Degree in Space Technology Applications from the Louis Pasteur University. Nicolas Peter is also a graduate from the International Space University's Master in Space Studies programme and holds a Master of International Science and Technology Policy from GWU's Elliott School of International Affairs.

**Serge Plattard** is currently consultant for the United Nations Office at Vienna, Austria. He formerly was Secretary General of the European Space Policy Institute (ESPI), located in Vienna, Austria, from September 2004 to August 2007. A nuclear Physicist by education, he earned his Doctorate from the University of Orsay (France) in 1973 and conducted scientific experimental work up to 1981 at the French Atomic Energy Commission (CEA) and in the U.S. in low energy nuclear fission, heavy ions reactions and low energy nuclear astrophysics. From 1981 to 1983, he was seconded to the French Ministry of Foreign Affairs at the Policy Planning Staff, and then returned to CEA until 1987 at the Directorate for Planning and Programs in charge of long term planning and of the EUREKA program. From 1987 to 1990, Serge Plattard served as Deputy Counsellor for Science and Cooperation at the French Embassy in New Delhi and in 1990 he was appointed as Assistant Director for scientific and technological cooperation at the French Ministry of Foreign Affairs, and moved the same year as Science and Technology Counsellor at the French Embassy in Tokyo. From 1994 to 1998 he was head of the French Mission on Science and Technology to the United States, as well as Counsellor for Science and Technology at the French Embassy in Washington DC. In 1998, Serge Plattard became the Director for International Relations of CNES, the French National Space Agency. He carried also duties as Deputy Director for Planning, Strategy, Programs and International Relations. Serge Plattard served too as Vice-Chair on the Committee for International Relations (IRC) of the European Space Agency (ESA) from 2002 to 2004. He is a former auditor of the French Institute for Higher Defense Studies (56th national session), IHEDN. Serge Plattard is also life member of the American Physical Society, founding member of Euroscience, member of the AIAA and corresponding member of the IAA. He is author/co-author of more than 30 publications and a book (Nucléaire, Merveille ou Menace?, Ed. Hatier, 1984). He was awarded in 1994 the Golden Rays in the Order of the Sacred Treasure (Japanese distinction) and in 1998 Knight in the Order of the Légion d'Honneur.

**Kai-Uwe Schrogl** is the Director of the European Space Policy Institute (ESPI) at Vienna, Austria since 1 September 2007. Before, he was Head Corporate Development and External Relations Department in the German Aerospace Center (DLR). In his previous career he worked with the German Ministry for Post and Telecommunications and the German Space Agency (DARA). He has been delegate to numerous international forums and recently served as the chairman of various European and global committees (ESA International Relations Committee, UNCOPUOS working groups). Kai-Uwe Schrogl has published seven books and more than 100 articles, reports and papers in the fields of space policy and law as well as telecommunications policy. He is Member of the Board of Directors of the International Institute of Space Law, Member of the International Academy of Astronautics (chairing its Commission on policy, economics and law) and the Russian Academy for Cosmonautics as well as member in editorial boards of international journals in the field of space policy and law (Acta Astronautica, Space Policy, Zeitschrift für Luft- und Weltraumrecht, Studies in Space Law/Nijhoff). He holds a doctorate degree in political science, lectures international relations at Tübingen University, Germany (as a Honorarprofessor) and has been a regular guest lecturer i.a. at the International Space University and the Summer Courses of the European Centre for Space Law.

**Kazuto Suzuki** is Associate Professor, Graduate School of Humanities and Social Sciences, University of Tsukuba, Japan. He graduated Department of International Relations, Ritsumeikan University, and received Ph.D. from Sussex European Institute, University of Sussex, England. He also acts as Academic Advisor to Japan Aerospace Exploration Agency (JAXA) and Society of Japanese Aerospace Companies (SJAC), as his capacity of the researcher on space and R&D policy. His current research agenda is the comparative study of strategic trade control, space technology and international security order, with particular focus on Europe and South East Asia. He has published number of articles and books, both in Japanese and English, including Policy Logics and Institutions of European Space Collaboration (Ashgate, 2003), 'Administrative Reforms and Policy Logics of Japanese Space Policy', Space Policy (vol.21 no.1, 2005), 'Transforming Japan's space policy-making' Space Policy (vol.23 no.2, 2007), and 'Foreign Perceptions of Chinese Space Activities: Japanese Perception' in European Space Policy Institute Report vol. 3 (2007).

**Tomas Valasek** is director of foreign policy and defence at the Centre for European Reform in London. He has written extensively on transatlantic relations, European foreign and security policy and on defence industrial issues. In his previous career he served as Policy Director and head of the Security and Defence Policy Division at the Slovak Ministry of Defence. He oversaw the ministry's defence analysis, planning & evaluation activities as well as its legislative efforts, military missions and international security initiatives. Prior to joining the ministry Tomas Valasek founded and directed the Brussels office of the World Security Institute (formerly the Center for Defense Information, CDI), a Washington, DC-based independent defence and security think-tank (2002–2006). From 1996 to 2002 he worked as Senior European Analyst in CDI's Washington, DC office. Mr. Valasek is the editor and co-author of "The 'Easternization' of Europe's Security Policy" (IVO-CDI, October 2004), and numerous articles appearing in newspapers and journals including Wall Street Journal, Jane's Defence Weekly, and the Cambridge Review of International Affairs. He is a holder of an M.A. in International Affairs from the George Washington University in Washington, DC, and a B.A. in journalism from the University of Georgia in Athens, Georgia.

# Index

## A



## B



## C

GPSR Compliance

*The European Union's (EU) General Product Safety Regulation (GPSR) is a set of rules that requires consumer products to be safe and our obligations to ensure this.*

*If you have any concerns about our products, you can contact us on ProductSafety@springernature.com*

In case Publisher is established outside the EU, the EU authorized representative is:

Springer Nature Customer Service Center GmbH
Europaplatz 3
69115 Heidelberg, Germany

**Batch number: 10786819**

Printed by Printforce, the Netherlands